D1750392

*Edited by*
*Sushil K. Misra*

**Multifrequency Electron Paramagnetic Resonance**

## Related Titles

Misra, S.K. (ed.)

**Multifrequency Electron Paramagnetic Resonance**

Theory and Applications

2011
Print ISBN: 978-3-527-40779-8
ISBN: 978-3-527-63353-1
Adobe PDF ISBN: 978-3-527-63354-8
ePub ISBN: 978-3-527-63355-5
eMobi ISBN: 978-3-527-64049-2

*Edited by Sushil K. Misra*

# Multifrequency Electron Paramagnetic Resonance

Data and Techniques

**WILEY-VCH**
Verlag GmbH & Co. KGaA

**The Editor**

**Sushil K. Misra**
Concordia University
Department of Physics
1455 de Maisonneuve Boulevard West
Montreal
Quebec H3G 1M8
Canada

All books published by **Wiley-VCH** are carefully produced. Nevertheless, authors, editors, and publisher do not warrant the information contained in these books, including this book, to be free of errors. Readers are advised to keep in mind that statements, data, illustrations, procedural details or other items may inadvertently be inaccurate.

**Library of Congress Card No.:** applied for

**British Library Cataloguing-in-Publication Data**
A catalogue record for this book is available from the British Library.

**Bibliographic information published by the Deutsche Nationalbibliothek**
The Deutsche Nationalbibliothek lists this publication in the Deutsche Nationalbibliografie; detailed bibliographic data are available on the Internet at <http://dnb.d-nb.de>.

© 2014 Wiley-VCH Verlag GmbH & Co. KGaA, Boschstr. 12, 69469 Weinheim, Germany

All rights reserved (including those of translation into other languages). No part of this book may be reproduced in any form – by photoprinting, microfilm, or any other means – nor transmitted or translated into a machine language without written permission from the publishers. Registered names, trademarks, etc. used in this book, even when not specifically marked as such, are not to be considered unprotected by law.

**Print ISBN:** 978-3-527-41222-8
**ePDF ISBN:** 978-3-527-67244-8
**ePub ISBN:** 978-3-527-67245-5
**Mobi ISBN:** 978-3-527-67246-2
**oBook ISBN:** 978-3-527-67243-1

**Cover Design**  Adam-Design, Weinheim, Germany
**Typesetting**  Laserwords Private Limited, Chennai, India
**Printing and Binding**  Markono Print Media Pte Ltd., Singapore

Printed on acid-free paper

## Contents

**Preface** *IX*
**List of Contributors** *XI*

**1**  **Introduction** *1*
*Sushil K. Misra*

**2**  **Rapid-Scan Electron Paramagnetic Resonance** *3*
*Sandra S. Eaton, Richard W. Quine, Mark Tseitlin, Deborah G. Mitchell, George A. Rinard, and Gareth R. Eaton*
2.1    Introduction *3*
2.1.1  Historical Background and Literature Survey *4*
2.1.2  Comparison of CW, Rapid-Scan, and Pulsed EPR: Advantages of Rapid Scan *6*
2.1.3  Scan Types *11*
2.1.4  Digital Rapid Scan *11*
2.1.5  Absolute Signal Quantitation *13*
2.1.6  Signal-to-Noise Advantage of Rapid Scan Relative to CW *14*
2.2    Post-Acquisition Treatment of Rapid-Scan Signals *17*
2.2.1  Deconvolution of Linear Scans *17*
2.2.2  Deconvolution of Sinusoidal Scans *19*
2.2.3  Low-Pass Filtering *20*
2.3    Simulation of Rapid-Scan Spectra *20*
2.4    Scan Coils *21*
2.4.1  Literature Background on Scan Coils *21*
2.4.2  Bruker Modulation Coils *22*
2.4.3  Coils for Rapid Scans *22*
2.4.4  Magnet Considerations *23*
2.5    Design of Scan Driver *25*
2.5.1  Linear Scan Drivers *25*
2.5.2  Sinusoidal Scan Drivers *27*
2.5.3  Integration into a Spectrometer System *28*
2.6    Use of ENDOR-Type Coils and RF Amplifiers for Very Fast Scans *30*

| | | |
|---|---|---|
| 2.7 | Resonator Design  32 | |
| 2.7.1 | X-Band  33 | |
| 2.7.2 | VHF (250 MHz)  35 | |
| 2.8 | Background Signals  37 | |
| 2.8.1 | Cause  37 | |
| 2.8.2 | Methods of Removing Background Signals from Rapid-Scan Spectra  37 | |
| 2.8.2.1 | Linear (Triangular) Scans  37 | |
| 2.8.2.2 | Sinusoidal Scans  40 | |
| 2.9 | Bridge Design  41 | |
| 2.10 | Selection of Acquisition Parameters  42 | |
| 2.10.1 | Resonator Bandwidth  43 | |
| 2.10.2 | Signal Bandwidth  43 | |
| 2.10.3 | Microwave Power  45 | |
| 2.10.4 | Electron-Spin Relaxation Times  45 | |
| 2.10.5 | Selection of Scan Rate  45 | |
| 2.11 | Multifrequency Rapid Scan  46 | |
| 2.12 | Examples of Applications  46 | |
| 2.12.1 | Comparison of Rapid-Scan Spectra Obtained with a Dielectric Resonator and Either Standard Modulation Coils or Larger Scan Coils  46 | |
| 2.12.1.1 | Standard Modulation Coils  47 | |
| 2.12.1.2 | Larger External Coils  48 | |
| 2.12.1.3 | Results of Comparison  49 | |
| 2.12.2 | S/N for Nitroxide Radicals  49 | |
| 2.12.3 | Estimation of Nitroxide $T_2$ at 250 MHz  50 | |
| 2.12.4 | Spin-Trapped Radicals  52 | |
| 2.12.5 | Improved S/N for Species with Long Electron-Spin Relaxation Times  54 | |
| 2.12.5.1 | E′ Center in Irradiated Fused Quartz  54 | |
| 2.12.5.2 | Amorphous Hydrogenated Silicon (a-Si:H)  56 | |
| 2.12.5.3 | N@C$_{60}$  57 | |
| 2.12.5.4 | Single Substitutional Nitrogen ($N_2^0$) in Diamond  58 | |
| 2.12.6 | Imaging  59 | |
| 2.13 | Extension of the Rapid-Scan Technology to Scans That Are Not Fast Relative to Relaxation Times  62 | |
| 2.14 | Summary  63 | |
| | Acknowledgments  64 | |
| | References  64 | |
| | | |
| **3** | **Computational Modeling and Least-Squares Fitting of EPR Spectra**  69 | |
| | *Stefan Stoll* | |
| 3.1 | Introduction  69 | |
| 3.2 | Software  70 | |
| 3.2.1 | EasySpin  70 | |

| | | |
|---|---|---|
| 3.2.2 | Other Software | 71 |
| 3.3 | General Principles | 72 |
| 3.3.1 | Spin Physics | 72 |
| 3.3.1.1 | Interactions | 72 |
| 3.3.1.2 | Quantum States and Spaces | 75 |
| 3.3.1.3 | Equations of Motion | 76 |
| 3.3.2 | Other Aspects | 77 |
| 3.3.2.1 | Isotopologues | 77 |
| 3.3.2.2 | Field Modulation for cw EPR | 78 |
| 3.3.2.3 | Frames and Orientations | 78 |
| 3.4 | Static cw EPR Spectra | 80 |
| 3.4.1 | Crystals and Powders | 80 |
| 3.4.1.1 | Crystals | 80 |
| 3.4.1.2 | Partially Ordered Samples | 81 |
| 3.4.1.3 | Disordered Systems and Spherical Grids | 81 |
| 3.4.2 | Field-Swept Spectra | 84 |
| 3.4.2.1 | Eigenfield Method | 84 |
| 3.4.2.2 | Matrix Diagonalization | 85 |
| 3.4.2.3 | Perturbation Theory | 87 |
| 3.4.2.4 | Hybrid Models | 88 |
| 3.4.3 | Transition Intensities | 88 |
| 3.4.4 | Isotropic Systems | 89 |
| 3.4.5 | Line Broadenings | 90 |
| 3.4.5.1 | Dipolar Broadening | 91 |
| 3.4.5.2 | Strains | 91 |
| 3.4.5.3 | Lineshapes | 92 |
| 3.4.6 | Frequency-Sweep Spectra | 93 |
| 3.4.7 | Simulation Artifacts | 93 |
| 3.5 | Dynamic cw EPR Spectra | 94 |
| 3.5.1 | Rotational Diffusion | 94 |
| 3.5.1.1 | Fast-Motion Limit | 94 |
| 3.5.1.2 | Fast-Tumbling Regime | 95 |
| 3.5.1.3 | Slow-Tumbling Regime | 95 |
| 3.5.2 | Chemical Exchange | 98 |
| 3.6 | Pulse EPR Spectra | 99 |
| 3.6.1 | Bloch Equations | 100 |
| 3.6.2 | Hilbert space | 100 |
| 3.6.2.1 | Scalar Equations | 100 |
| 3.6.2.2 | Matrix Equations | 101 |
| 3.6.3 | Liouville Space | 104 |
| 3.7 | Pulse and cw ENDOR Spectra | 104 |
| 3.7.1 | Transition Frequencies | 105 |
| 3.7.2 | Intensities | 105 |
| 3.7.3 | Broadenings | 107 |
| 3.8 | Pulse DEER Spectra | 108 |

| | | |
|---|---|---|
| 3.9 | Least-Squares Fitting | *108* |
| 3.9.1 | Objective Function | *109* |
| 3.9.2 | Search Range and Starting Point | *110* |
| 3.9.3 | Fitting Algorithms | *110* |
| 3.9.3.1 | Local Methods | *110* |
| 3.9.3.2 | Global Methods | *111* |
| 3.9.4 | Multicomponent and Multispectral Fits | *112* |
| 3.9.5 | Limits of Automatic Fitting | *112* |
| 3.9.6 | Error Analysis | *113* |
| 3.10 | Various Topics | *113* |
| 3.10.1 | Spin Quantitation | *113* |
| 3.10.2 | Smoothing and Filtering | *114* |
| 3.10.3 | Data Formats | *114* |
| 3.11 | Outlook | *115* |
| | References | *116* |
| | | |
| **4** | **Multifrequency Transition Ion Data Tabulation** | ***139*** |
| | *Sushil K. Misra, Sean Moncrieff, and Stefan Diehl* | |
| 4.1 | Introduction | *139* |
| 4.2 | Listing of Spin-Hamiltonian Parameters | *142* |
| | References | *270* |
| | | |
| **5** | **Compilation of Hyperfine Splittings and *g*-Factors for Aminoxyl (Nitroxide) Radicals** | ***287*** |
| | *Lawrence J. Berliner* | |
| 5.1 | Introduction | *287* |
| 5.2 | Tabulations | *288* |
| 5.3 | Concluding Remarks | *288* |
| | References | *294* |

**Index**  *297*

# Preface

Research into electron paramagnetic resonance (EPR), also known as *electron spin resonance* (*ESR*) and *electron magnetic resonance* (*EMR*), has been constantly expanding since the first article on this topic by Zavoisky in 1945. The field of EPR imaging, previously considered unachievable, is now well developed, complementing MRI (magnetic resonance imaging), a development of NMR (nuclear magnetic resonance), based on the resonance of protons possessing a nuclear magnetic moment. Moreover, EPR finds extensive applications in biology, medicine, chemistry, physics, and geology. It is, therefore, important to provide the scientific community with information on the latest developments in the field of EPR.

This volume is in continuation of the efforts put forward by my colleagues C. P. Poole, Jr. and H. A. Farach, who edited ESR Handbooks (ESRHB), Volume 1 (AIP Press, New York, 1994) and Volume 2 (Springer, AIP Press, New York, 1999), and by myself, editing the book "Multifrequency Electron Paramagnetic Resonance: Theory and Applications" (MFEPRTA; Wily-VCH, Weinheim, Germany, 2011). ESRHB Volume 1 dealt with the general aspects of the literature, with chapters on computer techniques, relaxation, and electron-nuclear double resonance (ENDOR), whereas Volume 2 contained chapters on sensitivity, resonators, lineshapes, electron-spin-echo envelope modulation (ESEEM), transition metal (TM) ion series, spin-Hamiltonian (SH) types and symmetries, evaluation of SH parameters from EPR data, EPR imaging, high-field EPR, and a thorough tabulation of TM ion data on SH parameters till 1993. Since the publication of these handbooks the technique of multifrequency EPR has been used extensively in EPR research. As for MFEPRTA book, it covered extensively the latest developments in theory and applications of multifrequency EPR. The present volume is aimed to provide chapters on the latest state-of-the-art information on EPR. It covers the technique of rapid-scan EPR and a thorough coverage of the literature on compilation of multifrequency EPR spectra and evaluation of SH parameters, as well as an exhaustive tabulation of TM ion SH parameters covering the period of 20 years (1993–2012, inclusive). In addition, a small chapter is devoted to a tabulation of hyperfine splittings and g-factors of some typical aminoxyl (nitroxide) ions published over the years. It is hoped that this volume will serve a useful and timely purpose to EPR researchers at large.

I am grateful to Professor C. P. Poole, Jr., University of South Carolina, for his constant mentoring throughout my efforts to put together this volume. Thanks are also due to Danielle Dennie and Katharine Hall, reference librarians at Concordia University, for their searches through the various databases to find the relevant articles.

Finally, I dedicate this book to my parents Mr. Rajendra Misra and (late) Mrs. Prakash Wati Misra, my daughters Manjula and Shivali, and my son Paraish.

# List of Contributors

**Lawrence J. Berliner**
Department of Chemistry and
Biochemistry
2190 E Iliff Ave
University of Denver
Denver
CO 80208
USA

**Stefan Diehl**
Justus Liebig University Griessen
2nd Physics Institute
Heinrich-Buff-Ring 16
35392 Griessen, Germany

**Gareth R. Eaton**
Department of Chemistry and
Biochemistry
University of Denver
Denver
CO 80208
USA

**Sandra S. Eaton**
Department of Chemistry and
Biochemistry
University of Denver
Denver
CO 80208
USA

**Sushil K. Misra**
Concordia University
Department of Physics
1455 de Maisonneuve Boulevard
West
Montreal
Quebec H3G 1M8
Canada

**Deborah G. Mitchell**
Department of Chemistry and
Biochemistry
University of Denver
Denver
CO 80208
USA

**Sean Moncrieff**
Concordia University
Physics Department
1455 de Maisonneuve Boulevard
West
Montreal
Quebec H3G 1M8
Canada

**Richard W. Quine**
Daniel Felix Ritchie School of
Engineering and Computer
Science
University of Denver
Denver
CO 80208
USA

**George A. Rinard**
Daniel Felix Ritchie School of
Engineering and Computer
Science
University of Denver
Denver
CO 80208
USA

**Stefan Stoll**
University of Washington
Department of Chemistry
Seattle
WA 98195
USA

**Mark Tseitlin**
Department of Chemistry and
Biochemistry
University of Denver
Denver
CO 80208
USA

# 1
# Introduction

*Sushil K. Misra*

This volume consists of five chapters including the introduction, covering the various aspects of the technique of EPR (electron paramagnetic resonance, also known as ESR – electron spin resonance and EMR – electron magnetic resonance), in a timely manner. The most notable feature in this context is the multifrequency aspect of the contents, which is now the practice in EPR research. The various chapters are briefly described here.

1) **Recording of EPR spectra using the recently developed techniques in rapid scan**. In this chapter (Chapter 2), the background, theory, instrumentation, and methodology of rapid-scan EPR, including the hardware and software required to implement rapid scans and analysis of the data, are described. Among other advantages, the ability of rapid scans to acquire data quickly permits higher temporal resolution for kinetics than can be achieved with CW (continuous wave) spectroscopy. In rapid-scan EPR, the magnetic field is scanned through resonance in a time that is short relative to electron spin-relaxation times. Direct detection of the EPR response yields the absorption and dispersion signals, instead of the derivatives that are recorded in the usual CW experiment. The rapid-scan signal provides the full amplitude of EPR absorption, and not just a small approximately linear segment as is recorded in field-modulated EPR. In a rapid scan, if the time on resonance is short relative to relaxation times, there is a scan-rate dependence response that can be deconvolved to yield the undistorted absorption signal. If the time on resonance is long enough that the signal is independent of scan rate, the deconvolution procedure does not change the spectrum; therefore, the data analysis method is general for any rate of passage through resonance. The signal-to-noise ratios obtained by rapid scan are higher by factors of as much as 20 to >250 than those obtained by CW EPR for samples ranging from spin-trapped superoxide and nitroxide radicals in fluid solution to paramagnetic centers in materials.

2) **Simulation of EPR spectra and evaluation of spin-Hamiltonian parameters (SHP)** as developed over the years. After summarizing the key aspects of available simulation software packages, the basic aspects of EPR simulations are discussed in Chapter 3. Thereafter, methods for simulation of static and

dynamic CW EPR, pulse EPR, ENDOR (electron nuclear double resonance), and DEER (double electron electron resonance) spectra are described. Subsequently, a section is dedicated to least-squares fitting. After a short section covering topics such as spin quantization and data formats, some of the challenges that still lie ahead are summarized in the conclusion. This chapter provides an expert overview of computational modeling and least-squares fitting of EPR spectra. A well-written summary of the theory and methods involved in EPR spectral simulation, covering a wide range of regimes (solids, liquids, slow motion, chemical exchange), experiments (cw and pulse EPR, ENDOR), and methods (Liouville space, Hilbert space, matrix diagonalization, perturbation theory) is provided. The discussion of least-squares fitting includes many aspects that are often only discussed in isolation. Apart from the author's very general and widely used software package EasySpin, many other existing programs are mentioned. The chapter concludes with an extensive list of over 500 references that encompasses not only the seminal high-impact papers from the last half century but also many less known contributions.

3) Chapter 4 is devoted to an exhaustive *tabulation of Spin-Hamiltonian parameters (SHPs) of transition metal ions*, as published in the last 20 years (1993–2012, inclusive). It supplements a similar data listing published in the *ESR Handbook*, Volume 2, edited by C. P. Poole, Jr. and H. A. Farach (Springer, AIP Press, New York, 1999), which covers the period from 1960s to 1992. Since then, in contrast, the technique of multifrequency EPR has been used extensively in EPR research. This information is useful for various purposes, which includes verification of new EPR results, planning of experiments, and finding what parameters have been reported in the literature, without having to do an extensive database search that may quite frequently involve journals that are not readily available.

4) Finally, Chapter 5 contains a tabulation of *hyperfine splitting and g-factors of some typical aminoxyl (nitroxide) radicals*. Since the first *Spin Labeling: Theory and Applications* volume in 1976, edited by L. Berliner, no compilation has been published on this topic. With the use of organic radicals such as spin labels, calibration agents, and so on, a complete reference listing of their physical parameters is useful. In particular, the aminoxyl (nitroxide) radicals are stable organic compounds that have found a plethora of uses in chemistry, biology, and physics. The compiled data, while not thorough, cover a range of these radical types at several frequencies, solvent environments, and hosts. In some selected cases, where the data were readily available, parameters in several host environments, solvents, and other states have been included, since polarity affects both the hyperfine and g-values.

# 2
# Rapid-Scan Electron Paramagnetic Resonance

*Sandra S. Eaton, Richard W. Quine, Mark Tseitlin, Deborah G. Mitchell, George A. Rinard, and Gareth R. Eaton*

## 2.1
## Introduction

The focus of this chapter is on the emerging and very powerful implementation of electron paramagnetic resonance (EPR) that is designated as rapid scan. Historically, most EPR instrumentation and methodology have been in one of two regimes: continuous wave (CW) [1, 2], or pulsed (saturation recovery, spin echo, and Fourier Transform) [3–6]. In a CW experiment the microwave power is constant, the magnetic field is scanned to achieve resonance, and the EPR signal is recorded by phase-sensitive detection at the frequency that is used for magnetic field modulation. Microwave powers and scan rates are selected such that spectra are independent of relaxation times. In pulse experiments, the microwave power is on only during excitation, signals are detected after the pulse(s), and differences in relaxation times are exploited to optimize information content. The rapid-scan regime is an intermediate case. As in CW experiments the microwave power is constant, but the magnetic field (or microwave frequency) is scanned through resonance in a time that is short relative to relaxation times, and phase-sensitive detection at a magnetic field modulation frequency is not used. Instead, the absorption and dispersion signals are recorded by direct detection with a double balanced mixer. Rapid scans and data analysis as discussed in the following paragraphs permit spectral acquisition with lineshapes that are not modulation broadened and have substantially improved signal-to-noise ratio ($S/N$) relative to CW spectroscopy. These experiments can be performed without the use of the high powers that are required for pulse experiments. In pulse experiments, data acquisition requires samples for which the decay time for a free induction decay (FID), $T_2^*$, is long relative to the instrument dead time. This is not a limitation for rapid scans. The combination of rapid scan with improvements in digital electronics provides opportunities to revolutionize the way that much EPR will be done in the future.

Historical development, analysis of data to recover the equivalent slow-scan spectrum, and hardware modifications of conventional spectrometers to implement

*Multifrequency Electron Paramagnetic Resonance: Data and Techniques,* First Edition.
Edited by Sushil K. Misra.
© 2014 Wiley-VCH Verlag GmbH & Co. KGaA. Published 2014 by Wiley-VCH Verlag GmbH & Co. KGaA.

rapid scans are discussed in this chapter. The technology and methodology to acquire, deconvolve, and interpret the transient responses are emphasized. In terms of instrumentation, the difference between CW and rapid scan is in the scan coils and drivers, optimized resonators, and detector bandwidth. The Hyde laboratory has developed segmental acquisition of spectra, scanning a few gauss at a time [7], while the Denver laboratory engineered faster and larger magnetic field sweeps to encompass spectra of most organic radicals [8, 9] to improve $S/N$ [10, 11], to enhance EPR imaging [12], and to measure relaxation times [13, 14]. Section 2.12 of this chapter provides examples of the dramatic improvements in $S/N$ that have been obtained by rapid scans of samples ranging from spin-trapped radicals to paramagnetic centers in materials. Most of the results surveyed are from the Denver laboratory.

2.1.1
**Historical Background and Literature Survey**

Rapid-scan EPR builds on prior work in NMR (nuclear magnetic resonance). Bloembergen *et al.* [15] observed a transient effect ("wiggles") after the magnetic field passed through resonance [16]. In 1974, it was shown that these transient effects could be deconvolved to obtain useful NMR spectra ("correlation NMR spectroscopy" or "rapid-scan Fourier transform NMR spectroscopy" (FT-NMR)) [16–19]. Rapid-scan NMR achieved almost as high an $S/N$ as pulsed FT-NMR, with the additional advantage that rapid-scan NMR could measure a portion of a spectrum, and hence avoid a strong solvent peak. Rapid-scan NMR was soon eclipsed by FT-NMR owing to the wide range of pulse sequences that became available. However, its use continued in a routine commercial NMR spectrometer.

Transient effects were also observed in the early days of EPR. The first observation of "wiggles" in EPR was by Beeler *et al.* [20, 21], using an approximately 25 mG wide line of sodium in liquid ammonia, and scan rates of $1.5 \times 10^4$ G s$^{-1}$. The EPR frequency was 23 MHz (8.2 G resonant field), the amplitude of the sinusoidal field scan was 1.1 G, and the scan frequency was 2 kHz. Effects of rapidly changing fields on signals from sugar char were reported by Gabillard [22]. Gabillard and Ponchel [23] showed that shapes of EPR spectra of DPPH (2,2-diphenyl-1-picrylhydrazyl) changed when the modulation period was comparable to $T_2$.

There is vast early literature on the effects of adiabatic passage on electron spins, of which a small portion is cited here to provide some background. Adiabatic rapid passage effects were observed in irradiated LiF by Portis [24] and by Hyde [25]. Feher and coworkers used rapid passage effects in electron-nuclear double resonance (ENDOR) experiments [26–28]. Multiple passage effects were described by Weger [29]. Although published 50 years ago, Weger's article remains the only extensive review of passage effects in magnetic resonance. Many cases were illustrated providing sketches of the expected effects on absorption and dispersion EPR spectra. It remains a guide to the range of phenomena that are potentially observable. In Weger's terminology [29], "rapid" refers to the regime in which

$$\frac{B_1}{\left[\left(\frac{dB}{dt}\right)(T_1 T_2)^{\frac{1}{2}}\right]} < 1 \tag{2.1}$$

where $B$ is the external magnetic field, $t$ is time, $T_1$ and $T_2$ are the electron spin relaxation times, and $B_1$ is the radio frequency (RF)/microwave magnetic field. Weger [29] defined adiabatic as the regime in which the signal is partially saturated and the scans are characterized by $(\omega_m H_m/\gamma B_1^2) << 1$ or $(dB_o/dt)/(\gamma B_1^2) << 1$. Thus, in his terminology, most of the scans discussed in this chapter are non-adiabatic. Slow scan is described by

$$\frac{dB}{dt} \ll \gamma(\delta B)^2 \tag{2.2}$$

where $\delta B$ is the relaxation-determined linewidth expressed in magnetic field units. $\gamma = -1.7609 \times 10^{11}$ rad s$^{-1}$ T$^{-1}$ = $-1.7609 \times 10^7$ rad s$^{-1}$ G$^{-1}$ is the free electron gyromagnetic ratio.

Czoch et al. [30] observed an EPR transient response for the TCNQ radical (N-methylpyridinium tetracyanoquinodimethane) with a sinusoidal magnetic field scan rate of $2 \times 10^5$ G s$^{-1}$. A low-Q helix was used as the resonator. Saturation transfer EPR exploits passage effects to estimate rotational correlation times of nitroxide radicals [31]. Other applications of rapid passage include $^{57}$Fe ENDOR obtained with dispersion derivative EPR under adiabatic rapid passage conditions [32]. Seamonds et al. [33] and Mailer and Taylor [34] used adiabatic rapid passage to enhance the intensity of EPR spectra of ferric hemoglobin and ferrocytochrome c, respectively. Signal enhancements in irradiated tooth samples were achieved by second harmonic out-of-phase signal detection at 77 K, using 100 kHz magnetic field modulation [35]. The out-of-phase response under adiabatic rapid passage conditions has been used to enhance intensities of the nitrogen signal in natural diamond [36]. The phase lag of the signal was used to estimate $T_1$ as $1.7 \pm 0.7$ ms. Periodic adiabatic passage with monitoring of the dispersion mode 90° out of phase with the modulation was used to measure $T_1$ of hydrogenated amorphous silicon and silicon carbide [37]. Rapid passage spectra also have been observed in rotational spectroscopy [38] and in infrared spectroscopy [39].

Hyde et al. [40] achieved rapid frequency scans, up to $1.8 \times 10^5$ GHz/s, in a 94 GHz EPR spectrometer. Rapid triangular and trapezoidal frequency sweeps through nitroxide lines resulted in transient responses ("wiggles"). The National Cancer Institute group of Murali Krishna and coworkers [41, 42] has incorporated rapid-scan and rotating magnetic field gradients into a fast CW EPR imaging method, and found a strong advantage of rapid scan for imaging nitroxide radicals. Direct-detected X-band (9.5 GHz) and L-band (1–2 GHz) absorption EPR spectra have been acquired using scans that are slow relative to relaxation times, that is, non-adiabatic rapid sweeps (NARS), as a replacement for the field-modulated, phase-sensitive-detected derivative CW EPR spectra [7, 43]. With scan rates that satisfy the non-adiabatic condition, pure absorption EPR spectra of nitroxide radicals were collected in magnetic field segments. By avoiding the line broadening that is inherent in any field-modulated CW spectrum, the L-band nitroxide center line was narrow enough to be sensitive to spin–spin interactions at distances of 18–30 Å [43].

The publications from the Eaton laboratory at the University of Denver (Denver laboratory) have emphasized the cases in which the magnetic field or frequency scan rate is rapid relative to relaxation rates as defined by Weger [29]. There is little or no saturation in most of these spectra, so the conditions are also non-adiabatic. If the time required to scan through the line is short relative to $T_2$, there are transient effects on the trailing edge of the rapid-scan signal, which damp out with time constant $T_2^*$. In the early literature those oscillations were called *wiggles*. If $T_2^* \ll T_2$, spectra may be in the rapid-scan regime even though oscillations are not observed. Rapid-scan EPR spectra from the Denver laboratory illustrate the impact of relaxation times and microwave power on signal amplitude and shape [10, 14, 44, 45], the impact of resonator parameters [46], the ability to deconvolve signals [47–49], and to reconstruct images from rapid-scan spectra [12]. Since the damping of wiggles depends on $T_2$ (as well as inhomogeneous broadening), $T_2$ can be determined from the rapid-scan response [14]. Rapid linear (triangular) [8] and sinusoidal [9] scan drivers have been described.

There also are cases in the literature where the term *rapid* has been used to describe magnetic field or frequency scans that are rapid relative to normal rates of scanning an iron-core magnet [50, 51], but are not rapid relative to relaxation rates. Since the scan rate that is rapid for one sample may be slow for another, similar technology applies in both regimes. The hardware and advances described in this chapter apply in that regime as well. If a spectrum includes lines with different relaxation times such that scan times are short relative to relaxation times for some lines, but not for others, the deconvolution methods described in Section 2.2 can still be applied. This chapter focuses on the observation of transient effects in the EPR spectra of organic radicals in solution and defect centers in solids at room temperature. This is a new regime for EPR, and new aspects of methodology and applications are being developed.

## 2.1.2
### Comparison of CW, Rapid-Scan, and Pulsed EPR: Advantages of Rapid Scan

Traditional CW EPR records the first-derivative of the absorption signal as a function of the slowly scanned Zeeman field, which often takes several minutes. Superimposed on the slowly scanned magnetic field is a sufficiently small, rapidly modulated, magnetic field that encodes the EPR response such that it can be distinguished from noise using phase-sensitive detection at the modulation frequency. The output is the first derivative of the absorption spectrum. In this chapter, this method is referred to as *slow-scan EPR* or *CW EPR*.

Prior to the use of magnetic field modulation, EPR spectra were recorded as the voltage output of a diode detector, often called a *crystal detector*, as the magnetic field was swept. The signal was usually displayed on an oscilloscope. This is detection at the resonance frequency, and is referred to as *direct detection*, as distinct from phase-sensitive detection at the field modulation frequency. The output is the absorption spectrum. Many such spectra are described in early papers [29, 52, 53]. This is also the detection method used in rapid-scan measurements – the signal is recorded directly with phase-sensitive detection at the resonance frequency using double-balanced mixer detection. The scans can be performed with magnetic field or microwave frequency sweeps such that the time on resonance is long or short relative to electron spin relaxation times. When the scan rate is slow relative to relaxation rates, and thus the time on resonance is long relative to the relaxation times, the absorption spectrum is detected. When the scan rate is fast and the time on resonance is short relative to relaxation times, oscillations may be observed on the trailing edge of the signal as shown in Figure 2.1. As the scan rate increases, the depth and number of oscillations increase and the signal broadens. Post-processing deconvolution can be used to remove both the broadening and the oscillations and obtain the undistorted absorption spectra [48]. Rapid scan is analogous to pulsed EPR in the sense that the microwaves are resonant with the spins for only a brief period during the scan. Consequently, much higher power can be used relative to a CW scan [10, 11, 54]. There is no magnetic field modulation broadening of the line. Examples in Section 2.12.5 show particularly large improvements in $S/N$ for rapid scan relative to CW for species with long electron spin relaxation times.

In CW spectroscopy, if the time to scan through the signal is short relative to relaxation times, passage effects distort the lineshape [29]. This can make it difficult to record spectra of samples with long electron spin relaxation times such as the E′ defect in irradiated quartz [55], or for many samples at low temperatures. For example, at about 5 K the spin-lattice relaxation time of vanadyl porphyrin is so long that reversing the direction of the magnetic field scan inverts the EPR spectrum, which then looks similar to an absorption (or emission) spectrum even though it is a field-modulated first-derivative spectrum (e.g., see Figure 6 in [56]). Rapid-scan spectroscopy takes advantage of the passage effects, which can be removed by deconvolution, to obtain the undistorted absorption lineshape with enhanced signal intensity.

If standard CW EPR is performed with small enough modulation amplitude to faithfully define the derivative of the absorption signal, the amplitude of the phase-sensitive detected signal is about 1/10 or less of the maximum possible signal. One can use modulation amplitudes less than or equal to the linewidth and recover the undistorted signal post-acquisition [57–62], but there are limitations to the corrections that can be made. In addition, these corrections cannot compensate for passage effects.

In a rapid-scan experiment, the microwave power is applied to the spin system for a short time during the scan of the magnetic field through resonance. The faster the scan, the shorter the time during which $B_1$ excites the spins; therefore, the balance between excitation and relaxation favors relaxation, and higher $B_1$ can be applied. The dependence of signal amplitude on scan rate for LiPc (lithium phthalocyanine) is shown in Figure 2.2 [54]. If $B_1$ is constant, signal amplitude decreases with increasing scan rate because the signal does not have time to respond to $B_1$ during

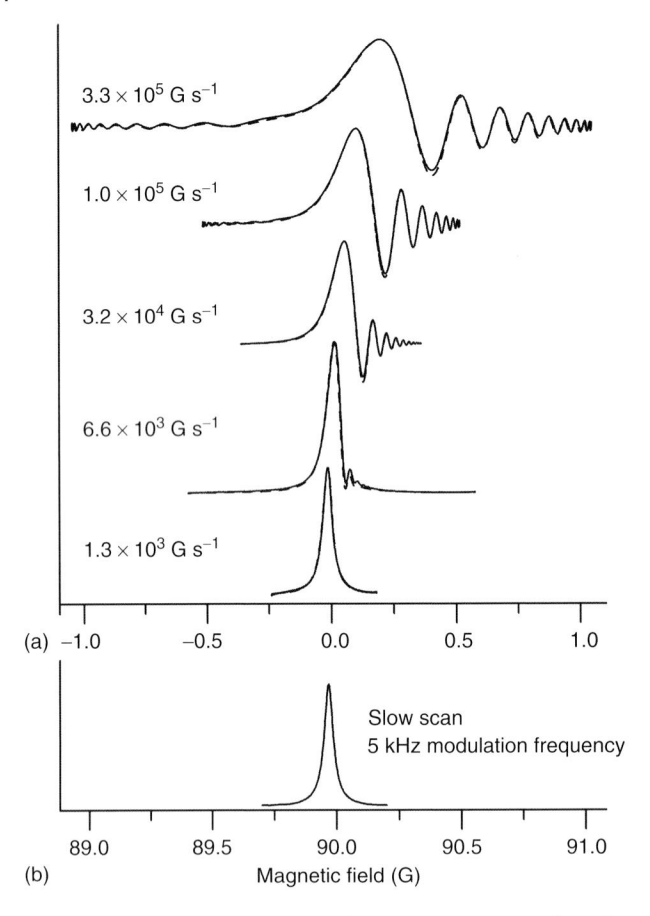

**Figure 2.1** EPR spectra of LiPc at about 252 MHz. (a) Sinusoidal rapid-scan spectra obtained at the scan rates shown. The x-axis is the offset from resonance (in gauss) and the spectral widths of the traces are the scan widths. The number of scans averaged for each scan rate was selected to give a total signal acquisition time of 84 s. The y-axis scales are arbitrary. The values of $B_1$, selected to be a factor of 2 below the values that gave the maximum signal intensity, were: $4.6 \times 10^{-2}$ G at $3.3 \times 10^5$ G s$^{-1}$, $2.6 \times 10^{-2}$ G at $1.0 \times 10^5$ G s$^{-1}$, $1.8 \times 10^{-2}$ G at $3.2 \times 10^4$ G s$^{-1}$, $1.15 \times 10^{-2}$ at $6.6 \times 10^3$ G s$^{-1}$, and $4.6 \times 10^{-3}$ G at $1.3 \times 10^3$ G s$^{-1}$. Simulated spectra (dashed lines) calculated using numerical integration of the Bloch equations are overlaid on the experimental data. (b) First integral of slow-scan CW spectrum obtained with 5 kHz modulation frequency, modulation amplitude of 10 mG, $B_1 = 4.6 \times 10^{-3}$ G, 84 s scan, 252.07 MHz. (Source: Stoner et al., 2004 [54]. Reproduced with permission of Elsevier Limited.)

the time on resonance. However, the maximum signal amplitude can be increased by increasing $B_1$ as the scan rate is increased. The transition from the regime where intensity is independent of scan rate to the regime where intensity is scan-rate dependent occurs when the scan rate satisfies the criterion shown in Eq. (2.1). Specific examples are shown in Section 2.12.

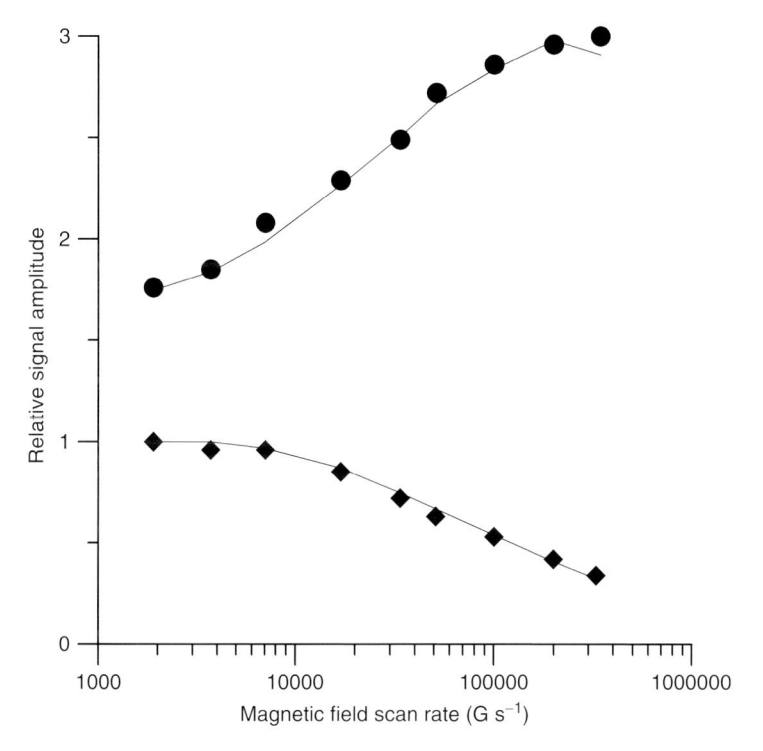

**Figure 2.2** Relative intensities for the LiPc signal at the center of the sinusoidal scan as a function of scan rate at constant $B_1$ of $6.5 \times 10^{-3}$ G (♦) and at the $B_1$ that gave the maximum signal amplitude (•). The number of scans was held constant. The relative signal amplitudes were scaled to 1.0 for the signal at constant $B_1$ of $6.5 \times 10^{-3}$ G collected at the scan rate of $1.3 \times 10^3$ G s$^{-1}$ (scan width of 0.42 G and scan frequency of 1 kHz). The solid lines connect points that were calculated by numerical integration of the Bloch equations (using the parameters for LiPc and the experimental scan widths and frequencies). (Source: Stoner et al., 2004 [54]. Reproduced with permission of Elsevier Limited.)

A power saturation curve is a plot of signal amplitude as a function of microwave $B_1$. (Recall that $B_1$ is proportional to the square root of incident power.) To have integrated signal intensity proportional to the number of spins in the sample as is required for spin quantitation, both rapid-scan and CW require operation in the regime where signal amplitude increases linearly with $B_1$. For CW spectra, the power saturation curve is independent of scan rate. Because of the scan-rate dependence of signal amplitude shown in Figure 2.2, rapid-scan power saturation curves depend on scan rate as shown in Figure 2.3 for a nitroxide radical [10]. As the scan rate is increased, higher powers can be used without saturating the signal, and the maximum signal amplitude increases. The dashed lines in the figure are calculated power saturation curves based on simulations using time-dependent Bloch equations.

If the spectrum can be fully excited by the microwave pulse, pulsed EPR detects the full signal. Standard X-band pulsed EPR spectrometers using amplifiers that

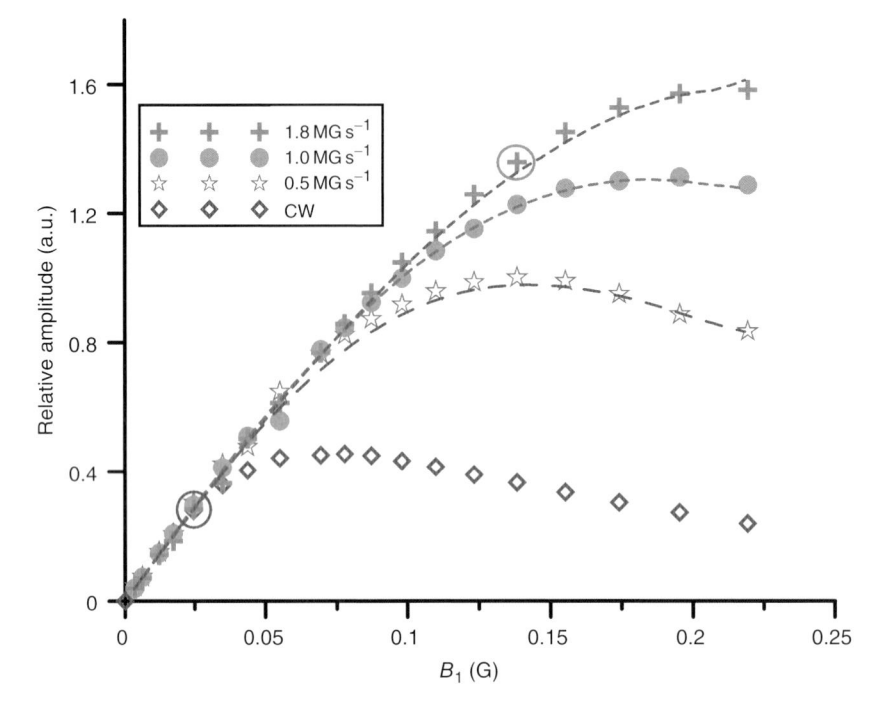

**Figure 2.3** Amplitude of CW and rapid-scan spectra of the low-field nitrogen hyperfine line of 0.1 mM $^{15}$N-mHCTPO solution as a function of microwave $B_1$. The scan widths were ~10 G and rapid-scan frequencies were 15.9, 31.5, or 57.4 kHz. Rapid-scan signals were 1024 averages, collected in less than 1 s. CW spectra were collected with a single scan acquired in ~82 s. The y-axis scale is the same for all of the rapid scans. The amplitude of the CW spectra is scaled to match that obtained for the rapid scans at low $B_1$. The dashed lines represent the calculated power saturation curves, which were simulated by numerical integration of the Bloch equations. The points selected for the acquisition conditions for rapid scan and CW spectra are circled. (Source: Mitchell et al., 2012 [10]. Reproduced with permission of Elsevier Limited.)

can deliver tens to hundreds of watts microwave power to the resonator can generate 20 ns 90° pulses, which excite about 50 MHz of spectral width, provided that resonator $Q$ is sufficiently low so that the bandwidth of the excitation is less than the resonator bandwidth. Some spectrometers can achieve shorter pulses with correspondingly larger spectral coverage. However, the relatively few spectra that are narrow enough to be fully excited with pulsed EPR can be studied with rapid-scan EPR with much lower power, and hence with instrumentation that is less expensive. Rapid scan can also be applied to spectra that are too wide to be fully excited by a 20 ns pulse. Pulsed EPR also is limited to signals with $T_2^*$ substantially longer than the dead time of the resonator, which is not a limitation for rapid scan.

In summary, the advantages of rapid scan relative to CW EPR arise from the detection of the full signal amplitude on every scan through the signal, the ability to use higher microwave powers, and the availability of deconvolution algorithms that take account of the known impact of passage effects on the signal.

## 2.1.3
## Scan Types

The emphasis in this chapter is on magnetic field scans. The rate at which the field of an iron-core magnet can be scanned is relatively slow; therefore, scan coils are used to generate the rapidly changing fields. Scans have been primarily triangular or sinusoidal, but many other shapes have potential applications. In a triangular scan, the rate of scan is constant across the spectrum and is given by

$$a_t = 2 f_s B_m \, (\text{G s}^{-1}) \tag{2.3}$$

where $a_t$ is the triangular scan rate, $f_s$ is the scan frequency, and $B_m$ is the scan width in gauss.

For a sinusoidal scan, the scan rate varies across the spectrum. The maximum rate at the center of the scan, $a_s$, is given by

$$a_s = \pi f_s B_m \, (\text{G s}^{-1}) \tag{2.4}$$

As discussed in Section 2.5.2, resonating the coils with the driver circuit reduces the voltage requirements for scan generation, so it is easier to generate faster and wider sinusoidal scans than triangular scans.

Alternatively, the microwave frequency can be swept at constant magnetic field. The primary limitation on frequency scans is that the reflected microwave power and phase follows the resonator $Q$ curve. The cavity reflection coefficient increases significantly at the extremes of the frequency sweep. The result is a frequency-dependent baseline. Since resonator $Q = \nu/\Delta\nu$, for the same resonator $Q$, the frequency bandwidth $\Delta\nu$ increases proportional to $\nu$. Consequently, frequency sweep is more feasible at the high microwave frequencies used in high-field EPR. Hyde and coworkers [40] exploited the 1 GHz bandwidth of a W-band (95 GHz) resonator to perform rapid frequency sweep EPR. Frequency sweep rates with the yttrium iron garnet (YIG) oscillator were up to $1.8 \times 10^5$ GHz s$^{-1}$. Absorption and dispersion signals were obtained for the nitroxide CTPO (3-carbamoyl-2,2,5,5-tetramethyl-3-pyrrolinyl-1-oxy) and for mixtures of $^{14}$N and $^{15}$N CTPO using triangular and trapezoidal waveforms (Figure 2.4).

The microwave frequency sweep can be very fast, as in a "chirp" of frequencies. Tseitlin et al. [63] demonstrated this capability with frequencies generated by an arbitrary waveform generator (AWG).

## 2.1.4
## Digital Rapid Scan

Recent advances in digital electronics make it possible to develop a fully digital EPR spectrometer. A major advantage of a digital spectrometer is its flexibility to be configured, with little cost or effort, to do rapid-scan and pulse experiments at multiple frequencies with a single spectrometer, rather than requiring multiple stand-alone systems. It is possible to do a variety of experiments more efficiently,

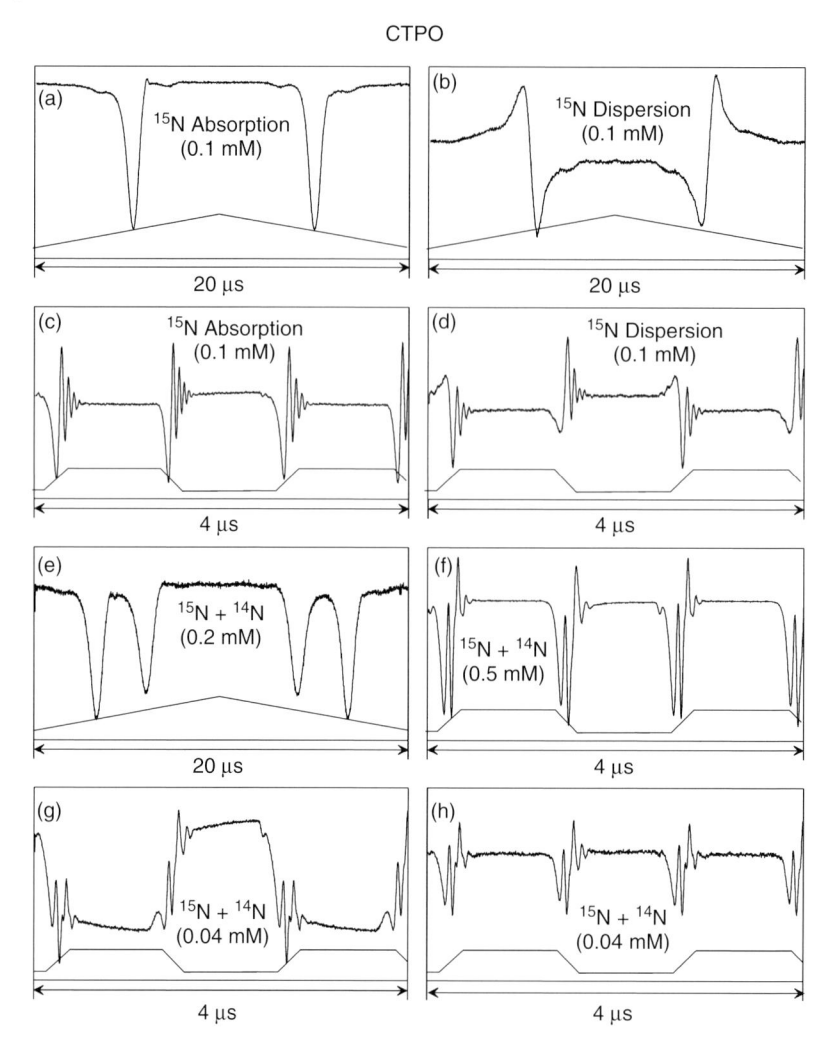

**Figure 2.4** Representative swept-frequency responses for the CTPO sample. Parts (a) and (b) used 50 kHz triangular waveforms of 44.7 MHz deviation centered on the low-field line. Part (e) used 50 kHz triangular waveforms of 45.3 MHz deviation centered between the two low-field lines of $^{14}$N and $^{15}$N CTPO. Parts (c), (d), (f), (g), and (h) used the trapezoidal waveform of 36.7 MHz deviation centered on the low-field or between the two low-field lines of the $^{14}$N and $^{15}$N CTPO. Sweep rates are 0.147 MHz ns$^{-1}$ (equivalent to 52.5 MG s$^{-1}$). Parts (c) and (h) are double-baseline corrected. (Source: Hyde et al., 2010 [40]. Reproduced with permission of Elsevier Limited.)

more accurately, with better S/N and at lower total cost. As a step in that direction, a comparison has been made of rapid-scan EPR spectra at 250 MHz with a traditional bridge and with a fully digital system. The 250 MHz excitation energy and the triangular rapid scan were generated with a Tektronix AWG.

The signal was amplified and digitally detected immediately after the resonator, without the use of a mixer. This approach eliminates the many complicated components of the "bridge" that are used in conventional EPR spectrometers. Tests were performed on samples with narrow lines including a trityl radical with well-resolved $^{13}$C hyperfine splittings (Figure 2.5a), the nitroxide radical mHCTPO (4-protio-3-carbamoyl-2,2,5,5-tetraperdeuteromethyl-3-pyrrolinyl-1-oxy), which is used for oximetry (Figure 2.5b), and semiquinones with well-resolved proton hyperfine splittings.

### 2.1.5
### Absolute Signal Quantitation

Quantitative EPR has long been recognized as difficult and subject to many confounding factors. Spin quantitation by EPR is typically done by comparison of the signal intensity of an unknown to that of a standard sample under comparable conditions. The goal of this study was to compare absolute experimental and theoretical signals and noise intensity for rapid scan at 250 MHz. The spectrometer, the resonator, and the sample were characterized in detail, which then permitted the measurement of any one of several free parameters, such as filling factor or $B_1$ for a particular incident power, which might otherwise be difficult to determine. These measurements are key to understanding the performance of the spectrometer and provide guidance in selecting the portions that are the most important targets for

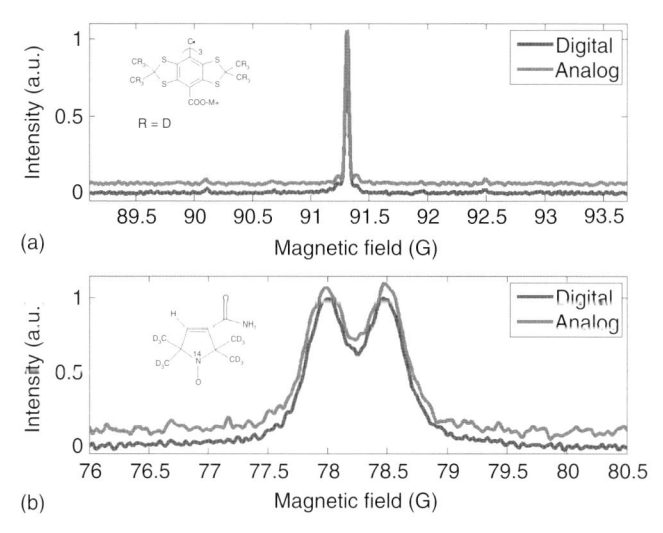

**Figure 2.5** Analog and digital triangular rapid-scan EPR spectra at 256 MHz with a field scan width of 4.8 G and scan frequency of 4 kHz. (a) 0.2 mM aqueous trityl-CD$_3$ and (b) low-field nitrogen hyperfine line of 0.25 mM aqueous $^{15}$N-mHCTPO. The doublet splitting is due to the single proton at position 4 of the ring. In each panel the upper scan is analog and the lower one is digital. (Source: Tseitlin et al., 2011 [63]. Reproduced with permission of Elsevier Limited.)

improvement [64]. This study extended the approach of prior studies by the Denver laboratory of spin echo amplitudes to rapid-scan EPR.

The calculation of signal and noise yielded an $S/N = 1.92$ for about $0.64\,\text{cm}^3$ of 0.43 mM aqueous tempone-$d_{16}$ in a 10 mm o.d. sample tube at 258.5 MHz. The experimentally measured value was $S/N = 2.07$, in excellent agreement. It should be noted that although the $S/N$ agreement is unexpectedly good, the experimental measurements of both signal and noise are about 4% lower than the theoretical values. This is attributed to errors in characterizing the bridge gain and noise figure and the fact that these values are interdependent. The experiment and calculation demonstrate the ability to fully characterize a spectrometer, resonator, and sample system [64].

### 2.1.6
### Signal-to-Noise Advantage of Rapid Scan Relative to CW

As discussed in Section 2.1.2, the rapid-scan method detects the absorption signal and CW spectroscopy detects the first derivative of the absorption. EPR spectroscopists are accustomed to viewing first-derivative displays of data. However, uncertainty analysis has shown that when spectral simulation is used to analyze data with the same $S/N$, (i) the linewidths can be obtained with the same uncertainties from absorption and first-derivative spectra, and (ii) the spin concentrations can be calculated more accurately from the absorption spectrum than from the first derivative [65]. Taking a derivative enhances high-frequency noise, and integration of the first derivative to obtain the absorption signal emphasizes low-frequency noise. Thus, interconversion between absorption and first-derivative signal displays changes the noise spectrum of the data. Therefore, in comparing rapid-scan and CW spectra it is appropriate to compare the $S/N$ in the original data acquisition forms, which is the first derivative for CW and absorption for rapid scan.

Comparison of $S/N$ for rapid-scan and CW spectra includes three factors: (i) differences in signal amplitudes detected with phase-sensitive detection at the modulation amplitude and direct detection, (ii) the signal-amplitude advantage from rapid scanning in the regime where increasing scan rate and increasing $B_1$ increases signal amplitude, and (iii) the noise in CW and rapid-scan spectra. These three factors are described in detail in the following paragraphs.

1) In CW with phase-sensitive detection at the magnetic field modulation frequency, when the magnetic field is set to the position with maximum signal, the time-dependent spin response, $R(t)$, is given by

$$R(t) = \frac{1}{2} r_m \cos(2\pi f_m t) \tag{2.5}$$

where $r_m$ is the peak-to-peak amplitude of the response and $f_m$ is the modulation frequency. Phase-sensitive detection, either in hardware or in software, consists of two steps [66]. Step 1 of phase-sensitive detection is multiplication of $R(t)$ by the reference signal $\cos(2\pi f_m t)$, which produces sum and difference frequencies, each of which has half of the original amplitude. The sum is at twice the modulation frequency and the difference is at the baseband. Step 2 is low-pass filtering, which eliminates signal at the second harmonic and suppresses high-frequency noise. The net result is a down-conversion of the spectrometer response to baseband. The resulting signal has an amplitude of $r_m/4$, which corresponds to a peak-to-peak amplitude of $r_m/2$ of the first derivative (Figure 2.6a).

In a rapid-scan experiment, the field is swept through resonance twice during the full scan cycle. The full amplitude of the absorption signal ($A_a$) is detected, which is larger than the maximum amplitude detected with field modulation by a factor of $A_a/r_m = D$ (Figure 2.6a). The ratio of the amplitudes of the rapid scan and CW spectra is, therefore, $(D\,r_m)/(r_m/2) = 2D$. For conservative choices of modulation amplitudes, $D$ may be as large as 5 or 10. For a strongly overmodulated signal, $D$ is still about 2. Thus, $2D$ is a substantial advantage in signal amplitude, which is due to direct detection, independent of whether the scan time is short or long relative to relaxation times.

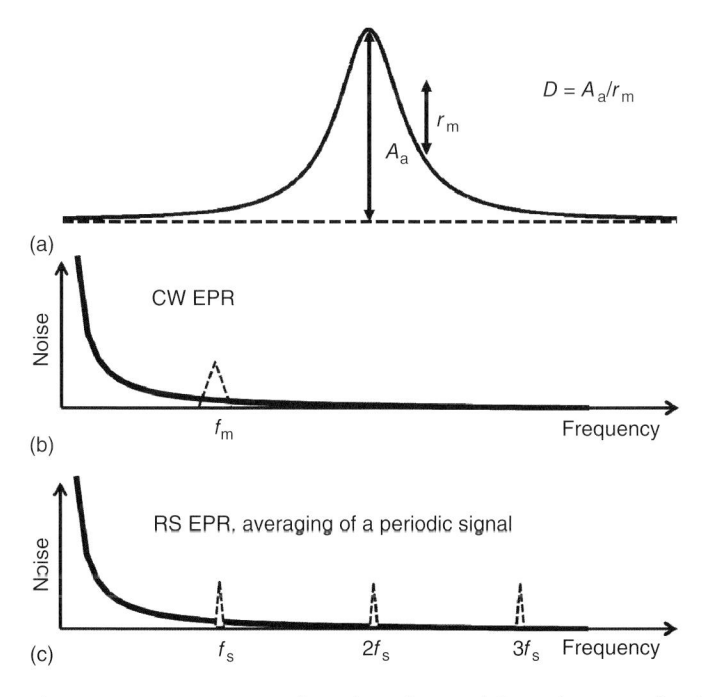

**Figure 2.6** (a) Comparison of signal amplitudes detected by rapid scan ($A_a$) and CW spectroscopy ($r_m$). (b) Comparison of noise reduction in CW and rapid-scan averaging. A hypothetical spectrometer noise profile relative to the carrier in the frequency domain is shown as a solid line. In CW the field modulation and phase-sensitive detection at the modulation frequency, $f_m$, shift the noise profile in the data away from that at the carrier by the offset $f_m$. (c) The averaging of a periodic signal in rapid scan shifts the noise profile to a comb of frequencies at multiples of the scan frequency, $f_s$. The noise bandwidths of the combs (Eq. (2.7)) are narrower than the noise bandwidths for CW EPR.

2) As discussed in Section 2.1.2, higher microwave powers can be used in rapid scans without saturating the signal (Figure 2.3), which produces higher signal amplitudes. This has been demonstrated for trityl radicals [54], LiPc [54], the E' signal in irradiated fused quartz [45], nitroxide radicals [10], and paramagnetic centers in materials [11]. The extent of signal enhancement, $F = A_a(\text{rapid scan})/A_a(\text{direct-detected slow scan})$, depends on relaxation times and scan rates. For the nitroxide example shown in Figure 2.3, $F$ is about 4. The combined signal enhancement that can be achieved by rapid scanning, by a factor of $F$, and direct detection, by a factor of $2D$, is then $2DF$.

3) The noise in the two types of experiments also needs to be considered. The use of field modulation and phase-sensitive detection at the modulation frequency reduces signal, but it also reduces noise. Noise in experimental data is not white. In addition to thermal noise (which is always present), there may be substantial contributions from source noise, which is highest near the carrier frequency. Use of field modulation offsets the EPR signal in the frequency domain by $f_m$ (Figure 2.6). If the modulation frequency is high enough, the detected signal is moved into the region where the noise spectrum is flat and noise is approximately white.

In rapid scans the source noise is reduced by averaging of a periodic signal. This time domain averaging is equivalent to applying a comb filter $H(f)$ in the frequency domain (Figure 2.6c). The absolute value of the filter function is

$$|H(f)| = \frac{\sin(N_{\text{aver}})\frac{\pi f}{f_s}}{\sin\left(\frac{\pi f}{f_s}\right)} \qquad (2.6)$$

where $f_s$ is the scan frequency. Noise components that are not coherent with the periodic signal are efficiently attenuated. The width of each comb is equal to the inverse of the total averaging time, $Time$, in Eq. (2.7).

$$\Delta f = \frac{f_s}{N_{\text{aver}}} = \frac{1}{Time}, \qquad (2.7)$$

where $N_{\text{aver}}$ is the number of scans averaged. This is a much narrower filter bandwidth than that used in CW experiments. Coherent averaging of the periodic rapid-scan signal offsets the signal in the frequency domain to the region of the noise spectrum where the noise spectrum is approximately frequency independent (Figure 2.6). The quantitative impact of the narrower comb filter has not been estimated, so it is not included in the overall comparison of CW and rapid-scan spectra. This principle was demonstrated by Klein and Barton in 1963 [67] for NMR spectra that included "wiggles" and for CW EPR spectra.

If CW and rapid-scan experiments are performed in the regime where both $f_m$ and $f_s$ are high enough to shift the spectrum into the regime where the noise is approximately independent of frequency, the white noise approximation can be used to compare two experiments performed on the same spectrometer. Before detection, both CW and rapid scan have noise with the same standard deviation, $\sigma$. In the phase-sensitive detection of the CW spectrum, multiplication by $\cos(2\pi f_m t)$

reduces $\sigma$ by a factor of $\sqrt{2}$. The cut-off frequency for the low-pass filter is selected such that high-frequency noise is attenuated, while the lineshape is preserved; thus, the cut-off frequency depends on the scan rate. In a rapid-scan experiment, only the low-pass filter is used, which also has a cut-off frequency that depends on scan rate. Noise bandwidth is proportional to the square root of the cut-off frequency. If the scan rate in the rapid-scan experiment is higher than that of the CW spectrum by a factor of $M$, the cutoff frequency in the low-pass filter would be higher by a factor of $M$, and the noise in a single rapid scan would be higher than that for the CW spectrum by $\sqrt{M}$. However, if the data-acquisition time is the same for the two experiments, the rapid-scan signal can be averaged $M$ times, reducing the noise by $\sqrt{M}$. As a result, the noise in the rapid-scan signal would be higher than for CW by only a factor of $\sqrt{2}$.

The net effect of the three factors is that the $S/N$ for rapid scan is better than that for CW by $\sqrt{2DF}$. $S/N$ improvements have been achieved for a variety of samples, including BDPA ($\alpha,\gamma$-bisdiphenylene-$\beta$-phenylallyl) [13], irradiated quartz [45], trityl radicals [9], nitroxide radicals [10], defects in diamonds [11], and hydrogenated amorphous silicon [11]. The longer the relaxation times the greater the $S/N$ advantage of rapid scan. Examples for real samples are discussed in Section 2.12.

## 2.2
### Post-Acquisition Treatment of Rapid-Scan Signals

As shown by Dadok and Sprecher [17] and Gupta and coworkers [18] for rapid-scan NMR, the transient response from linear scans can be deconvolved, and the absorption spectrum recovered. If the time on resonance is long relative to electron-spin relaxation, oscillations are not observed, and there is no need to deconvolve the driving function. However, no information is lost by applying the deconvolution procedure to a slow-scan spectrum, so one does not have to distinguish between overlapping species characterized by different relaxation times during post-acquisition processing.

### 2.2.1
### Deconvolution of Linear Scans

Deconvolution of linear scans is straightforward over the portion of the magnetic field scan that is linear within the uncertainties of experimental measurement [47]. If the system is designed with feedback linearization, then about 95% of the scan is accurately linear [47].

The time-dependent driving function $a'(t)$ is [47]

$$a'(t) = \exp\left(\frac{ib_t t^2}{2}\right) \qquad (2.8)$$

where $i$ is $\sqrt{-1}$, and $b_t$ is the scan rate in frequency units.

After Fourier transformation of the experimental data, the convolution by the driving function becomes simple multiplication. If the rapid-scan spectrum is acquired in the regime where signal increases linearly with the square root of microwave power, the effect of the driving function can be deconvolved by dividing by

$$A(\omega) = \exp\left(\frac{-i\omega^2}{2b_t}\right) \qquad (2.9)$$

The Fourier transform of the resultant is the slow-scan spectrum. Deconvolution gives undistorted absorption spectra provided $B_1$ is in the linear response regime. Examples of deconvolution of triangular rapid scans for LiPc are shown in Figure 2.7 [47].

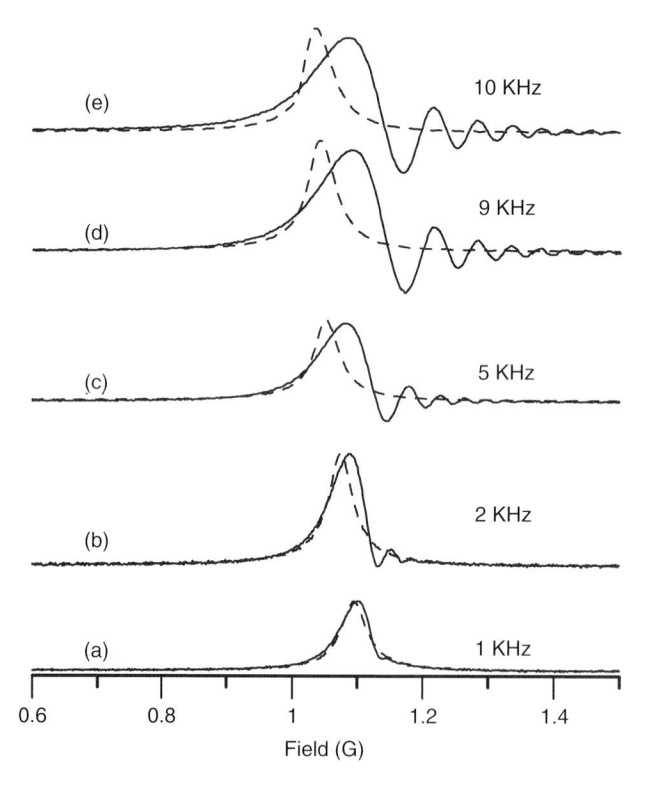

**Figure 2.7** Triangular rapid-scan signals (solid lines) for a LiPc sample, obtained at RF = 248 MHz and $B_1$ = 3.6 mG at various scan frequencies and the signals obtained by deconvolving these signals (dashed lines). The signals are obtained at various scan frequencies and a scan width of 2.16 G. The central segment of each scan is plotted. The corresponding scan rates (a–e) are 4.32, 8.64, 21.6, 38.8, and 43.2 kG s$^{-1}$. (Source: Joshi et al., 2005 [47]. Reproduced with permission of Elsevier Limited.)

## 2.2.2
### Deconvolution of Sinusoidal Scans

As discussed in Section 2.5.2, resonating the driver and coil circuit permits sinusoidal scans to be faster than triangular scans for the same applied voltage. Sinusoidal deconvolution requires an approach different than triangular deconvolution because the scan rate changes throughout the scan. Tseitlin et al. [48] described a general approach to Fourier deconvolution of rapid scans and demonstrated its use to recover the slow-scan lineshape from sinusoidal rapid scans. Since an analytical expression for the Fourier transform of the driving function for a sinusoidal scan was not readily apparent, a numerical method was developed to do the deconvolution. The signals from the up-field and down-field half-cycles were deconvolved separately and the resulting spectra were combined. The slow-scan EPR absorption lineshapes recovered from sinusoidal scans were in excellent agreement with slow-scan spectra for a wide variety of samples. This method was used to deconvolve all of the sinusoidal scans shown in figures in this chapter. An example for tempone-$d_{16}$ (2,2,6,6-tetramethyl-4-piperidone-1-oxyl-$d_{16}$) at 250 MHz is shown in Figure 2.8. The experimental data for the down-field half-cycle is shown in Figure 2.8a. The sum of deconvolved up-field and down-field scans is in excellent agreement with the integral of the CW spectrum (Figure 2.8b). The amplitudes

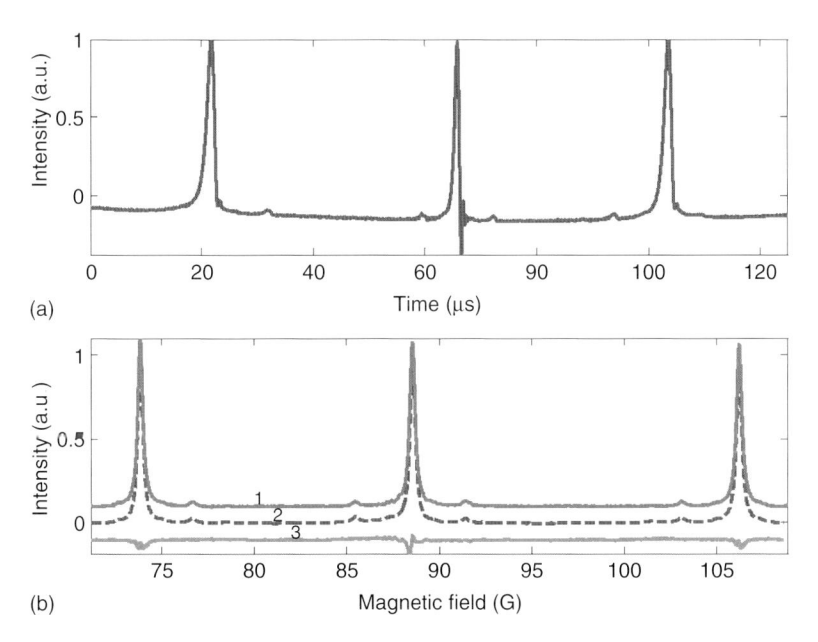

**Figure 2.8** EPR spectra of aqueous tempone-$d_{16}$. (a) Down-field half-cycle of the sinusoidal rapid scan. (b) Comparison of the sum of spectra obtained by deconvolution of up-field and down-field signals (trace 1) with CW spectrum (trace 2), and the difference between traces 1 and 2 (trace 3). (Source: Tseitlin et al., 2011 [63]. Reproduced with permission of Elsevier Limited.)

of the $^{13}$C hyperfine lines and spacings between the hyperfine lines are accurately represented [48].

### 2.2.3
### Low-Pass Filtering

As discussed in Section 2.10.2, rapid-scan spectra are acquired with large resonator and detection system bandwidths to preserve lineshape features. After deconvolution of the rapid-scan response, the signals are low-pass filtered to decrease high-frequency noise. As in CW spectroscopy, there are tradeoffs inherent in the selection of filter parameters. If the filter bandwidth is too small, the S/N is improved, but spectral lineshapes are broadened. The tradeoff between reducing high-frequency noise with little line broadening is particularly good for the fourth-order Butterworth filter, and hence it is the filter of choice in the Denver laboratory [65]. Although pseudomodulation can be used to obtain the derivative of the absorption signals [68], numerical differentiation, followed by a Butterworth filter, produces a less noisy derivative [65]. The bandwidth of the first-derivative spectrum is larger than that for the absorption spectrum of the same signal [65]. To calculate the first derivative, the deconvolved rapid-scan spectrum before low-pass filtering is used, with subsequent application of low-pass filtering.

## 2.3
## Simulation of Rapid-Scan Spectra

The Bloch equations, with numerical integration, are used to simulate rapid-scan spectra by adding a term for the scanning magnetic field [54] (Eq. (2.10)).

$$\frac{dM_u}{dt} = \frac{-M_u}{T_2} - (\Delta\omega + \Omega_m \cos(\omega_m t))M_v$$

$$\frac{dM_v}{dt} = (\Delta\omega + \Omega_m \cos(\omega_m t))M_u - \frac{M_v}{T_2} - \gamma B_1 M_z$$

$$\frac{dM_z}{dt} = \frac{M_0}{T_1} + \gamma B_1 M_v - \frac{M_z}{T_1} \quad (2.10)$$

The Bloch equation notation is conventional, with the addition that $\Omega_m$ is the amplitude of the field scan in angular units, and equals $0.5\gamma B_m$, where $B_m$ is the peak-to-peak scan amplitude in Gauss. $\gamma = -1.7608 \times 10^7$ rad s$^{-1}$ G$^{-1}$. $\Delta\omega = \omega_0 - \omega$ is the offset of a relaxation-determined spin packet from the center of the scan in angular units. $f_m$ is the scan frequency in hertz. $\omega_m = 2\pi f_m$ is the angular scan frequency. $B_1$ is the RF or microwave magnetic field in Gauss (peak-to-peak) [54].

If the EPR signal is a homogeneously broadened Lorentzian line, as in Fremy's salt and some organic radicals such as semiquinones, spectral simulation can be performed for a single spin packet. Some magnetically concentrated species of general interest, such as the solids LiPc, DPPH, and BDPA, also have approximately

Lorentzian lines if the spin concentration is high enough to have strong exchange interactions. More commonly, there are unresolved or partially resolved hyperfine couplings. If these couplings are known, they can be included explicitly in the computations. Some unresolved and unknown couplings have to be treated as a distribution. The description of the distribution for the inhomogeneous broadening can be determined from the slow-scan spectrum and by iteratively estimating $T_2$ from rapid-scan spectra at a series of scan rates. With sufficient knowledge of the hyperfine couplings, the rapid-scan response can be simulated well enough to measure the $T_2$ relaxation time [14, 54].

## 2.4
## Scan Coils

The magnetic field scan coils have to provide the rapidly swept magnetic field with a specified uniformity over the volume of the sample to be measured and are designed for the size and shape of the sample to be studied. It is convenient to use a simple Helmholtz arrangement of two coils, with the diameter chosen on the basis of two criteria: The coil spacing may be determined (i) by the size of the resonator and (ii) by the size of the homogeneous region desired for the sample. If the sample is very small, small scan coils can be used. Even the modulation coils built into standard commercial resonators, or ENDOR coils, can be used for some rapid-scan experiments. For *in vivo* imaging, larger resonators and thus larger scan coils are required. As discussed in Section 2.5 concerning scan drivers, the properties of the driver must match the properties of the coils. A four-coil assembly can produce a larger homogeneous volume for the scanning field than can a two-coil assembly, as in the case for an air-core magnet [8, 69].

### 2.4.1
### Literature Background on Scan Coils

The initial paper by Beeler *et al.* [20] used 1.1 G sinusoidal field modulation at 2 kHz. Rengan *et al.* [70] fastened scan coils on the quartz variable temperature Dewar inside a rectangular $TE_{102}$ cavity. A circuit was designed (Figure 10 in [70]) to linearly scan the current from 0 to 400 mA within about 6 μs, resulting in a scan rate of about $10^6$ G/s. Hirasawa *et al.* [51] used 16.5 cm diameter coils consisting of 50 turns of 1 mm copper wire, with a low impedance (3–20 Ω) source to create 0–100 G linear scans in 1–50 ms. Hsi *et al.* [71] used 5 cm radius Helmholtz coils constructed with 200 turns of 26 AWG copper wire. A trapezoidal driving voltage was used to achieve a linear current ramp. The Hall probe of the Varian 30 cm electromagnet sensed the rapid-scan field, causing nonlinearities in the sweep. Sinusoidal magnetic field sweep was implemented using modulation coils by Czoch *et al.* [30]. An example of using standard modulation coils with an external driver included 100 G sweeps in 200 ms [72]. A locally built voltage-controlled current-source amplifier provided 50 G triangular scans at 2.6 kHz [7, 43].

## 2.4.2
### Bruker Modulation Coils

Some rapid-scan experiments can be performed with the standard modulation coils of a Bruker spectrometer. The parameters for the standard modulation coils for 40 $G_{pp}$ at 100 kHz in the Bruker ER4118-MD5 dielectric resonator are as follows: coil constant ~35 G/A, $R \approx 8\,\Omega$ (includes both DC and AC components at typical scan frequencies), $I_{pp} = 40G_{pp}/35G/A = 1.14\ A_{pp}$, $I_{rms} = 0.40\ A$, Power $\approx 1.3$ W. There are the following limitations to the use of standard Bruker modulation coils and drivers for rapid-scan EPR. (i) *Limitation due to small coil size*: Standard modulation coils are about 25 mm in diameter. This limits the homogeneous-field region produced by the coils. Considerable distortion in rapid-scan spectra taken with the standard modulation coils has been observed for extended samples (see Section 2.12.1.1). When using the standard modulation coils, lineshapes are more accurate for samples that are small relative to the dimensions of the coils. (ii) *Limitation due to maximum scan rate*: Taking the maximum modulation field available from the standard system to be 40 $G_{pp}$ at 100 kHz the maximum scan rate is about 12.5 MG/s. For narrow lines and long $T_2$ such as for BPDA, LiPc, or the E' center in irradiated fused quartz this is fully adequate for rapid scan, if the directly detected signal were accessible to the operator. However, for shorter $T_2$ faster scan rates are needed. For samples with $T_2^* \sim T_2$ the observation of oscillations on the trailing edge of the signal is a convenient way to estimate whether a spectrum is in the rapid-scan regime. Typically, oscillations are observed when $(aT_2^*/LW)$ is greater than about 2, where $a$ = scan rate and $LW$ = linewidth of the absorption line. Although $T_2^*$ and linewidth are not independent, this simple expression is a convenient way to use readily available information about the spin system. For example, approximate parameters for recording one line of a nitroxide radical could be 10 Gpp, 40 kHz sinusoidal scan, $T_2 \sim 0.5\ \mu s$, and $LW \sim 0.3$ G. This gives an $aT_2^*/LW$ value approximately equal to 2, so the onset of an oscillatory response should be observed. (iii) *Power limitations in the coils*: The standard modulation coils themselves are robust enough to handle 40 $G_{pp}$ continuously, but their proximity to the resonator can sometimes cause heating and resultant RF tuning drift when run continuously. Since the power is proportional to the square of the sweep width, this limitation only occurs at high sweep widths.

## 2.4.3
### Coils for Rapid Scans

In the Denver laboratory, for samples in resonators not exceeding about 25 mm (1 in.) dimensions, coils with average diameter of about 3 in. (76 mm), spaced about 1.5 in. (38 mm) apart have been found to be convenient. These coils fit the Bruker FlexLine resonator modules as well as several locally built resonators. The coils have been constructed of AWG 20 solid copper wire and of Litz wire. Approximately 60 turns of wire in each coil yields a practical coil set. These coils produce about

14 G/A. The magnetic field generated with 3 in. diameter Helmholtz coils is linear to better than 1% in both the radial and axial directions out to about 1 cm. Deviations from linearity increase at larger offsets from the center of the coils. At a distance of 1.5 cm the deviation from linearity is about 2.5% in the axial direction and about 1% in the radial direction.

As the scan frequency increases, the AC resistance of the coils increases, and there are distinct advantages in replacing normal solid copper wire with Litz wire. The AC resistance of Litz wire is much less than that of solid wire, for the same number of turns. Litz wire was used by Sloop et al. [73] to achieve very rapid magnetic field changes. It was used in Varian modulation coils, and also in many NMR systems (e.g., Doty Scientific). In the Denver laboratory several sizes of Litz wire, including sizes 255/44 (255 strands of #44 wire), 270/44, and 330/46 have been used.

### 2.4.4
### Magnet Considerations

Air-core magnets are convenient for measurements at 1 GHz and below, and have the advantage that there is space for scan coils. However, most labs have iron-core magnets. Those designed for EPR usually have air gaps between the pole faces of about 2.5 in. (about 62 mm); an air gap of about 4 in. (100 mm) is large for an iron-core electromagnet, which limits the space available for scan coils. The currents induced in the poles of the magnet pose a problem if the coils are placed in the gap of an iron-core electromagnet. The magnet poles are usually made of moderately resistive metal alloys. Less loss in the magnet poles can be achieved by placing a highly conductive metal (e.g., aluminum or copper) plate against the pole face.

Table 2.1 presents comparisons of performance of scan coils constructed with AWG20 solid copper wire and coils of the same size (3 in. average diameter and 1.5 in. average spacing) constructed with Litz wire (255 strands of AWG44 wire). In the example described in Table 2.1, a 1/8″ thick aluminum plate was placed between the scan coils and the magnet pole faces. In air, the coils provided a field of about 14 G/A. Of particular note is that the AC coil resistance for the Litz wire coils is less than half that of the solid wire coils even without the aluminum plates, and with the aluminum plates against the pole faces, the AC resistance of the Litz wire coils is about 28% that of the solid wire coils. Although our focus in this section is on the effect of Litz wire and aluminum plates, the tables also provide examples of practical limits in a coil and driver system: the maximum scan rate can be limited by the power dissipation in the driver amplifier, by the power dissipation and consequent heating acceptable for the coils, or by the maximum design current. For a particular resonated coil driver, the design limits were a sinusoidal sweep of 6 A and an amplifier temperature of 60 °C. These limits are functions both of the coils and the driver as discussed in the following section.

**Table 2.1** Comparison of scan coil performance for sinusoidal scans.

| Coil wire type | AC resistance (Ω) at about 80 kHz[a] | Inductance (mH) | Max sweep width limited by coil driver (Gpp), (limit type)[b] | Power in each amplifier at the driver limit (W) | Power in coils at max sweep width limited by coil driver (W) | Power in magnet pole faces at max sweep width limited by coil driver (W) | Max sweep width limited by 5 W in coils (Gpp) |
|---|---|---|---|---|---|---|---|
| Solid wire in air | 19 | 1.0 | 84 (I) | 44 | 85.7 | | 20 |
| In magnet w/o Al plates | 27.7 | 0.96 | 52 (V) | 30 | 38 | 17 | 19 |
| In magnet with 1/8 in. Al plates | 22 | 0.93 | 66.5 (V) | 38.5 | 66 | 9.8 | 18 |
| Litz in air | 2.9 | 1.0 | 62 (T) | 39.5 | 6.7 | | 53 |
| In magnet w/o Al plates | 12.8 | 0.96 | 65 (T) | 31.2 | 9.0 | 31 | 48 |
| In magnet with 1/8 in. Al plates | 6.25 | 0.93 | 71 (T) | 45 | 11 | 13 | 46.9 |

[a] With the Al plates against the poles of the iron-core magnet, the AC resistance was linear with scan frequency.
[b] The limit types are (I) maximum design current, (V) maximum power dissipation in the driver amplifier, or (T) maximum heating.

## 2.5 Design of Scan Driver

The fundamental issue in designing a magnetic-field scan driver is that a specified time dependence of the current is desired, but readily available amplifiers are voltage amplifiers, and not current amplifiers. The required voltage waveform is

$$V(t) = L\frac{di}{dt} + Ri(t) \tag{2.11}$$

where $V(t)$ is the time-varying voltage applied across the coils to achieve the desired current waveform, $i(t)$, $L$ is the total inductance of the coils, and $R$ is the total resistance of the coils, including the AC resistance that is frequency dependent. The scan driver generates the voltage waveform that is required to produce a desired current waveform, $i(t)$. The driver design has to meet the specifications for the $L$ and $R$ of the coil and the coil constant, which is the magnetic field produced per ampere of current through the coils. Note that $R$ will change with temperature and with the frequency of the varying current. A high-gain feedback loop is needed to produce the required drive voltage defined by Eq. (2.11).

### 2.5.1 Linear Scan Drivers

The linear (triangular) scan driver described by Quine et al. [8] uses a high-gain feedback system. It is important to emphasize that the driver does not produce a linear voltage ramp – it produces a voltage output (Eq. (2.11)) that creates a linear current ramp in the scan coils. There is also a secondary effect (sometimes called *compensation*) that relates to the tuning of the feedback loop. The tuning network is adjusted for a particular set of coils. If the coils are replaced, the network is retuned to match the new inductance. The tuning is necessary to compensate for the amplifier response to the time response of the coils. The maximum magnitude of the scan field was increased by using two amplifiers in a push–pull configuration.

For linear magnetic field scans, the system has to support at least the seventh harmonic of the nominal drive frequency. That is, a 10 kHz triangular scan requires a driver that can achieve at least 70 kHz bandwidth. Even with seven harmonics, only about 95% of the scan will be linear enough to use. In addition, the voltage slew rate has to be as high as possible to support the instantaneous voltage sign reversal that occurs in Eq. (2.11) in the $di/dt$ term. An alternate to a triangular scan is a sawtooth. The flyback of the sawtooth could be very nonlinear if it exceeds roughly the factor of seven criterion given above. There may also be some field stabilization time required at the beginning of the next scan. However, even ignoring the affected time regions, using sawtooth current waveforms, one can approach the time efficiency of triangular scans.

Linear drivers have been built in several versions. The linear driver described in detail in [8] produced a waveform whose frequency and amplitude were each controlled in a computer with 12 bit resolution. A block diagram is shown in

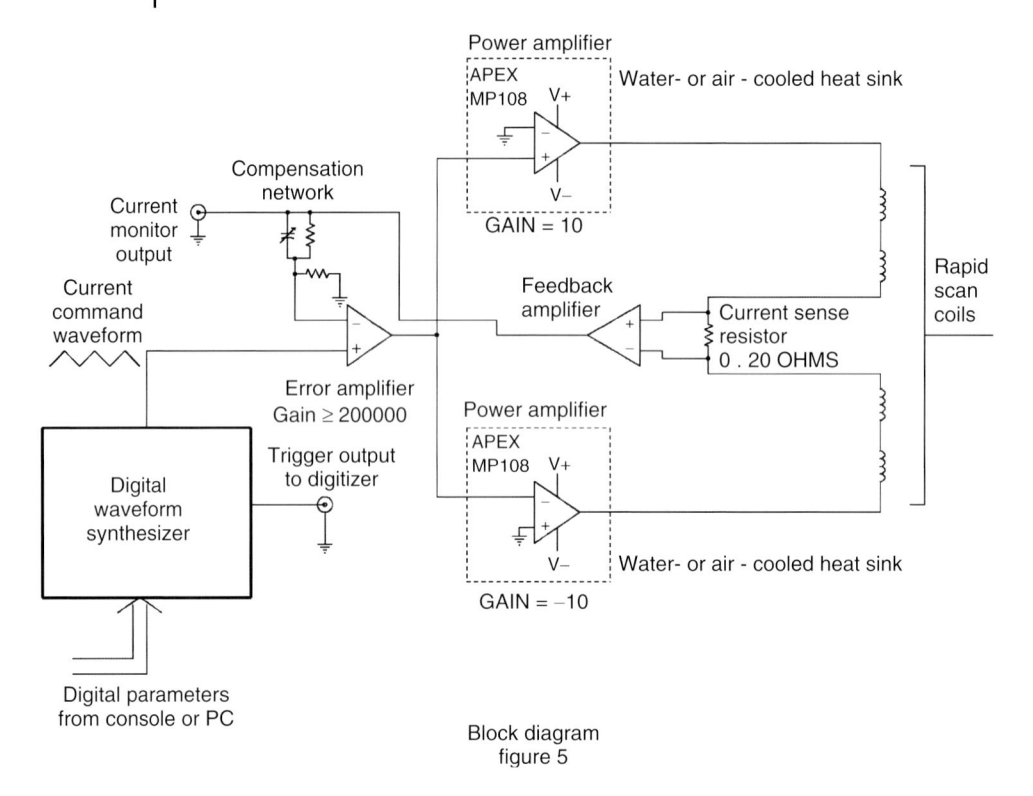

**Figure 2.9** Block diagram of the linear scan coil driver using two amplifiers in a push–pull configuration.

Figure 2.9. The digital waveform synthesizer voltage output is compared with the output of a current monitor, a $0.2\,\Omega$ resistor that produces 0.2 V/A. The difference between the synthesized voltage waveform and the current waveform (the error voltage) is shown in Figure 2.10 for the particular case of an 8 kHz, 20 $G_{pp}$ sweep width using scan coils with a coil constant of 38.6 G/A and having inductance of 1.2 mH and resistance of $1.3\,\Omega$. This error waveform, after amplification by $4 \times 10^6$, is the voltage drive to the coils. Frequencies ranged from 500 to 20 000 Hz, and amplitudes were up to 80 $G_{pp}$. For the highest scan rates the power amplifier is water cooled. The best power driver amplifier module that is currently available for the frequency range desired is the PAD-35 or the PAD-108 manufactured by Power Amp Design, Inc Tucson, AZ. Two of these are used in a push–pull arrangement (the amplifiers run 180° out of phase to each other) to achieve twice the output voltage obtainable with one amplifier [8]. The linearity of the scan is indicated by the excellent agreement in lineshapes and line positions for the three lines of a nitroxide signal at 250 MHz [8].

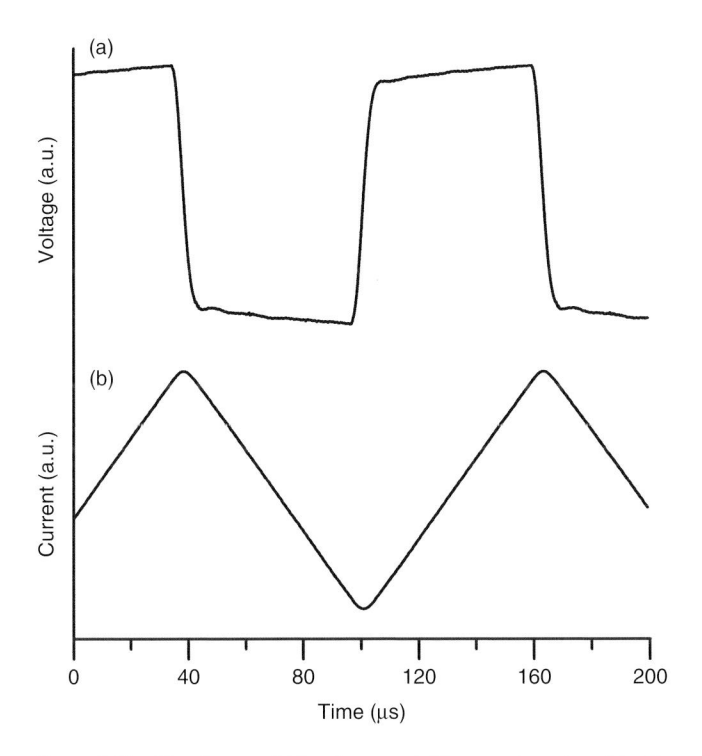

**Figure 2.10** Waveforms of (a) voltage and (b) current produced by the coil driver at 8 kHz and 20 Gpp.

## 2.5.2
### Sinusoidal Scan Drivers

The advantages of sinusoidal magnetic field scans are that (i) they can achieve the highest scan rate for a given driver amplifier power; (ii) they are accurately a single frequency, in contrast to linear scans, which as described above have to consist of a Fourier series of many frequencies to operate reasonably linearly; (iii) much higher scan rates are feasible with resonated systems than with linear driven systems; (iv) they dissipate very little power in the scan coils, other than owing to AC resistance, which increases with scan frequency; and (v) as shown in Section 2.8.2, it is easier to remove the background for sinusoidal scans than for linear scans [49]. The disadvantages of sinusoidal scans are that (i) it is difficult to change scan frequencies quickly because resonating capacitors have to be changed; (ii) the scan rate is changing continually through the scan, so interpretation of the spin response is more complicated; and (iii) the scan rate cannot be varied as easily as in a linear system because the circuit has to be resonated at each scan frequency. Scan rates can be varied by changing the sweep width (Eq. (2.2)), but changing the scan frequency requires changing the resonating capacitor.

The frequency of a resonated coil, $\omega_{coil}$, is expressed as

$$\omega_{coil} = \frac{1}{\sqrt{LC}} \tag{2.12}$$

The reactive component of the coil is negligible at resonance. If the frequency is high enough, the AC resistance becomes important. The resistance does not change the resonant frequency, but it changes the $Q$ of the circuit, and thus the amplitude of the maximum scan field that can be achieved by the circuit for a given voltage output from the coil driver. Hence, decreasing coil resistance without decreasing the coil constant is a design goal. Section 2.4.3 describes ways to decrease the AC resistance by using Litz wire coils.

How fast can one scan the magnetic field? Scan coils can be resonated at selected frequencies to be able to scan the magnetic field faster than one can with a linear (triangle) scan. There are several physical limits to the scan rate. Specific examples for a particular set of coils are shown in Table 2.1. A scan driver and its power supply have to be designed on the basis of the inductance and resistance of the coils, or, conversely, the coils have to be designed to match available power amplifiers. Scan rate limitations include available voltage and current (6 $A_{pp}$ maximum for the present driver) to drive the coils, heat dissipated in the scan coils, and heat dissipated in the driver amplifier. A temperature limit of 60 °C was established as the amplifier heat sink limit. For small samples, very rapid magnetic field scans are possible, using ENDOR coils as described in Section 2.6.

Resonated coil drivers have been designed to operate with air-cooled amplifiers. For applications where very high scan rates are needed and operation with solid wire coils is required, a system was developed to run the amplifiers at reduced duty cycle to keep their heat sink temperatures within the required limits. This system provides rapid scan at high scan rates but in discrete bursts. The number of rapid-scan cycles per burst and the number of bursts per second are adjustable. In this way, the average power in the amplifiers can be reduced to the heat transfer levels available from air-cooled heat sinks. When solid wire scan coils were replaced with Litz wire coils, duty-cycled operation was not required. This greatly improves the efficiency of data collection because nearly 100% of the available time can be utilized. Figure 2.11 presents the simplified block diagram of a scan coil driver designed to provide a resonated, thus sinusoidal, magnetic field scan.

### 2.5.3
**Integration into a Spectrometer System**

Integration of a rapid-scan system into an EPR spectrometer system involves considerable engineering to minimize interactions between subtle contributors to the signal and the background. In a CW EPR spectrometer, eddy currents caused by magnetic field modulation result in limitations on modulation amplitude, increases in noise at various modulation frequencies, and overall baseline drift. In addition, the magnetic field modulation frequency has to avoid the AFC modulation frequency. Lock-in detection at the magnetic field modulation frequency partially

## 2.5 Design of Scan Driver | 29

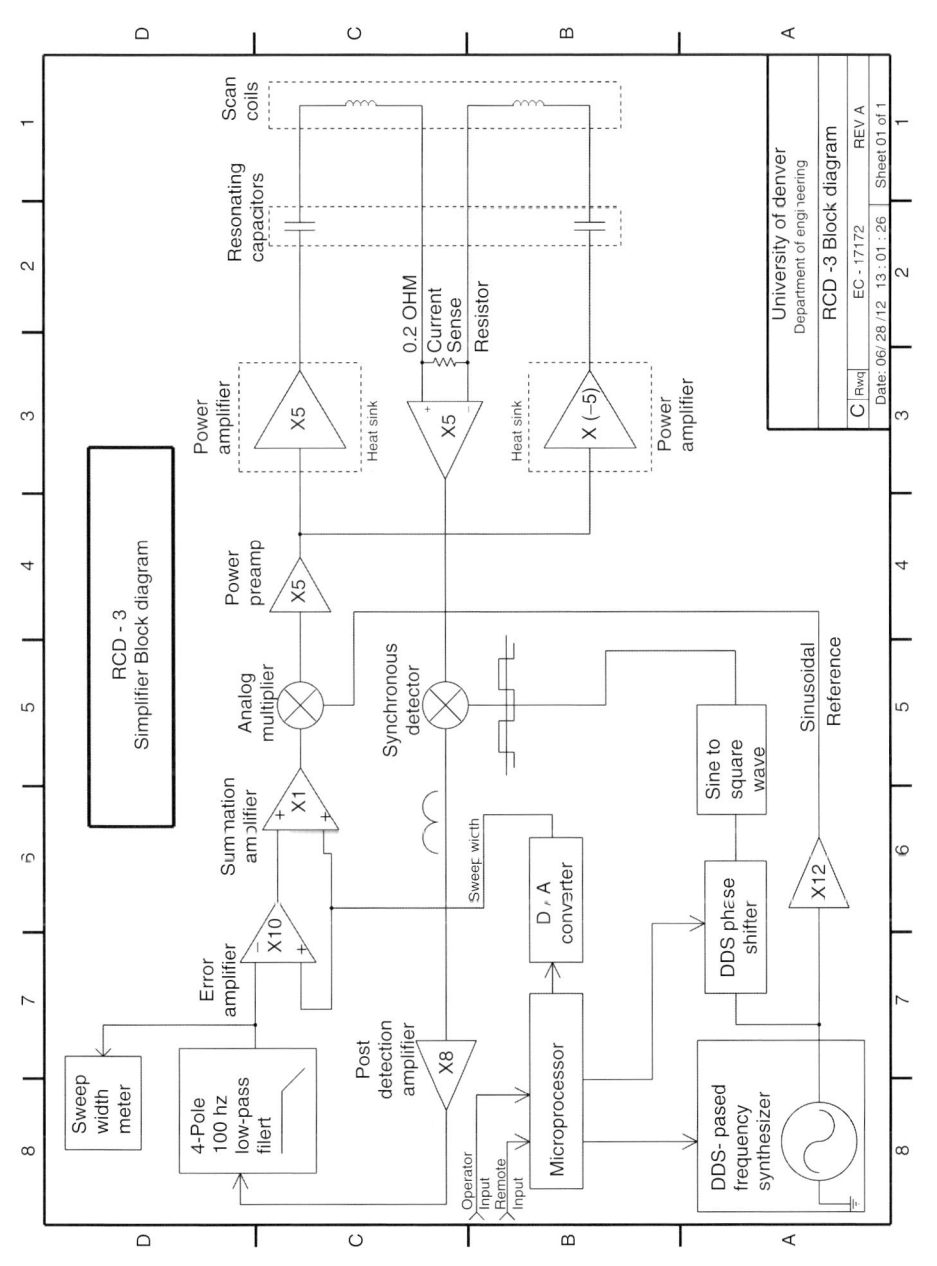

**Figure 2.11** Simplified block diagram of a resonated coil driver [9]. The capacitors are selected to resonate the circuit.

compensates for these problems and provides some baseline stability. In rapid-scan EPR, as described in this chapter, the eddy currents produced by the scan field are larger because the scan amplitudes are larger. Section 2.7 discusses resonator design to minimize eddy current problems. Mechanical aspects of the resonator and scan coil assembly have to be arranged to minimize metal in the scanning field, and to be as rigidly supported as tuning mechanisms permit. Positioning of coaxial cables can be critical. Usually, there are one or more mechanical vibrational frequencies that are excited by the eddy currents. Operation at these frequencies should be avoided. The more rigid the structure, the higher the mechanical resonances. Vibration damping material in unexpected places may reduce the background also.

Shielding the RF of the resonator adequately from the scan coils is a major challenge. The interaction between the RF and the coils introduces a background signal in the data at the rapid-scan frequency (and its harmonics for a linear scan). Three methods have been used so far: enclosing the coils in a shield, enclosing the resonator in a shield, and putting an RF trap (choke) between the scan driver and the coils. This is particularly important at the low frequencies used for *in vivo* imaging because the scan frequencies are closer to the Zeeman frequency than, for example, at the X-band.

The primary remedy to RF energy absorption in the scan coils is to insert resonant choke coils in series with the drive cables to the scan coils. The chokes allow the low frequency scan current to flow unimpeded while the RF current is blocked by the high impedance that the choke coils have at the RF. The resonant chokes are parallel resonant $L-C$ circuits tuned to the operating RF. A parallel resonant circuit has high impedance at the resonant frequency. This type of choke filter has been effective when the resonator and scan coils are placed inside a larger shielding box, with the filter effectively at the penetration point of the scan coil wires. The coils are driven with a four-wire drive system so that there are four chokes, one on each wire. Shielding the scan coils is about as effective as using the resonant choke coils. The shield can be local – for example, conductive tape wound around the coils – or between the coils and the resonator.

Grounding is also very important. Normal cables may not provide adequate shielding between the drive and RF voltages; so a well-grounded shield covering the cables from the scan driver to the scan coils reduces background signals, as do additional ground straps between the resonator, first-stage amplifier, and bridge. An inner/outer DC block strategically placed may contribute to suppressing grounding problems. The weaker the signals of interest, the more critical these steps become.

## 2.6
### Use of ENDOR-Type Coils and RF Amplifiers for Very Fast Scans

ENDOR is a well-developed method of measuring nuclear spin couplings to electron spins, with commercially available resonators, amplifiers, and so on.

It uses simultaneous excitation of nuclear spins and electron spins. Since their resonant frequencies at the same magnetic field differ by a factor of roughly 658, wire coils for the nuclear resonant frequency are usually placed within a cavity resonator or wrapped around a dielectric resonator, as in the Bruker FlexLine series of resonators. The nuclear RF magnetic field vector is commonly called $B_2$ by EPR spectroscopists. The usual experimental arrangement is to position the EPR resonator with the microwave $B_1$ perpendicular to the external field, $B_0$, and $B_2$ perpendicular to both $B_1$ and $B_0$. Since the $B_2$ vector is a magnetic field, rotating the ENDOR resonator assembly, such that $B_2$ becomes parallel with $B_0$, provides a sinusoidal sweep of the magnetic field at the frequencies (tens of megahertz) used for ENDOR. ENDOR coils are small, so the magnetic field generated using the ENDOR coils as scan coils is homogeneous only over a relatively small volume.

The feasibility of performing rapid-scan EPR with this arrangement was demonstrated using BDPA crystals [13]. In this study, CW linewidths, direct measurements of $T_2$ by pulse techniques, and rapid scan were used together to demonstrate the ability to simulate the rapid-scan response (Figure 2.12) with a value of $T_2$ that agreed with the pulse and CW results. The ENDOR coils in a Bruker ER4118X-MD4 pulse ENDOR resonator were driven with a sine wave created by a Tektronix AWG2021 arbitrary waveform generator and a locally built amplifier. The AWG

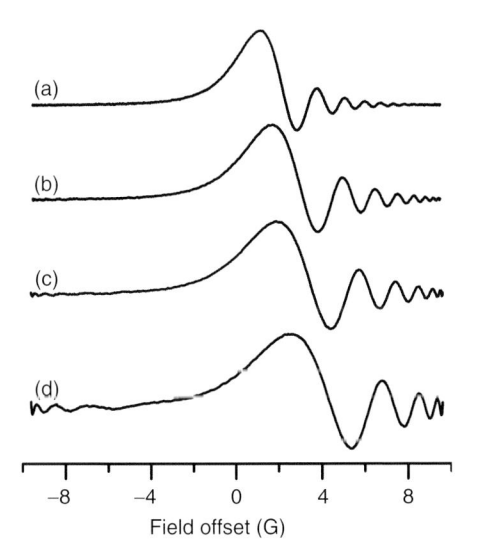

**Figure 2.12** Sinusoidal rapid-scan spectra of a BDPA particle with $T_2 = 88 \pm 3$ ns, obtained with constant 19 G scan width, and different scan frequencies generated with a Bruker ENDOR accessory. 10240 averages were collected with resonator $Q \sim 200$ and 0.2 mW power. (a) 300 kHz scan frequency (18 MG s$^{-1}$), recorded in ~30 s. (b) 500 kHz scan frequency (30 MG s$^{-1}$), recorded in ~20 s. (c) 700 kHz scan frequency (42 MG s$^{-1}$), recorded in ~15 s. (d) 1 MHz scan frequency (60 MG s$^{-1}$), recorded in ~10 s. (Source: Mitchell et al., 2011 [13]. Reproduced with permission of American Chemical Society.)

was operated in pulse mode with 1% duty cycle to avoid overheating the ENDOR coils. Scan frequencies ranged from 300 kHz to 1.5 MHz, and scan widths were from 17 to 60 G, corresponding to scan rates at the center of the sinusoidal scan of 16–280 MG/s. The ENDOR coils can be used to generate sweep widths up to 70 G peak-to-peak at scan frequencies up to 5 MHz, which corresponds to scan rates in excess of 1 GG/s. However, these rates are higher than those needed to characterize $T_2$ of BDPA and would have required a significant decrease in resonator $Q$ to record spectra with sufficient signal bandwidth (see Section 2.10.1). The parameters for the modulation coils in the ENDOR resonator are coil constant $\approx 20$ G/A, R $\approx 3\,\Omega$ (includes both DC and AC components), $I_{pp} = 40\,G_{pp}/20\,G/A = 2\,A_{pp}$, $I_{rms} = 0.7$ A, power $\approx 1.5$ W.

Schweiger and coworkers [74, 75] similarly rotated an ENDOR resonator to obtain absorption mode EPR spectra of broad lines. They used amplitude modulation of the RF field longitudinal to the $B_0$ field, with a constant transverse $B_1$ field. Phase-sensitive detection at the amplitude modulation frequency was performed. The technique requires partial saturation of the spin system, and consequently reduces intensity of forbidden transitions. Broad lines are emphasized by the amplitude modulation method. See, for example, Figure 7 in [74]. Passage effects were demonstrated to be important [75].

## 2.7
**Resonator Design**

Rapid-scan EPR can be performed to some extent using any of the standard EPR resonators (see, e.g., [46]), but optimizing a resonator to maximize the signal relative to background requires several tradeoffs, as outlined in this section.

Because the rapidly changing magnetic field induces eddy currents in metal parts and the eddy currents distort the scanned field, reducing the amount of metal between the sample and the scan coils reduces distortions of the scanned field. Eddy currents are the reason that modulation coils in CW EPR resonators are small relative to the main metal support of the resonator, for example, the brass box that is the X-band $TE_{102}$ cavity resonator, and the reason that there are very thin conductors forming the cavity walls between the modulation coil and the sample. In many types of resonators, the walls are helical coils of wire, which also pass the modulation field with minimal attenuation. Even with these approaches, eddy currents limit the modulation amplitude that is feasible to use with many commercial EPR resonators. For example, in a commercial spectrometer one might see a suggested limit of 4 G modulation at 100 kHz, above which excessive noise and resonator heating may be observed. Field modulation and phase-sensitive detection result in baseline flattening, but rapid scan with direct detection of the EPR signal inherently has a background signal coherent with the magnetic field scan. For a sinusoidal scan, the background usually will be accurately sinusoidal and can be removed mathematically [49]. A linear scan can be viewed as a superposition of multiple harmonics of the scan frequency. For example, the frequency response of the driving circuit has to be at least seven times the fundamental frequency to approximate a triangular scan. Consequently, there are multiple harmonics in the background response also.

Every resonator assembly has some mechanical vibration frequencies that can be within the bandwidth of the EPR experiment. Minimizing and avoiding "microphonics" (mechanical resonances) is a major part of the effort of refining an EPR resonator, whether for CW or for rapid-scan EPR. For example, in one wire-wound cross loop resonator (CLR) built for 250 MHz *in vivo* imaging, there was a mechanical resonance at 1.18 kHz, but a clean background at 1.4 kHz. Comparison of mechanical resonances for two CLRs is shown in Table 2.2 [76]. These mechanical resonances may impact resonator performance over a narrow frequency range. After characterizing the resonator, one simply avoids operating at frequencies close to known mechanical resonances. Open-style resonators can also minimize the metal near the sample [76].

A CLR has the advantage that source noise at the detector is decreased by the isolation between the excitation and detection resonators. Reflected power also does not get to the detector. The resonator $Q$ has to be designed for the information desired from the rapid-scan measurements. The higher the $Q$ of the mechanical resonance, the narrower the frequency range of interfering background signals. The following paragraphs outline some of the rapid-scan resonators used in the Denver laboratory, with examples taken from X-band and VHF (250 MHz) EPR spectrometers.

## 2.7.1
### X-Band

Bruker split-ring and dielectric FlexLine resonators and a rectangular cavity resonator were used to illustrate the impact of the resonator on X-band rapid-scan spectra [46]. Sinusoidal scans were generated with the modulation driver in the Bruker console. The small bandwidth of higher $Q$ resonators attenuates the rapid-scan response at fast scan rates. The closer proximity of metal to the sample in split-ring and dielectric resonators than in the cavity resonator contributes to more inhomogeneous broadening in the former than in the cavity resonator, as shown in Figure 2.13. Eddy currents are not created in dielectric resonators, so

**Table 2.2** Cross loop resonator performance – two examples.

| CLR resonator | Normalized S/N obtained under same conditions with no significant background | Mechanical resonance frequencies (kHz) | Normalized background amplitudes at worst mechanical resonance | Normalized ratio of standard signal to worst background |
|---|---|---|---|---|
| CLR-1 | 1 | 1.04, 1.18, 2.25 | 1.62 | 1 |
| CLR-2 | 1.67 | 1.07 | 1 | 2.73 |

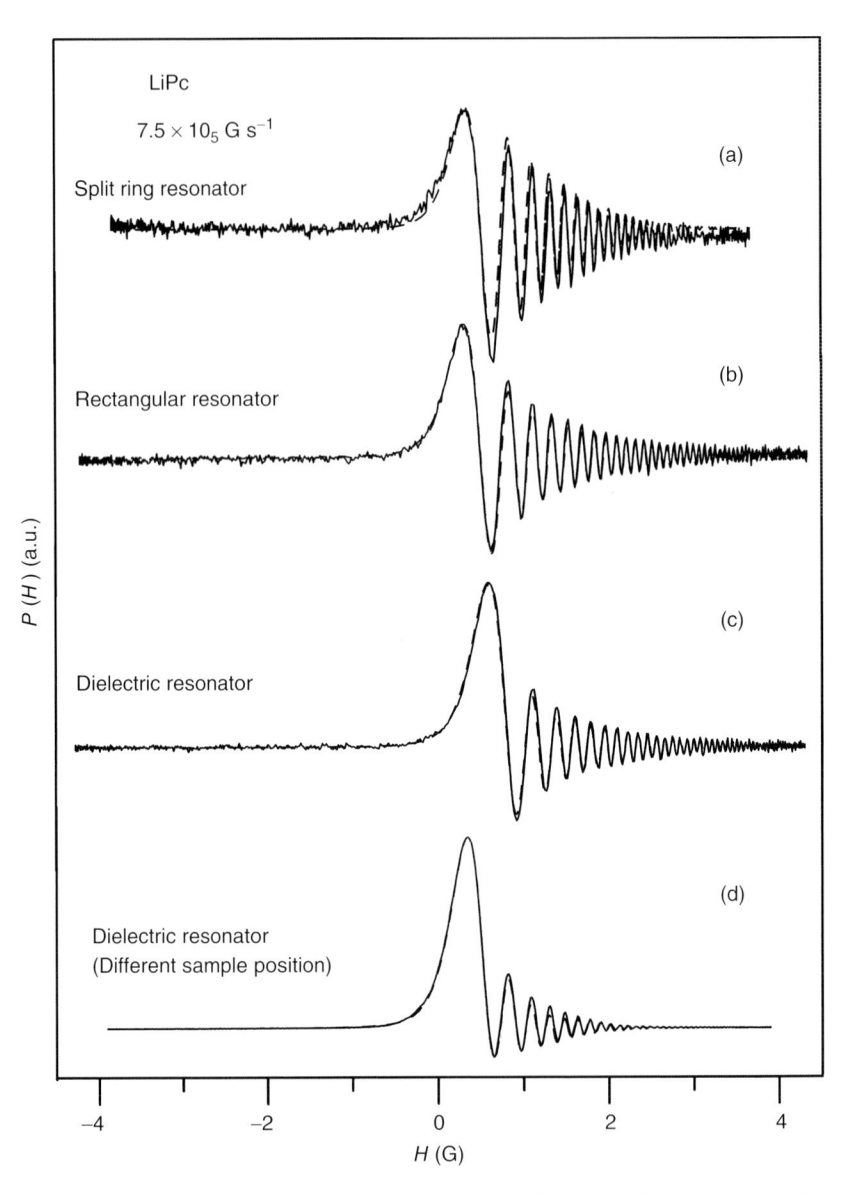

**Figure 2.13** Sinusoidal rapid-scan spectra from an LiPc crystal recorded at 9.8 GHz, at a scan frequency of 30 kHz and scan width of 8 G (scan rate $\sim 7.5 \times 10^5$ G s$^{-1}$). The x-axis is the offset from the resonance field in gauss. The solid lines show the experimental spectrum obtained in (a) split-ring resonator, (b) rectangular resonator, (c) the dielectric resonator, and (d) the dielectric resonator with a different sample position in the resonator. The spectrum in (d) shows the effects of higher $Q$ and larger inhomogeneous broadening. The dashed lines show the simulations.

dielectric resonators, including those in the Bruker FlexLine series, can be used effectively for rapid-scan EPR. However, the position of transmission line that is used to couple microwaves to the resonator may result in eddy currents that affect the spectrum [46].

Following the studies described in [46], X-band rapid scan has been performed primarily with a Bruker ER4118X-MD5 dielectric resonator [10]. This resonator has a $Q$ of about 11 000, which would restrict scan rates for narrow-line spectra to very slow scans. In practice, the $Q$ was lowered to about 300 by putting tubes of water or other lossy solvents into the resonator along with the sample, if the sample was not lossy.

Because of the impact of $Q$ on the rapid-scan signal (Section 2.10.1), it is useful to have a method of measuring resonator $Q$. A circuit for measuring $Q$ on a CW or rapid scan spectrometer, stimulated by the methods used routinely on pulsed EPR spectrometers, is described in [77].

## 2.7.2
## VHF (250 MHz)

Since the most highly developed rapid-scan resonators have been built for VHF studies, they will be described in detail here. As discussed in Section 2.4.2, for small samples, standard EPR resonators and their modulation coils can be adapted for rapid-scan EPR. However, for larger samples, such as mice at VHF or L-band, it is necessary to have resonators with much better access to the sample. In addition, as discussed in Section 2.10, the sample linewidth, resonator $Q$, and scan rate are interrelated. For many applications, to avoid line broadening a lower resonator $Q$ is used for rapid-scan EPR than for CW, similarly to the low $Q$ for pulsed EPR. When using CLRs, the resonator can be designed to be inherently low $Q$ or $Q$ can be lowered by overcoupling, as with pulsed EPR. Fortunately, if the design criteria are both an open design and a low $Q$, both criteria can be achieved by using fine wire to construct the resonators. The smaller the diameter of the wire, the higher the resistance and the lower the resonator $Q$ ($Q = \omega L/R$). Eddy currents pose another problem, as mentioned in Section 2.7. Eddy currents in a conductor close to the sample can distort the magnetic field at the sample. Smaller conductors have less eddy currents, so making a resonator of fine wire also reduces eddy current problems.

The wire CLR resonator sketched in Figure 2.14 consists of two sets of coils. The sample resonator is 16 mm in diameter and 15 mm long. The excitation resonator consists of two coils 32 mm in diameter and 20 mm apart. Capacitors are needed in the coils to keep wire segments less than about 1/4 wavelength in circumference. Each loop of the excitation resonator consists of two series turns of AWG 38 wire (0.1 mm, 0.004 in., diameter). The sample resonator consists of six turns of this fine wire with a chip capacitor between each 1 1/2 turns and also at the two ends of the winding for a total of five capacitors. The characteristic parameters for the resonator are summarized in Table 2.3. The fine wire that was used to construct

## 2 Rapid-Scan Electron Paramagnetic Resonance

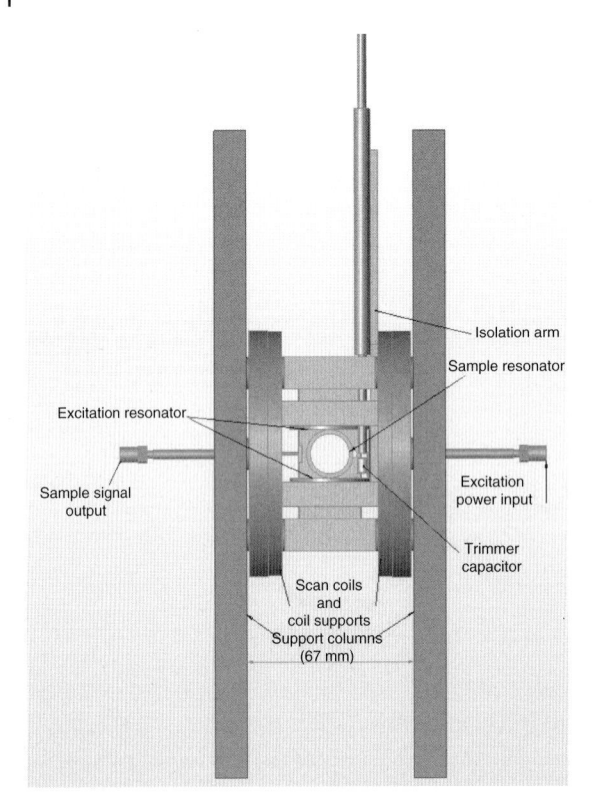

**Figure 2.14** Sketch of VHF CLR, showing the arrangement of the two resonators and scan coils.

**Table 2.3** Measured parameters for the 250 MHz wire CLRs and scan coils [78].

| CLR resonator | $Q_{source}$ | $Q_{detector}$ | Best isolation (dB) | Scan coil inductance (mH) | Scan coil DC resistance ($\Omega$) | Scan coil field constant (G A$^{-1}$) |
|---|---|---|---|---|---|---|
| CLR-1 | 35 | 30 | 47 | 1.2 | 1.8 | 10.1 |
| CLR-2 | 66 | 54 | 44 | 1.2 | 1.8 | 10.1 |

this resonator resulted in $Q < 100$ (Table 2.3), which is much lower than the ~1000 for the 25 mm sample diameter solid copper 250 MHz CLR reported in [78]. Source and detector refer here to the two resonators of the CLR, which sometimes also called the *excitation* and *signal resonators*. The scan coils, whose $L$ and $R$ are listed in Table 2.3, are 89 mm diameter coils consisting of 50 turns of AWG 20 enameled wire, encapsulated in epoxy. To use very high incident RF/microwave power, the fine wire would have to be replaced by a more robust conductor.

## 2.8
## Background Signals

### 2.8.1
### Cause

Background signals are ubiquitous in CW and pulse EPR, as well as in rapid-scan EPR. Sometimes data collected off resonance can be subtracted from the on-resonance data. This commonly works in pulsed EPR. For CW EPR, the background signal often is due to paramagnetic impurities in the resonator and sample tube. Therefore, it is usually necessary to replace the sample with one that has the same dielectric effect on the resonator and to collect a background data set to subtract from the data of interest. Background subtraction can be omitted only if the sample of interest produces a signal that is strong enough that the impurity signals can be ignored. In addition to the resonator and sample tube signals, rapid-scan spectra usually are superimposed on a scan-related signal because of periodic eddy currents that vibrate the resonator, and to the leakage from the scan voltage into the RF/microwave signal.

Rapidly changing magnetic fields induce eddy currents in conductive materials in the resonator and shield. The resultant forces cause slight changes in the frequency and/or match of the resonator that are coherent with the magnetic field scan. This results in a change in the RF signal that is detected along with the EPR signal, but it is independent of the EPR signal. The dependence on the external magnetic field makes it difficult, or even impossible, to fully subtract the signal by recording an off-resonance signal at a magnetic field that is sufficiently far off resonance. For nonsinusoidal scans, the background can have many component frequencies and harmonics that confound the field dependence. In some cases, the background signal can look similar to an EPR signal, because it is magnetic field dependent. These interactions are major causes of background signals that have the same frequency as the magnetic field scan.

Methods of minimizing the eddy currents and leakage are discussed in Sections 2.4 and 2.7, which describe the scan coils and resonators.

### 2.8.2
### Methods of Removing Background Signals from Rapid-Scan Spectra

#### 2.8.2.1 Linear (Triangular) Scans
A method of correcting for background signals in triangular rapid scans was developed on the basis of data collection at two center fields that are offset by much less than the sweep width [79]. This approach yields two signals with offset EPR lines and very similar background. The background is a superposition of a few harmonics of the scanning frequency with unknown amplitudes and phases. The goal of the background removal algorithm is to calculate these amplitudes on the basis of the experimental background. The background subtraction process is outlined in Figure 2.15. The positions of the EPR lines depend on the center

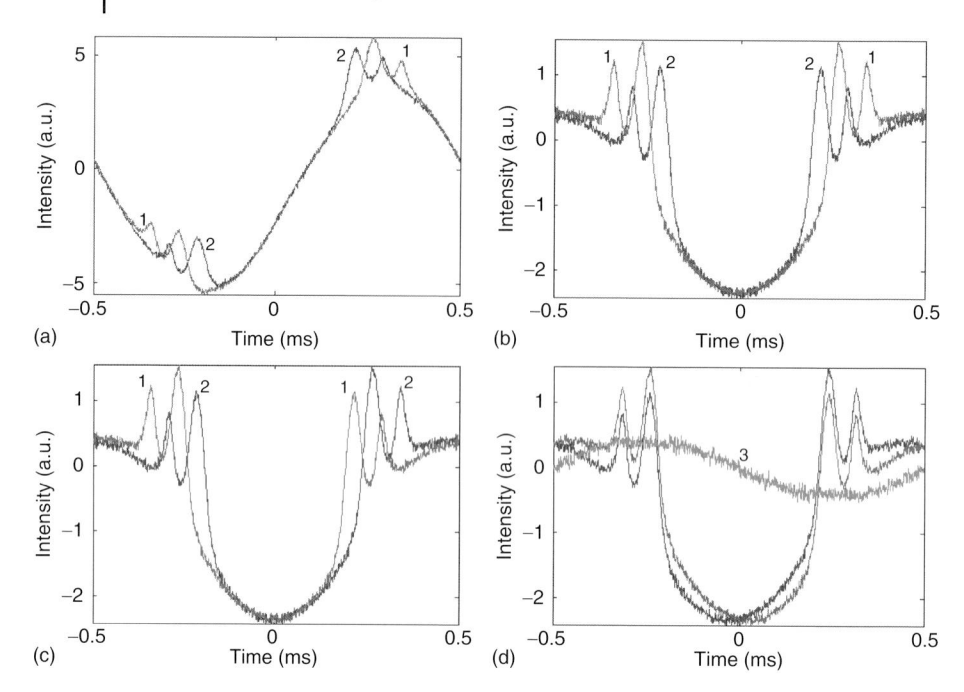

**Figure 2.15** Background removal procedure for triangular scans. The data were obtained for two tubes containing BDPA in the presence of a magnetic field gradient, with scan widths of 20 G. (a) Absorption signals recorded with center fields offset by 2.0 G; (b) symmetric parts of the two signals obtained by combining the up and down half-cycles (Eq. (2.3)); (c) interchange of the second half-cycles for traces 1 and 2; and (d) shift of the signal toward each other by half the center field offset. Trace 3 is the background signal calculated by subtracting the two shifted traces. (Source: Tseitlin et al., 2009 [79]. Reproduced with permission of Elsevier Limited.)

field. Appropriately changing the center field preserves mirror symmetry relative to the midpoint between two half-cycles (Figure 2.15a). After deconvolution the EPR lines for the first half-cycle (up-scan) and the second half-cycle (down-scan) have mirror symmetry relative to the midpoint ($t=0$) in the scan and such that $EPR(t) = EPR(-t)$. By combining signal $(t)$ and signal $(-t)$ one can obtain the symmetrical part, which eliminates the asymmetric component of the baseline and also decreases random noise by a factor of $\sqrt{2}$ (Figure 2.15b). To permit separation of the background from the EPR signal, the data points at times 0 to $T/2$ ($T$ is the period of the rapid-scan cycle) in trace 1 are swapped with the corresponding points in trace 2. The result is shown in Figure 2.15c. This procedure does not affect the background, but changes the positions of EPR lines. The two signals are then cyclically shifted toward each other by $\tau/2$, so that the positions of the EPR lines coincide (Figure 2.15d). Data points removed from one end of the signal are moved to the other end of the cycle, so that the array remains the same length. This procedure is mathematically valid because of the cyclic nature of the signal. The difference between the two traces contains only the background information,

## 2.8 Background Signals | 39

which can be modeled as the sum of sinusoids and then subtracted from the experimental data. This approach solves three problems: (i) As the signal is present in both scans, no time is lost in acquiring background scans that contain no signal. (ii) The time delay between two "offset" spectra is small enough to minimize the effects of time dependence. (iii) The field offset is small enough so that there is negligible impact on the background [79].

As shown in Figure 2.16a, the reconstructed background signal is a better match to the background signal in the experimental data than a background signal collected

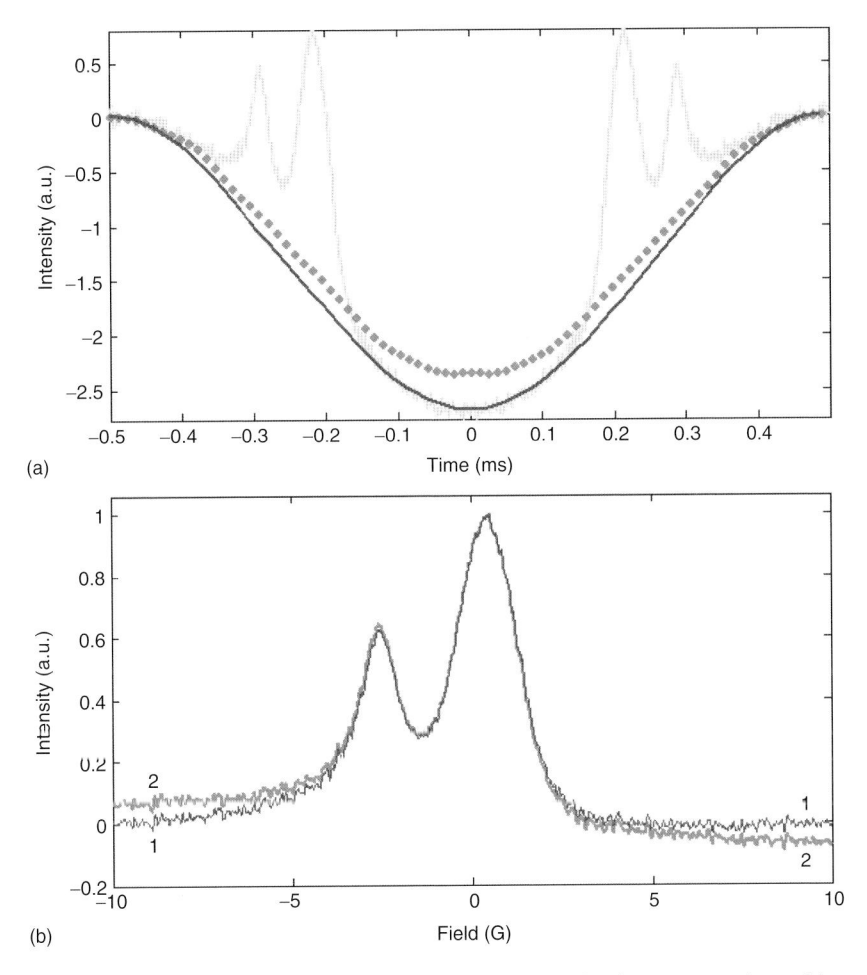

**Figure 2.16** (a) Comparison of the experimental signal with the reconstructed background (solid), and the off-resonance background (dotted). (b) The baseline is flatter for the spectrum obtained by subtraction of the reconstructed background (trace 1) than for the spectrum obtained by subtraction of the off-resonance background (trace 2). (Source: Tseitlin et al., 2009 [79]. Reproduced with permission of Elsevier Limited.)

off resonance. The resulting spectrum after subtraction of the reconstructed background possesses less residual background than the one obtained when an off-resonance background was subtracted (Figure 2.16b).

### 2.8.2.2 Sinusoidal Scans

Sinusoidal scans provide a special opportunity for accurate background subtraction. For data acquired in quadrature, up-field and down-field signals can be separated in the frequency domain. For each scan direction, the background oscillation can be calculated by fitting to the half-cycle that does not contain the EPR signal. The fit function is then extrapolated into the other half-cycle and subtracted from the half-cycle that contains the EPR signal. By zeroing the array for the half-cycles that do not contain the EPR signal, the S/N is improved and the data are corrected for non-orthogonality of the quadrature channels [49].

The procedure is outlined in Figure 2.17 and consists of the following steps: (i) On the basis of the known scan frequency and time per data point, calculate the number of data points for each sinusoidal scan cycle. (ii) Identify a point in the data array that corresponds to the beginning of a sinusoidal cycle. This can be done by deconvolving the data, and checking that the positions of the peaks in the up-field and down-field scans coincide. Steps (i) and (ii) define the first and last elements of the array $s(t)$, as in Figure 2.17a. (iii) Perform a complex Fourier transform to obtain $S(\omega)$, as in Figure 2.17b. (iv) Separate $S(\omega)$ into arrays $S^\uparrow(\omega)$ and $S^\downarrow(\omega)$ that contain the positive and negative frequencies, respectively, as in Figure 2.17c,d. (v) Perform reverse Fourier transforms independently for $S^\uparrow(\omega)$ and $S^\downarrow(\omega)$, to obtain $s^\uparrow(t)$ and $s^\downarrow(t)$, respectively, as in Figure 2.17e,f. (vi) Fit the half-sinusoid in the second half of $s^\uparrow(t)$, extrapolate into the first half-cycle, and subtract from $s^\uparrow(t)$ to obtain the baseline corrected signal for the up-field half-cycle of the scan, as in Figure 2.17g. (vii) Perform the analogous fitting procedure for the first half of $s^\downarrow(t)$, extrapolate, and subtract from the second half to obtain the baseline-corrected signal for the down-field half-cycle of the scan, as in Figure 2.17h. (viii) Perform sinusoidal deconvolution [48] to obtain the slow-scan spectrum. (viii) Combine the up-field and down-field scans to obtain the final spectrum. (ix) To correct the phase, multiply the original data by $e^{j\varphi}$ and repeat the deconvolution and background subtraction process. The steps are implemented with the array index for first point of a cycle and the phase correction as adjustable parameters.

Application of the subtraction procedure to the X-band spectrum of the spin-trapped superoxide radical, BMPO-OOH (BMPO is 5-tert-butoxycarbonyl 5-methyl-1-pyrroline N-oxide), is shown in Figure 2.18. Since both absorption and dispersion signals are used in the analysis, it is important that data are acquired in the linear-response regime, without signal saturation, because the absorption saturates more readily than the dispersion. In performing the corrections, it is important to know the scan frequency precisely. This background correction procedure was used for a variety of samples, and it consistently reduced the sinusoidal background to less than the noise level [49].

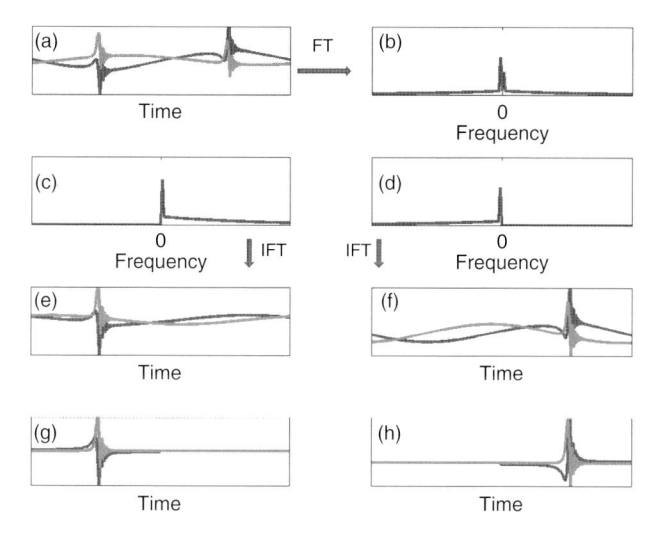

**Figure 2.17** Separation of the sinusoidal background from the rapid-scan EPR signal. (a) Full cycle of a rapid-scan signal, $I(t)$ and $Q(t)$ including the sinusoidal background, which may be out of phase with the scan. The first half of the rapid-scan signal corresponds to the up-field scan and the second half corresponds to the down-field scan. (b) The Fourier transform (magnitude spectrum) of the signal in (a) is $S(\omega)$. The sinusoidal background in the time domain is transformed into two spikes in the frequency domain, at the positive and negative scan frequencies, respectively. The intensities of these two spikes are much larger than the intensity of the rapid-scan signal. (c, d) The frequency domain magnitude signal in (b) is divided into two halves, one for positive frequencies $S^{\uparrow}(\omega)$ (c) and the other for negative frequencies $S^{\downarrow}(\omega)$ (d). (e, f) The signals in (c) and (d) are inverse Fourier transformed to produce time domain signals, (e, f), respectively. In (e), $s^{\uparrow}(t)$, the first half of the signal contains up-field rapid scan + background, but the second half contains only background. As the signal is a full cycle of the sinusoid, the background in the second half can be fitted, extrapolated into the first half, and subtracted. Similarly, the first half of the signal in (f), $s^{\downarrow}(t)$, does not contain a rapid-scan signal, but the second half contains the down-field half-cycle + background. The background in the first half can be fitted, extrapolated into the second half and subtracted. Alternatively, the full cycle backgrounds from (e) and (f) can be summed and subtracted from the data in (a). (g, h) Background subtracted up-field (g) and down-field (h) scans. (Source: Mitchell et al., 2012 [49]. Reproduced with permission of Elsevier Limited.)

## 2.9 Bridge Design

Several features distinguish a rapid-scan bridge from a standard CW bridge. (i) The narrow-band detection and amplification system that directs the EPR signal to the phase-sensitive detection and filtering system is not used. Instead, the signal from the resonator goes to a high-gain low-noise first-stage amplifier and then to the detector. (ii) The diode detector is replaced by a quadrature-mixer detector, so that both in-phase and quadrature components of the signal are obtained. (iii) The bandwidth of the bridge is comparable to that of a pulsed EPR bridge, and the

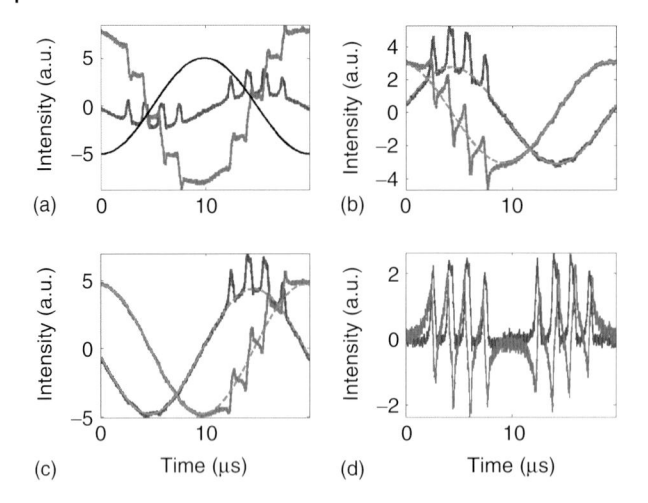

**Figure 2.18** Application of background subtraction procedure to spectra of spin-trapped superoxide, BMPO-OOH at X-band. The signals in the two channels are shown. (a) Experimental data for a full cycle of sinusoidal scan overlaid on the magnetic field scan waveform. (b) Up-field scan, $s^{\uparrow}(t)$, (c) Down-field scan $s^{\downarrow}(t)$. For both the up and down scans, the fitted background (solid) was extrapolated into the half-cycle that includes the EPR signal (dashed). (d) Results after background subtraction. (Source: Mitchell et al., 2012 [49]. Reproduced with permission of Elsevier Limited.)

output of the bridge goes to a fast digitizer, such as the Bruker SpecJet or SPU (signal processing unit). If a cross-loop resonator is used, tuning paths are provided for both resonators.

Quadrature signal acquisition provides an opportunity to further improve $S/N$ in the final spectrum by combining the absorption and dispersion signals [80]. The use of a CLR results in similar $S/N$ in both channels. The dispersion signal can be converted to an equivalent absorption signal by means of the Kramers–Kronig relations [80]. The converted signal is added to the directly measured absorption signal. Since the noise in the two channels is not correlated, this procedure increases the $S/N$ of the resultant absorption signal by up to a factor of $\sqrt{2}$. Tseitlin et al. [80] demonstrated the $S/N$ improvement for spectral-spatial oximetric imaging. The sinusoidal background procedure discussed in Section 2.8.2.2 combines the two channels via the data work-up procedure [49].

## 2.10
### Selection of Acquisition Parameters

As in the case of CW and pulsed EPR, one can find a rapid-scan EPR signal with parameters far removed from optimum values [5, 46, 66]. Similarly, the optimization depends on the information desired, and the amount of post-acquisition processing acceptable. In CW EPR, if a lineshape with less than 1% distortion is desired, one would have to be very conservative in the selection of microwave power to avoid power saturation, in modulation amplitude to avoid broadening of the line, in the choice of scan time and filter time constant [66]. Higher powers can be used for rapid scan than for CW without saturating the signal. The scan rate should be chosen consistent with the bandwidth of the EPR signal relative to the bandwidths of the resonator and the detection system.

### 2.10.1
### Resonator Bandwidth

A resonator acts as a filter for the rapid-scan EPR signal as was shown experimentally at X-band for scans in a high Q resonator [46]. Resonator Q can be expressed in many ways, but one that is convenient for discussion of rapid-scan EPR is in terms of the frequency bandwidth of the resonator, $\Delta \nu$.

$$Q = \frac{\nu}{\Delta \nu} = \frac{\nu}{BW_{res}} \quad (2.13)$$

Since each half-cycle of a rapid-scan experiment is recorded with either increasing or decreasing field/frequency, the relevant bandwidth that is available for a rapid-scan signal ($BW_{RS}$) is only half of the resonator bandwidth.

$$BW_{RS} = \frac{\nu}{2Q} \quad (2.14)$$

For an X-band resonator with $\nu = 9.3 \times 10^9$ Hz and $Q = 300$, $\Delta \nu = 31$ MHz, the resonator bandwidth available for selecting rapid-scan parameters ($BW_{RS}$) is 15.5 MHz (Eq. (2.14)). To take advantage of this bandwidth the spectrometer detection and amplification bandwidths should be at least 15.5 MHz, preferably greater.

The Q of a critically coupled X-band resonator is often too high for rapid-scan EPR. In the experiments in the Denver laboratory the Q has been lowered by various means, usually by introducing water into the sample area of the resonator in one or more separate tubes or by using water as the solvent for the sample. Since rapid scan is a continuously driven experiment, lowering the Q by over-coupling, as is done in pulsed EPR, is not an option for a reflection resonator, because of the resulting increase in reflected power. However, overcoupling could be used to lower the Q of a cross-loop resonator, since high power is incident on one resonator and signal detection is from the second resonator.

### 2.10.2
### Signal Bandwidth

The bandwidth of a rapid-scan signal is given by

$$BW_{sig} = \frac{N\gamma}{2\pi} a T_2^* \quad (2.15)$$

where the value of $N$ depends on the acceptable extent of lineshape broadening, and the scan rate $a$ is either $a_t$ (Eq. (2.3), triangular) or $a_s$ (Eq. (2.4), sinusoidal). $T_2^*$ is the time constant for decay of the oscillations on the trailing edge of the signal, and is also the decay time constant that would be observed for an FID.

Although the definition of the rapid-scan regime is that the magnetic field scans through the signal in a time that is less than $T_2$, the frequency bandwidth of the spectrum depends on $T_2^*$. A value of $N$ between 5 and 10 in Eq. (2.15) is a relatively conservative starting point for selecting scan parameters. The bandwidth required depends on the $T_2^*/T_2$ ratio for the sample and the linewidth accuracy desired in an experiment. The larger the value of $N$, the less the signal is broadened.

For a Lorentzian lineshape,

$$T_2^* = T_2 = \frac{2}{\sqrt{3}\Delta B_{pp}} \qquad (2.16)$$

where $\Delta B_{pp}$ is the peak-to-peak first-derivative linewidth. The bandwidth for the rapid-scan signal from a Lorentzian line is then

$$BW_{sig} = \frac{Na}{\sqrt{3}\pi\Delta B_{pp}} \qquad (2.17)$$

The relationship between $T_2^*$ and $\Delta B_{pp}$ depends on the lineshape. Unresolved hyperfine structure results in EPR line broadening that is approximately Gaussian, and decreases the signal bandwidth, which therefore changes the selection of parameters for rapid scan. For example, $^{15}$N perdeutero tempone has $\Delta B_{pp} \sim 175$ mG with Gaussian broadening due to unresolved deuterium hyperfine coupling. Use of Eq. (2.16), which assumes a Lorentzian shape, gives $T_2^* = 380$ ns, but the experimental $T_2^*$ by pulse methods is about 430 ns. Equation (2.17) is a useful starting point for estimating signal bandwidth and selecting scan rates consistent with a particular resonator $Q$.

Consider calculations of these requirements for two specific cases:

1) For anoxic deuterated trityl radical $\Delta B_{pp} \sim 20$ mG. A 20 kHz sinusoidal scan with 5 G sweep width ($a_s = 3.4 \times 10^5$ G/s) results in a signal bandwidth of $\sim 15$ MHz, at $N = 5$. At 9.6 GHz, resonator $Q$ (Eq. (2.14)) would have to be less than about 320.

2) A nitroxide with $\Delta B_{pp} \sim 150$ mG, recorded with a 80 kHz sinusoidal scan and 40 G sweep width ($a_s = 1.0 \times 10^7$ G/s), has a signal bandwidth of $\sim 62$ MHz at $N = 5$. At 9.6 GHz, resonator $Q$ would have to be less than about 78.

Case (2) could be collected into a Bruker SpecJet, which has a bandwidth of 150 MHz, but the same signals would be distorted by the 20 MHz bandwidth of a Bruker SPU. In each case, the video amplifier bandwidth in the bridge would need to be set to greater than the resonator and signal bandwidths.

For immobilized or slowly tumbling species, anisotropies are not fully averaged and $T_2^*$ may be much smaller than $T_2$, which makes the signal bandwidth requirement smaller than that determined by $T_2$.

In practice, the scan rate is initially set to a conservatively slow value. The linewidth is monitored as the scan rate is increased, until the line broadening approaches the maximum acceptable value. The S/N can be improved at the expense of line broadening by acquiring data with smaller bandwidth, which permit use of higher resonator $Q$ and increased post-processing filtering.

## 2.10.3
### Microwave Power

As discussed in Section 2.1.2 the power saturation curves for rapid-scan experiments depend on the scan rate. The regime in which the signal intensity increases with increasing $B_1$ ($\sqrt{P}$) extends to higher power for increasing scan rate. Thus, the power to be used depends on scan rate and on acceptable power distortion.

## 2.10.4
### Electron-Spin Relaxation Times

If the relaxation times are long relative to the time to scan through the line, there are transient effects on the trailing edge of the signal, which damp out with the time constant $T_2^*$. If the time between successive passes through resonance is less than about 5 $T_2^*$, oscillations have not fully damped out before the next spin excitation, which complicates interpretation of the data. This constraint requires that $1/f_s > \sim 20\ T_2^*$, and depends on the position of the signal in the scan. Since $T_2^*$ usually is $< T_1$ this limitation on the scan frequency is less restrictive than the limitation on pulse repetition rates for pulsed EPR. In a pulsed experiment with 90° pulses, it is necessary to have the pulse repetition time longer than about 5 $T_1$ to permit the magnetization to return to equilibrium between pulses. Alternatively, smaller tip angles can be used with faster repetition times [81]. Rapid-scan experiments typically are performed in the linear region of the power saturation curve, which means that the power is low enough to cause a small perturbation of the spin state populations. The repetition time, $1/(2\ f_s)$, can, therefore, be much shorter than $T_1$.

## 2.10.5
### Selection of Scan Rate

Multiple factors influence the selection of the scan rate for a particular sample. As discussed in Section 2.1.6, in order to decrease the impact of source noise it is advantageous to have $f_s$ large enough to approach the regime in which the noise is approximately white (Figure 2.6). There is additional S/N advantage in scanning at rates in the range where Eq. (2.1) is satisfied. However, if the scan rate causes the signal bandwidth to be too large relative to the bandwidth set by resonator Q, the signal is broadened. Analogous to CW spectroscopy, the tradeoffs between lineshape broadening and improving S/N are dependent on the goals of the experiment. To obtain the most accurate lineshapes, slower scans are needed. However, if the goals are, for example, increasing S/N, precision of spin counting, or acquiring transient signals as quickly as possible, greater signal broadening is acceptable. Wider spectra require wider scans, which may result in faster scan rates (Eqs. (2.3) and (2.4)).

There may also be instrumental constraints on the selection of parameters. As described in Section 2.6, which discusses the scan driver, various combinations of scan amplitude and scan frequency may result in exceeding the heat specification of the power amplifier, or exceeding the power deposition criterion for the scan coils, which causes over-heating.

## 2.11
## Multifrequency Rapid Scan

Examples of rapid-scan spectra shown in Section 2.12 were obtained at frequencies ranging from 250 MHz to 9.5 GHz. Lower magnetic fields/frequencies are important for *in vivo* experiments. Higher magnetic fields/frequencies are predicted to be advantageous for rapid scans for several reasons. At higher frequency, the same resonator Q corresponds to a higher bandwidth, which permits faster scans without signal broadening. Greater g value dispersion often results in broader lines, which increases $T_2^*$ and permits the use of higher Q and/or faster scans. The smaller size of higher frequency resonators means that sample sizes are smaller, so that the size of the scan coils can be decreased, which decreases the power required for a scan and facilitates wider scans. The higher bandwidth for the same Q at higher frequencies facilitates rapid frequency scans [40]. If $T_1$ and $T_2$ are frequency dependent [82, 83], the parameter selection may be frequency dependent. Background signals due to mixing of scan frequencies with resonance frequencies are a larger problem at lower frequencies.

## 2.12
## Examples of Applications

Rapid-scan spectroscopy has been performed in the Denver laboratory at 9.5 GHz, 1.0 GHz, or 250 MHz. Many of the examples shown in this section are at X-band because of its widespread use for EPR. The impact of the resonator on rapid-scan signals, reported in 2005, used a Bruker E580 spectrometer [46]. More recent experiments, including many of those described in this section, used a Bruker E500T, which was designed for rapid-scan EPR.

### 2.12.1
### Comparison of Rapid-Scan Spectra Obtained with a Dielectric Resonator and Either Standard Modulation Coils or Larger Scan Coils

One question that arises with respect to the hardware requirements is how well rapid scan can be performed with the standard Bruker modulation coils, compared with results obtained with larger scan coils. To address that question, the experiments described in Sections 2.12.1.1 and 2.12.1.2 were performed with a Bruker dielectric resonator, for which the height of the active volume is about 10 mm.

### 2.12.1.1 Standard Modulation Coils

The sinusoidal rapid-scan spectrum (Figure 2.19a) of the low field line of a 0.1 mM solution of the nitroxide $^{15}$N-mHCTPO in 80/20 EtOH/water solution was obtained using the ER4118X-MD5 dielectric resonator with modulation coils (Sections 2.4.2 and 2.4.3) resonated at ∼29 kHz with ∼30 G scan width ($a_s = 2.7 \times 10^6$ G/s). The sample was degassed by performing six freeze–pump–thaw cycles and sealing the tube. The sample, in a 4 mm o.d., 3 mm i.d. quartz tube, had a height of 3 mm, resulting in an approximately $3 \times 3$ mm cylindrical shape. The sample loss resulted in a resonator $Q \sim 150$. Oscillations were observed on the trailing edge of the signal (Figure 2.19a). Deconvolution gave the absorption spectrum (Figure 2.19b) in which the $^{13}$C hyperfine lines and the coupling to the unique ring proton were well resolved. There was good agreement between the first derivative of the deconvolved rapid scan and the CW spectrum. In this experiment, the signal was centered in the 30 G sinusoidal scan window.

The experiment was repeated with a 55 G scan that encompassed both lines of the $^{15}$N nitroxide ($a_s = 5.0 \times 10^6$ G/s), which positioned the signals toward the ends of the scan (Figure 2.20a). For this spectrum, the lines were broadened, and the proton hyperfine splitting were not as well resolved as the CW spectrum

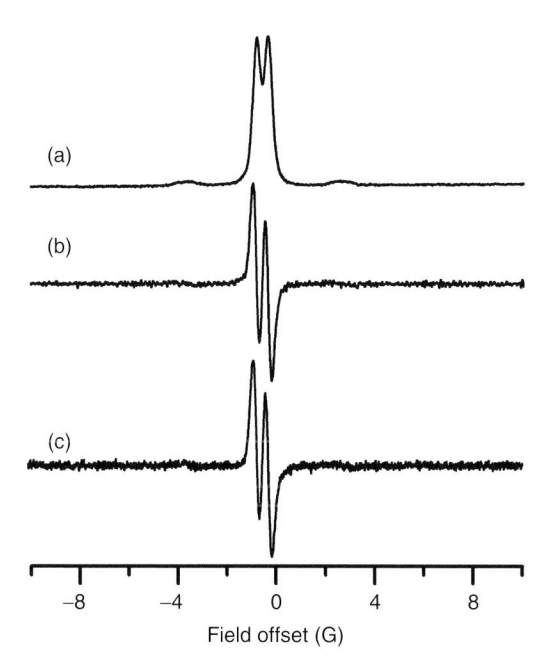

**Figure 2.19** Comparison of deconvolved rapid scan and CW spectra of the low-field line for $^{15}$N-mHCTPO, obtained with standard modulation coils. (a) Slow-scan absorption spectrum obtained by deconvolution of sinusoidal rapid scan. (b) First-derivative spectrum obtained by pseudomodulation of the signal in a. (c) Single scan of a field-modulated first-derivative CW EPR spectrum of the same sample. (D. G. Mitchell 2012, unpublished results.)

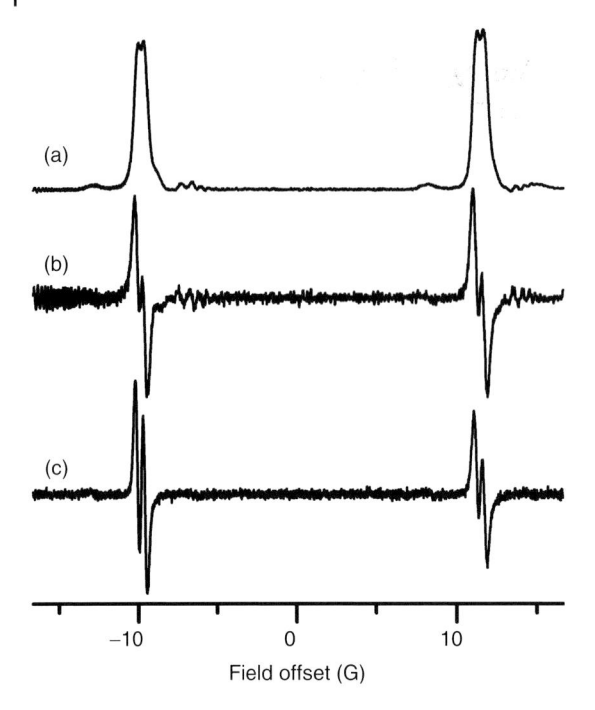

**Figure 2.20** Comparison of deconvolved rapid scan and CW of full $^{15}$N-mHCTPO, obtained with standard modulation coils. (a) Slow-scan absorption spectrum obtained by deconvolution of sinusoidal rapid-scan signal. (b) First-derivative spectrum obtained by pseudomodulation of the signal in (a). (c) Single scan of a field-modulated first-derivative CW EPR spectrum of the same sample. (D. G. Mitchell 2012, unpublished results.)

(Figure 2.20b,c). The broadening observed in Figure 2.20a is attributed to the significant size of the sample, relative to the size of the modulation coils. The modulation field, $B_m$, may not be homogeneous over the sample. The effect of inhomogeneous $B_m$ is larger if the EPR line is near the extremes of the sinusoidal magnetic field scan (Figure 2.20). Inhomogeneity of the scan field has less impact on the signal if the line is near the center of the scan (Figure 2.19).

#### 2.12.1.2 Larger External Coils

The rapid-scan spectrum of the low-field line for a 0.1 mM solution of $^{15}$N-mHCTPO in 80/20 EtOH/water solution in the same dielectric resonator with external, circular 9.5 cm coils, separated by 4.5 cm, resonated at ~60 kHz with ~10 G scan width is shown in Figure 2.21. The sample was the same as that for the spectra shown in Figures 2.19 and 2.20. The power saturation curves in Figure 2.3 (Section 2.1.3) are of this sample. The larger coils provide greater homogeneity of the scan field over the dimensions of the sample. Good agreement is observed between the first derivative of the deconvolved rapid scan and CW spectra both for the low-field line (data not shown) and for the full spectrum (Figure 2.21a,b).

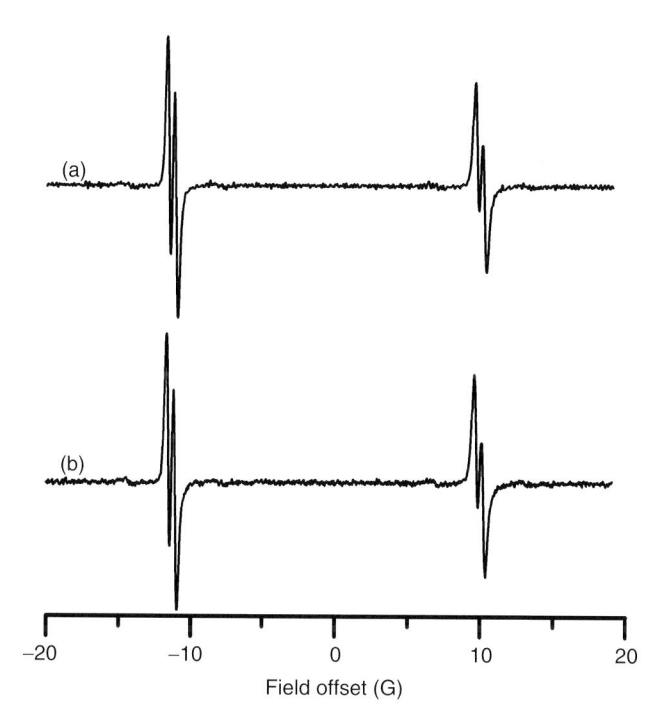

**Figure 2.21** (a) CW spectrum of degassed 0.2 mM mHCTPO solution. 40 G sweep width, 0.05 G modulation amplitude. (b) Pseudomodulated, deconvolved rapid-scan spectra of a degassed 0.1 mM mHCTPO solution with 55 G scan width, and 29.7 kHz scan frequency (~5.1 MG s$^{-1}$). 1024 averages were collected with resonator $Q \sim 150$ and 2 mW power ($B_1 = 0.02$ G). (D. G. Mitchell 2012, unpublished results.)

### 2.12.1.3 Results of Comparison

The experiments that produced the spectra shown in Figures 2.20 and 2.21 had approximately the same scan frequency and width. The difference was in the size of the coils used to create the rapid magnetic field scans. The improvement in resolution observed with the external coils is attributed to the larger size of the coils. The 9.5 cm external circular coils are large relative to the sample size, creating a more homogeneous field over the sample.

### 2.12.2
### S/N for Nitroxide Radicals

The spectra in Figure 2.22 demonstrate the S/N advantage of rapid scan for nitroxide radicals in solution at ambient temperature. The sample is the same as was used for Figures 2.19–2.21. Sinusoidal rapid-scan data published in [10], obtained with 4% duty cycle for the rapid scans, showed significant improvement in S/N relative to CW, for the same data acquisition time. Improvements in the scan coil driver [9] and coils now permit 100% duty cycle for these scan widths and substantially larger improvement in S/N and S/N per unit time. For the same data

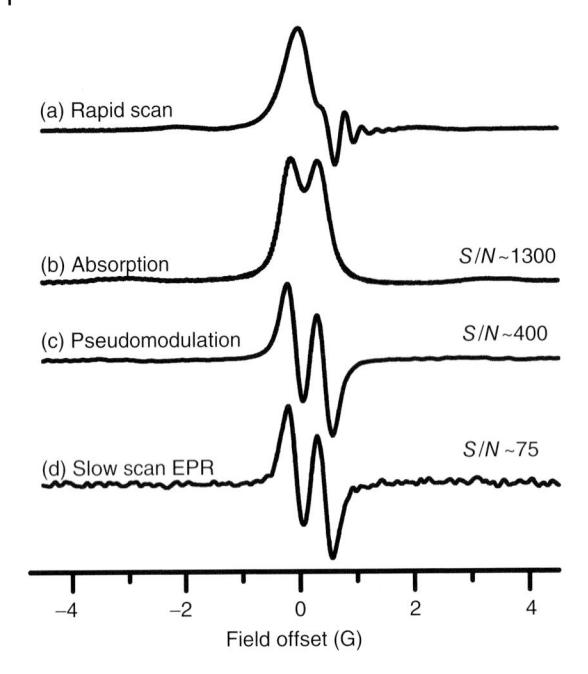

**Figure 2.22** Comparison of rapid scan and CW EPR spectra of mHCTPO. Magnetic field scans were from low field to high field using 9.5 cm diameter external coils. (a) As-recorded sinusoidal rapid-scan signal obtained with a scan rate of 1.8 MG s$^{-1}$. 1024 averages were recorded in about 0.9 s using SpecJet II. The incident microwave power was about 80 mW ($B_1 = 0.14$ G). (b) Slow-scan absorption spectrum obtained by deconvolution of signal in (a). (c) First-derivative spectrum obtained by pseudomodulation of the signal in (b). First-derivative spectrum was filtered using a fourth-order Butterworth filter allowing less than 2% broadening of the linewidth. (d) Single scan of a field-modulated first-derivative CW EPR spectrum of the same sample, obtained in 0.9 s using about 5 mW incident microwave power, 10 kHz modulation frequency, 0.9 ms conversion time, 1024 points, 0.13 G modulation amplitude. Modulation amplitude, power, and fourth-order Butterworth filter were chosen to maximize signal-to-noise while allowing less than 2% broadening of the linewidth.

acquisition time, and incident powers selected to give less than 2% broadening, the S/N for the absorption spectrum acquired by rapid scan with 100% duty cycle was ∼1300 compared with the S/N for the first-derivative spectrum obtained by CW of ∼75 (Figure 2.22). These results are representative of performance for other rapidly tumbling nitroxides.

### 2.12.3
### Estimation of Nitroxide $T_2$ at 250 MHz

The first direct measurement of $T_2$ of a nitroxide radical (tempone-d$_{16}$) at low frequency (250 MHz) was made using rapid scan [14]. The frequency dependence of nitroxide $T_1$ reported in a highly cited science paper [57], convinced many people that nitroxides would not be useful relaxation probes at low frequencies. However, a CW EPR study by Lloyd and Pake [84] reported $T_1 \sim 0.5\,\mu s$ for the nitroxide Fremy's salt at 60 MHz ($B_o = 32\,G$). In the rapid tumbling regime, $T_1 \sim T_2$ is expected. If this generalization were valid, the $T_2$ for nitroxides would be expected to be $\sim 0.5\,\mu s$ at 250 MHz. When other researchers reported in lectures that it was not possible to measure pulsed EPR of nitroxides at 300 MHz, it appeared that there was some unknown $T_2$ relaxation mechanism applicable to nitroxides at low frequency. The rapid-scan transient response for tempone-$d_{16}$ was analyzed by simulation, as shown in Figure 2.23. The damping of the rapid-scan oscillations depends on $T_2^*$. Simulations of slow-scan spectra define the inhomogeneous broadening, which is then used iteratively with simulations of the oscillations as a function of scan rate to determine $T_2$ [14]. The values of $T_2$ obtained from the rapid-scan experiments are shown in Table 2.4. These values are in good agreement with the results obtained subsequently by two-pulse spin echo [83]. These estimates of $T_2$ at 250 MHz, and subsequent demonstration at the National Cancer Institute of pulsed EPR imaging

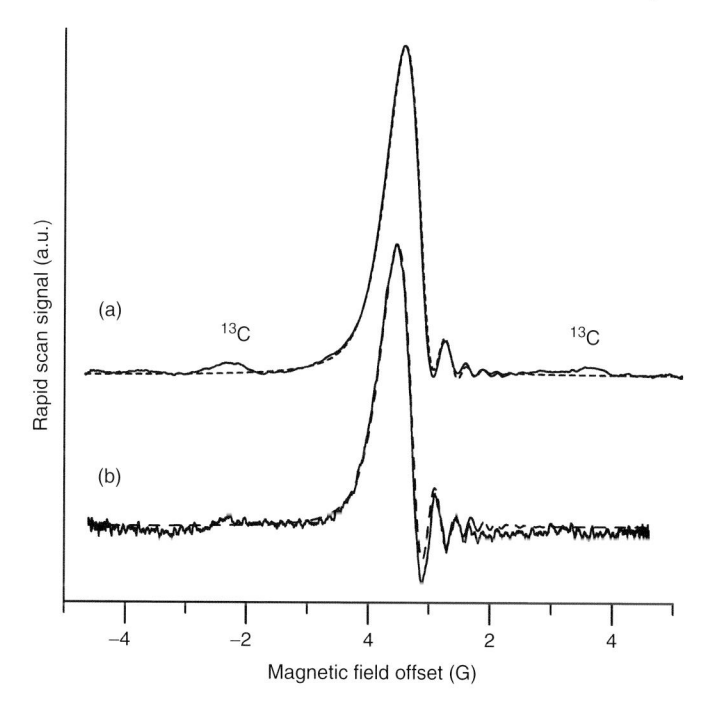

**Figure 2.23** Rapid-scan signals for the low-field line of tempone-$d_{16}$ in water obtained with 40 kHz sinusoidal field sweeps and RF of 245 MHz. (a) Signal for 0.5 mM solution with a sweep width of 10.0 G. The dashed line is a simulation obtained with $T_2 = 0.38\,\mu s$ and an inhomogeneous broadening of 15 mG. (B) Signal for 0.1 mM solution with sweep width of 9.5 G. The dashed line is a simulation obtained with $T_2 = 0.56\,\mu s$ and an inhomogeneous broadening of 15 mG. The $^{13}C$ sidebands were not included in the simulations. (Source: Tseitlin et al., 2006 [14]. Reproduced with permission of Springer-Verlag.)

**Table 2.4** $T_2$ values[a] (in microseconds) at 250 MHz for tempone-d$_{16}$ in aqueous solution at room temperature [14].

| Concentration (mM) | $m_I = +1$ | $m_I = 0$ | $m_I = -1$ |
|---|---|---|---|
| 0.5 | 0.41 | 0.41 | 0.35 |
| 0.2 | 0.50 | 0.50 | 0.43 |
| 0.1 | 0.53 | — | 0.42 |

[a]Uncertainties are about ±0.03 μs.

of nitroxides stimulated by our work [85], have given new impetus for development of nitroxide radicals for *in vivo* imaging.

### 2.12.4
### Spin-Trapped Radicals

The short lifetime of superoxide, $O_2^{\bullet-}$, and low rates of formation expected *in vivo*, make detection by standard CW EPR challenging. The rapid scan methodology offers improved sensitivity for these types of samples. To validate the application of

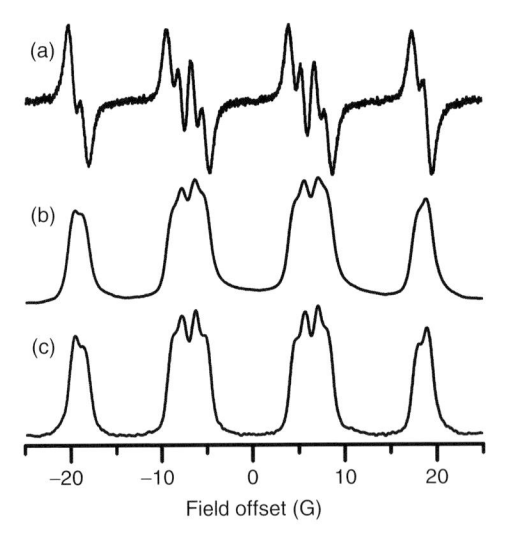

**Figure 2.24** Comparison of CW and rapid-scan spectra of BMPO-OOH in a solution with a $O_2^{\bullet-}$ production rate of 6 μM min$^{-1}$ $O_2^{\bullet-}$, recorded 10 min after mixing reagents. The $O_2^{\bullet-}$ was produced by a hypoxanthine/xanthine oxidase mixture. (a) CW spectrum, obtained with 55 G sweep width, 0.75 G modulation amplitude, single 42 s, and 20 mW microwave power ($B_1 = 170$ mG). (b) The first integral of (a). (c) Deconvolved rapid-scan spectrum obtained with 55 G scan width, 51 kHz scan frequency, 20 mW ($B_1 = 170$ mG) microwave power, 100 K averages, and a total time of ~4 s. Reproduced with permission from [87].

rapid scan to spin trapping, $O_2^{•-}$ was generated by the reaction of xanthine oxidase and hypoxanthine with rates of 0.1–6.0 μM/min and trapped with BMPO [86] at pH = 7.4 [87]. Spin trapping converts the very short-lived superoxide radical, $O_2^{•-}$, into a more stable spin adduct. The half-life of BMPO-OOH at ambient temperature is reported to be about 23 min [88]. CW spectra were recorded with a Bruker X-band (9.5 GHz) EMX Plus and SHQ resonator. Rapid-scan spectra were recorded on a Bruker X-band E500T with a dielectric resonator. CW and rapid-scan spectra for BMPO-OOH that were observed for 6 μM/min generation of $O_2^{•-}$ exhibited the characteristic 12-line spectrum [88] (Figure 2.24). The data acquisition time for the deconvolved rapid-scan spectrum (Figure 2.24c) was 10% of the time that was used to acquire the CW spectrum (Figure 2.24a). The hyperfine splittings observed in the CW and rapid-scan spectra are in good agreement. The small hyperfine splittings are better resolved in the rapid-scan spectrum than those in the first integral of the CW spectrum because the high-modulation amplitude and power used to obtain the CW spectrum broadened the lines.

CW and rapid-scan spectra in Figure 2.25 were obtained in 30 s of data acquisition time for samples with formation rates of 0.1 μM/min $O_2^{•-}$, which is 60 times lower than the rate that was used to acquire the spectra in Figure 2.24. In the CW

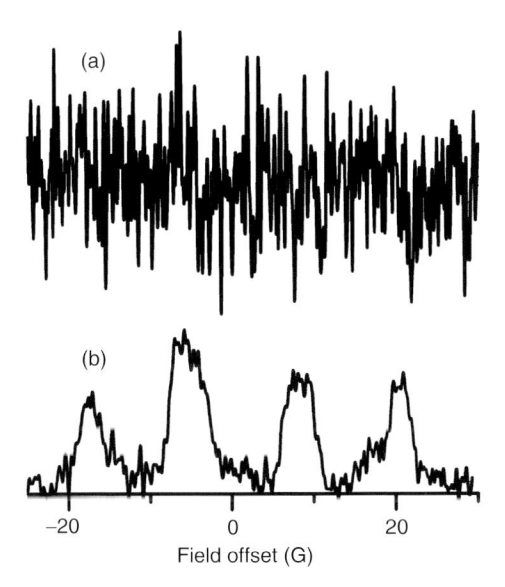

**Figure 2.25** Comparison of CW and rapid-scan spectra of ~0.3 μM BMPO-OOH detected in solution with a $O_2^{•-}$ production rate of 0.1 μM min$^{-1}$, recorded 10 min after mixing reagents. The $O_2^{•-}$ was produced by a hypoxanthine/xanthine oxidase mixture. (a) CW spectrum obtained with 55 G sweep width, 0.75 G modulation amplitude, single 30 s scan, 15 ms conversion time, 10 ms time constant, 2048 points and 20 mW microwave power ($B_1$ = 170 mG). (b) Deconvolved rapid-scan spectrum obtained with 55 G scan width, 51 kHz scan frequency, 53 mW (6 dB) microwave power ($B_1$ = 250 mG), segments consisting of 12 sinusoidal cycles were averaged 100 k times with a total data acquisition time of about 30 s. Reproduced with permission from [87].

spectrum (Figure 2.25a) there is barely a hint of the BMPO-OOH signal. The rapid-scan spectrum recorded in the same 30 s of data acquisition time has an S/N of about 10 (Figure 2.25b). From the comparison in Figure 2.25 it is evident that rapid-scan EPR permits detection of BMPO-OOH with good lineshape fidelity at low production rates that are too low for detection by CW EPR for the same data acquisition time. Rapid-scan EPR was also applied to a bacterial system, the extracellular production of $O_2^{\bullet-}$ by *Enterococcus faecalis*, at a superoxide production rate of 0.1 nmol/min per $1.0 \times 10^6$ CFU (Figure 3 in Ref. [87]). At this rate of $O_2^{\bullet-}$ production it was difficult to determine whether the EPR spectrum of BMPO-OOH was present in a CW spectrum with 30 s acquisition time. By contrast, the characteristic BMPO-OOH signal in a rapid-scan spectrum with a 30 s acquisition time has an S/N of about 42, which is thus an improvement by a factor of more than 40 over CW EPR. These data demonstrate the improved sensitivity of rapid-scan relative to CW EPR in a living system and demonstrates that rapid-scan can detect superoxide produced *by E. faecalis* at rates that are too low for detection by CW EPR. The ability of rapid-scan EPR to acquire higher S/N data in a shorter time than CW will improve temporal resolution of spin trapping experiments greatly and be crucial for *in vivo* imaging.

### 2.12.5
### Improved S/N for Species with Long Electron-Spin Relaxation Times

The electron spin relaxation times for many paramagnetic centers in solids are very long at ambient temperature, which makes it difficult to obtain CW spectra that are free of passage effects. Rapid scan is advantageous for these samples because the deconvolution procedures correct for the passage effects, and there are substantial enhancements in S/N. In the following comparisons the CW spectra were obtained on Bruker E580 or EMX spectrometers and the rapid-scan spectra were obtained on a Bruker E500T spectrometer with a dielectric resonator. These combinations of spectrometers and resonators provide the best currently available performance for the respective methods. As discussed in Section 2.1.2, higher power and microwave $B_1$ can be used to acquire the rapid-scan spectra than those used to acquire the CW spectra (Table 2.5). For the same data acquisition time, and parameters selected to keep line broadening at less than about 2%, rapid scan gives substantially higher S/N than CW, or field-swept echo detected, as discussed below for individual samples. The advantages relative to Fourier transform electron paramagnetic resonance (FT-EPR) are sample dependent.

#### 2.12.5.1  E' Center in Irradiated Fused Quartz
The $T_1$ for the E' center in irradiated fused quartz is about 200 μs and depends on position in the spectrum [55, 44]. $T_2$ depends on the concentration of paramagnetic centers [44]. The X-band data shown in Figure 2.26 were obtained for a sample with $T_2$ about 20 μs [45]. These relaxation times are so long that to obtain an undistorted CW spectrum requires the use of very low microwave power and low modulation frequency, which decreases the S/N. For the rapid-scan experiments, the Q of the

## 2.12 Examples of Applications

**Table 2.5** Electron relaxation times, linewidths, and microwave $B_1$ for CW and rapid scan for paramagnetic centers in materials at ~20 °C.

| Sample | $T_1$ (μs)[a] | $T_2$ (μs)[a] | $\Delta B_{pp}$ (G)[a] | $B_1$ for CW[b] (mG) | $B_1$ for rapid scan[b] (mG) | Rapid-scan rate (MG s$^{-1}$) |
|---|---|---|---|---|---|---|
| E′ in irradiated fused quartz | 200 | 20 | ~1[c] | 17 | 220 | 4.7 |
| a-Si:H | 11 | 3.3 | 6 | 35 | 200 | 3.9 |
| N@C$_{60}$ | 120–160 | 2.8 | 0.25 | 6 | 53 | 1.5 |
| $N_S^0$ in diamond[d] | 2300 | 230 | 0.045 | 0.03 | 5.8 | 0.14 |

[a] Uncertainties are about ±5% for relaxation times and ±2% for linewidths.
[b] Selected to give less than 2% power broadening.
[c] Lineshape is anisotropic.
[d] Parameters are for nitrogen $m_I = 0$ line.

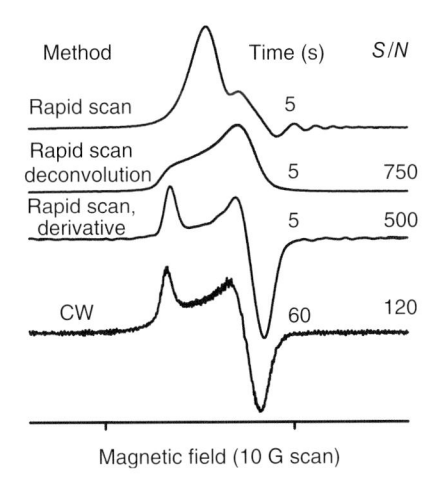

| Method | Time (s) | S/N |
|---|---|---|
| Rapid scan | 5 | |
| Rapid scan deconvolution | 5 | 750 |
| Rapid scan, derivative | 5 | 500 |
| CW | 60 | 120 |

Magnetic field (10 G scan)

**Figure 2.26** Comparison of rapid scan and CW EPR spectra of E′ center in irradiated fused quartz. (a) As-recorded sinusoidal rapid-scan signal obtained with a scan rate of 4.7 MG s$^{-1}$. 1024 averages were recorded in about 5 s. The incident microwave power was about 3.3 mW. (b) Slow-scan absorption spectrum obtained by deconvolution of the signal in part (a). (c) First derivative spectrum obtained by pseudomodulation of the signal in part (b). (d) Single scan of a field-modulated first-derivative CW EPR spectrum of the same sample, obtained in 1 min using about 0.02 mW incident microwave power, 10 kHz modulation frequency and 0.05 G modulation amplitude. (Source: Mitchell et al., 2011 [45]. Reproduced with permission of Elsevier Limited.)

dielectric resonator was reduced to about 300 by placing water around the sample (2 mm diameter by 10 mm long rod) in a 4 mm OD quartz tube. The sinusoidal rapid-scan signal (Figure 2.26a) was deconvolved to give the absorption spectrum (Figure 2.26b). The lineshape of the first derivative of the absorption spectrum (Figure 2.26c) is in good agreement with that for the CW spectrum (Figure 2.26d).

Parameters for both experiments were selected to minimize line broadening. The S/N for the spectra obtained by rapid scan in 5 s is substantially better than that for the CW spectrum obtained in 1 min.

### 2.12.5.2 Amorphous Hydrogenated Silicon (a-Si:H)

The paramagnetic centers in undoped a-Si:H are three-coordinated silicon atoms that usually are referred to as *dangling bonds*. X-band EPR spectra for a-Si:H obtained by several methods are shown in Figure 2.27 [11]. For each experiment, the data acquisition and processing parameters were selected to keep line broadening at less than 2%. For data acquired in about the same time, the S/N is much higher for rapid scan than that for CW. Owing to instrumental limitations, the time required for a field-swept echo-detected spectrum was longer than that for CW or rapid scan, but the S/N still was substantially lower than that for rapid scan (Figure 2.27b). Although $T_1$ and $T_2$ are 11 and 3.3 µs, respectively (Table 2.5), $\Delta B_{pp}$ (G) of the CW spectrum (Figure 2.27d) is about 6 Gauss, which means that $T_2^*$ is too short to perform FT-EPR. The short $T_2^*$ corresponds to a small signal bandwidth for rapid

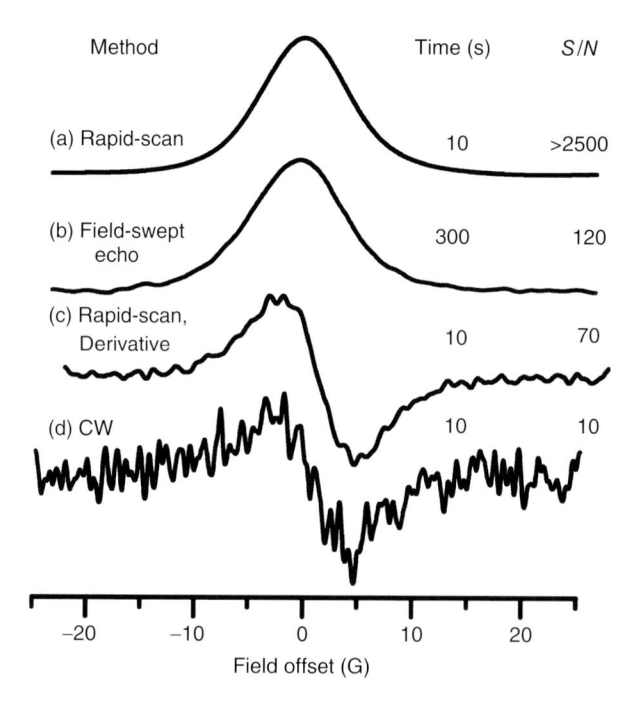

| Method | Time (s) | S/N |
|---|---|---|
| (a) Rapid-scan | 10 | >2500 |
| (b) Field-swept echo | 300 | 120 |
| (c) Rapid-scan, Derivative | 10 | 70 |
| (d) CW | 10 | 10 |

Field offset (G)

**Figure 2.27** Comparison of X-band spectra of a-Si:H. (a) Slow-scan absorption spectrum obtained by deconvolution of sinusoidal rapid-scan signal acquired with a scan rate of 3.9 MG s$^{-1}$, 102 400 averages, and $B_1 = 200$ mG. (b) Field-swept echo-detected spectrum obtained with constant 500 ns spacing between pulses, SRT = 100 µs, 1024 shot/point, 10 scans. (c) Derivative of deconvolved rapid-scan spectrum. (d) Field-modulated first-derivative CW EPR spectrum acquired with 2 G modulation amplitude at 30 kHz, and $B_1 = 35$ mG. (Source: Mitchell et al., 2013 [11]. Reproduced with permission of Taylor and Francis.)

scan, which permits acquisition of the rapid-scan and CW spectra with the same resonator $Q$, and contributes to the improvement in $S/N$ by $> 250$ for rapid scan relative to CW. Thus, the short $T_2^*$ that prevents FT-EPR is an advantage for rapid scan.

### 2.12.5.3 N@C$_{60}$

A 0.2% N@C$_{60}$ in solid C$_{60}$ sample [89] was placed in a 0.8 mm capillary tube supported in a 4 mm OD quartz tube. X-band EPR spectra obtained by several methods are shown in Figure 2.28 [11]. All three nitrogen hyperfine lines were included in the scans. For each experiment, the data acquisition and processing parameters were selected to keep line broadening at less than 2%. For the rapid-scan experiments, the space in the 4 mm OD tube surrounding the 0.8 mm capillary contained a 50/50 ethanol/water mixture, which lowered the $Q$ to ~250 [11]. For comparable data acquisition times the $S/N$ is substantially higher for rapid scan (Figure 2.28a,d) than that for CW (Figure 2.28e) or pulse methods (Figure 2.28c,d).

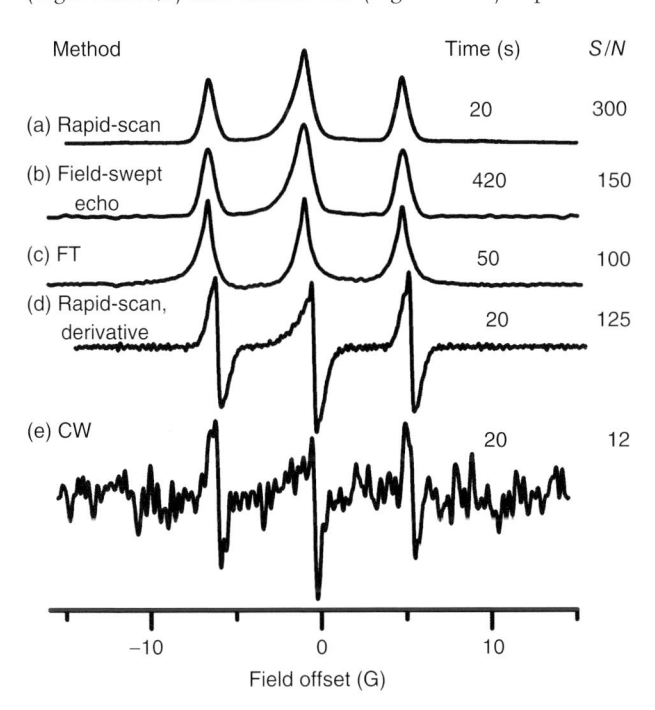

| Method | Time (s) | S/N |
|---|---|---|
| (a) Rapid-scan | 20 | 300 |
| (b) Field-swept echo | 420 | 150 |
| (c) FT | 50 | 100 |
| (d) Rapid-scan, derivative | 20 | 125 |
| (e) CW | 20 | 12 |

Field offset (G): −10, 0, 10

**Figure 2.28** Comparison of X-band spectra for 0.2% N@C$_{60}$. (a) Slow-scan absorption spectrum obtained by deconvolution of sinusoidal rapid-scan signal acquired with a scan rate of 1.5 MG s$^{-1}$, 102 400 averages, and $B_1 = 53$ mG. (b) Field-swept echo-detected spectrum obtained with constant $\tau = 600$ ns spacing between pulses, SRT = 200 μs, 1024 shots/point, 2 scans. (c) FT-EPR of data obtained with SRT = 200 μs, 90° tip angle, and 20 480 averages. (d) Derivative of deconvolved rapid-scan spectrum. (e) CW spectrum acquired with 0.1 G modulation amplitude at 30 kHz and $B_1 = 6$ mG. (Source: Mitchell et al., 2013 [11]. Reproduced with permission of Taylor and Francis.)

## 2.12.5.4 Single Substitutional Nitrogen ($N_S^0$) in Diamond

A diamond sample with 20 ppb $N_S^0$ defects grown by vapor deposition had dimensions of $4 \times 4 \times 2$ mm and was wedged in a 4 mm OD Teflon tube [11]. For each experiment, the data acquisition and processing parameters were selected to keep line broadening at less than 2%. For the rapid-scan experiments a 3 mm OD quartz tube filled with a 25/75 EtOH/$H_2O$ mixture to a height of ~5 mm was placed above the sample, which lowered the $Q$ to ~400. X-band EPR spectra of the central line of the nitrogen hyperfine triplet are shown in Figure 2.29. As for the other samples with long electron spin-relaxation times, for comparable data acquisition times the S/N is substantially higher for rapid scan than for CW or field-swept echo-detected EPR. Shorter shot repetition time (SRT) and smaller tip angles calculated using the Ernst equation [90] give S/N for FT EPR similar to that for rapid scan.

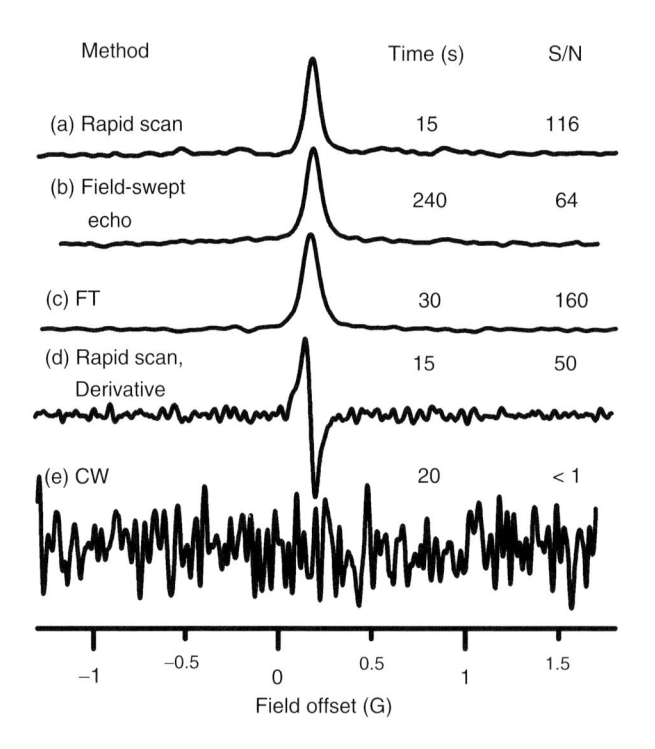

| Method | Time (s) | S/N |
|---|---|---|
| (a) Rapid scan | 15 | 116 |
| (b) Field-swept echo | 240 | 64 |
| (c) FT | 30 | 160 |
| (d) Rapid scan, Derivative | 15 | 50 |
| (e) CW | 20 | < 1 |

Field offset (G)

**Figure 2.29** Comparison of X-band spectra for the center line of $N_S^0$ in diamond. (a) Slow-scan absorption spectrum obtained by deconvolution of triangular rapid-scan signal acquired with a scan rate of 0.14 MG/s, 102 400 averages, $B_1 = 4$ mG. (b) Field-swept echo-detected spectrum with a constant 600 ns spacing between pulses, SRT = 3 ms, 64 shots/pt, 1 scan. (c) FT-EPR of data obtained with SRT = 200 μs, 24° tip angle, and 40 960 averages. (d) Derivative of deconvolved rapid-scan spectrum. (e) CW spectrum acquired with 0.05 G modulation amplitude at 6 kHz and $B_1 = 0.25$ mG, one scan. (Source: Mitchell et al., 2013 [11]. Reproduced with permission of Taylor and Francis.)

## 2.12.6
### Imaging

The acquisition of the absorption signal, rather than the first derivative, by rapid scan is particularly advantageous for imaging [47]. In the presence of the magnetic

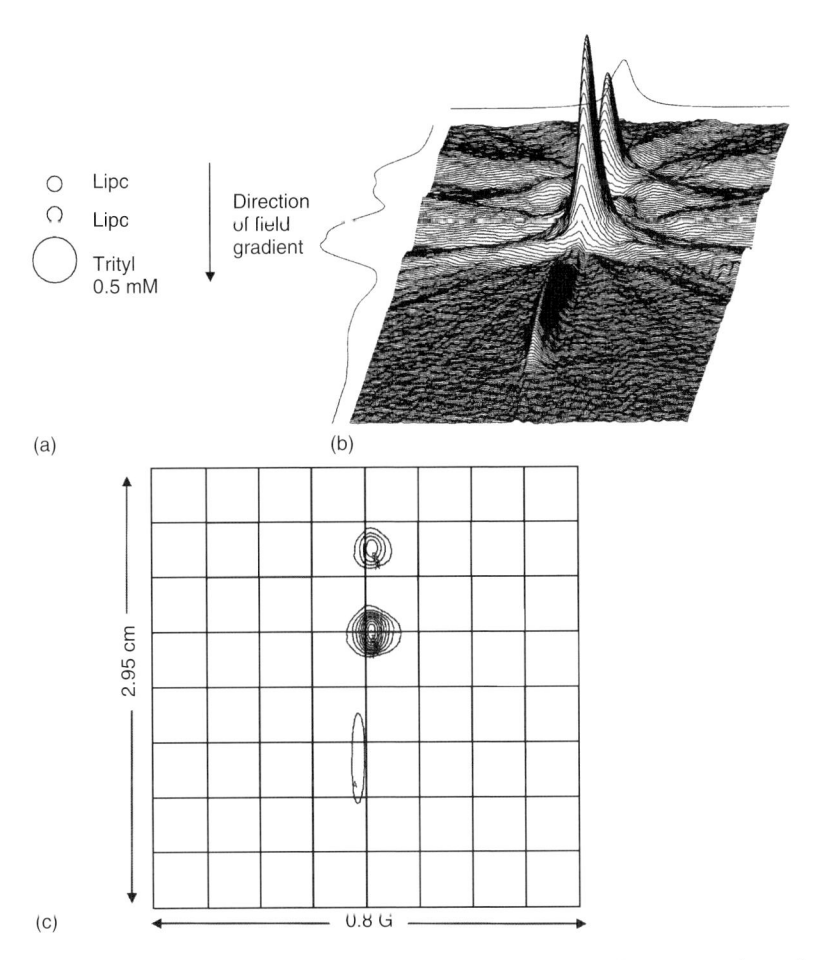

**Figure 2.30** (a) A cartoon of the phantom consisting of two LiPc samples and a 0.5 mM trityl sample. The separations between centers of tubes are ~5 and ~8 mm, respectively. The direction of the field gradients is shown by the arrow. (b) Perspective plot of the 2-D spectral-spatial image of the phantom obtained by rapid-scan at 8 kHz. The RF was 248 MHz and $B_1$ was 3.6 mG. (c) Contour plot of the image showing circularly symmetric contours corresponding to the LiPc samples and a much less intense elongated contour due to the trityl sample. On the spectral axis, the contour due to the trityl sample is displaced relative to the contours due to the LiPc samples by ~0.03 G, consistent with the difference in g values. The lowest contour is 15% of maximum. (Source: Joshi et al., 2005 [47]. Reproduced with permission of Elsevier Limited.)

field gradients that are used to encode spatial information, the amplitude of the CW first-derivative spectrum decreases approximately quadratically with the magnitude of the gradient. The amplitude of the rapid-scan absorption spectrum decreases only linearly with the magnitude of the gradient [47], which is a large advantage at the high gradients that define spatial resolution. The first rapid-scan imaging demonstration was a proof-of-principle experiment using two LiPc samples and a solution of trityl radical in a spectrometer operating at 248 MHz (Figure 2.30) [47].

An improved image reconstruction method was developed using the maximum entropy method (MEM) and compared with filtered back projection (FBP) (Figure 2.31) [12]. Rapid-scan signals were recorded using triangular scans with frequencies in the range 1–8 kHz and scan widths in the range 0.85–6.93 G. Each approach was found to have advantages and disadvantages. For FBP the advantages are that (i) less computation time is required and (ii) the relative intensities of features in the image are more accurate. The disadvantages of FBP are that (i) the "star" effect is observed when the number of projections is small, (ii) projections must be equally spaced, (iii) imperfections in a small number of projections adversely impact the whole image, and (iv) both noise and streak-like artifacts in the image distort spin concentration profile along the spatial axis; For MEM the advantages are that (i) even with a small number of projections, there is no "star" effect, (ii) less noise in baseline regions of the image permits recognition of weak signals, (iii) non-negativity is implicit in the algorithm, (iv) projections do not need

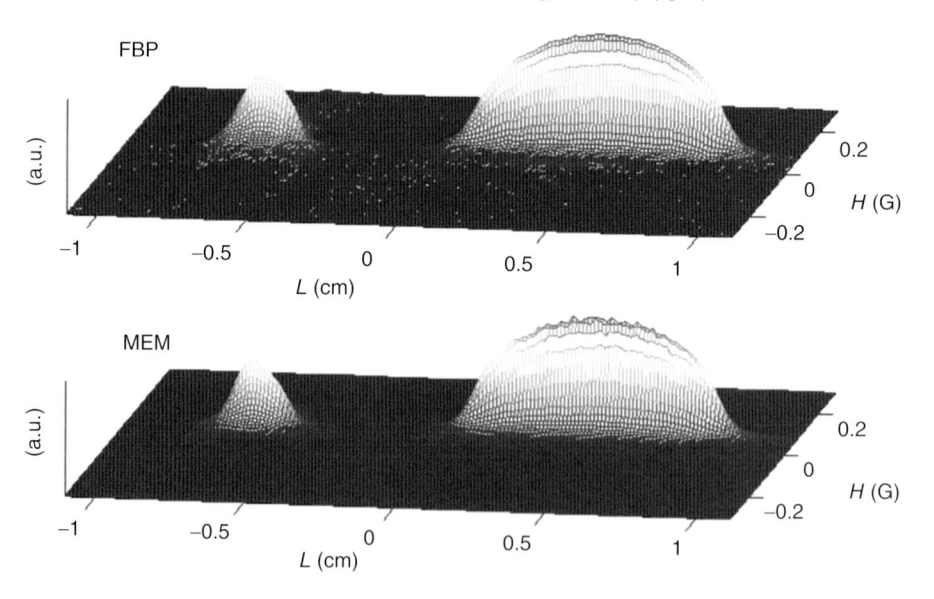

**Figure 2.31** (a, b) 2-D spectral-spatial EPR image of a small tube containing solid LiPc and a larger tube containing an aqueous solution of trityl-CD$_3$. The centers of tubes were separated by 11 mm. Each of the 60 projections was averaged 50 000 times with scan frequencies of 1–8 kHz and a scan rate of 13.9 kG/s. A Hamming filter was used in conjunction with FBP. (Source: Tseitlin et al., 2007 [12]. Reproduced with permission of Elsevier Limited.)

to be at equally spaced angular increments, (v) the shape of the spatial profile is more accurate, and (vi) there is better matching with experimental projections. The disadvantages of MEM are that (i) it is computationally intensive, (ii) the amplitude scale is nonlinear and amplitudes of weak peaks are underestimated, and (iii) noise superimposed on peaks is higher than that with FBP. The overall pattern is that FBP works well when the number of projections is large enough that the star effect is negligible and $S/N$ is higher. MEM has advantages when projections are not equally spaced, when there are fewer projections, and/or when $S/N$ is not as good.

Another image reconstruction method that has been applied to rapid-scan imaging is regularized optimization (RO) (Figure 2.32) [91]. The phantom tested consisted of tubes containing LiPc in different oxygen concentrations. Instead of creating a 2D spectral-spatial image, two one-dimensional profiles were reconstructed: the concentration of the radical and the corresponding oxygen concentration, which reduced the dimensionality of the problem. The algorithm seeks to minimize the discrepancy between experimental data and projections calculated from the profiles, and uses Tikhonov regularization [92] to constrain the smoothness of the results. This approach controllably smoothens profiles rather than the data, while preserving sharp features. The spatial distribution of oxygen can be defined more precisely than by FBP [91].

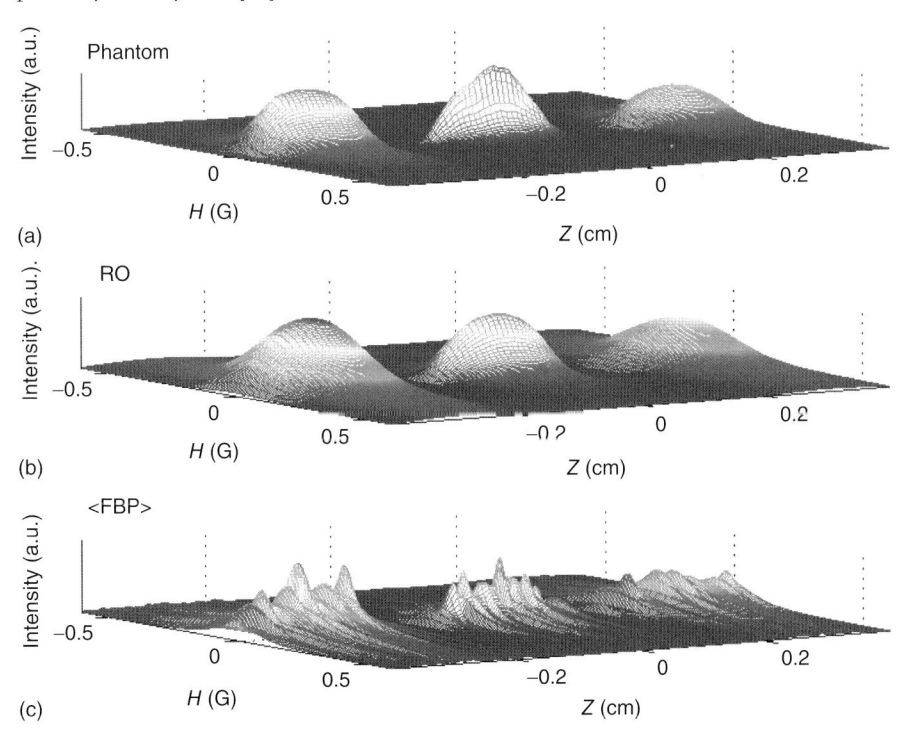

**Figure 2.32** (a–c) Comparison of the images reconstructed by RO and FBP with the phantom image that consisted of three tubes with LiPc at different oxygen concentrations. (Source: Tseitlin et al., 2008 [91]. Reproduced with permission of Elsevier Limited.)

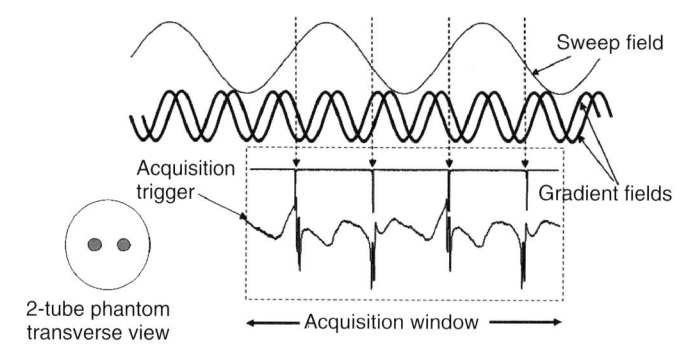

**Figure 2.33** Direct-detected rapid-scan absorption EPR signals of a phantom consisting of two capillary tubes containing 1 mg each of particulate TCNQ. The top sinusoid represents the rapid-scan sweep, and the two higher frequency sinusoids represent the X- and Z-gradients, which together provided the rotating gradients in the XZ plane. The left side of the dotted rectangle represents the start of the trigger. The field scan frequency was 333.33 Hz and the gradient rotation frequency was 1 kHz. The sampling frequency was 4 Ms/s and 25 000 points were collected, giving rise to two downfield scans (first and third spectra) and two upfield scans (second and fourth spectra). (Source: Subramanian et al., 2007 [42]. Reproduced with permission of Elsevier Limited.)

Subramanian et al. [42] performed rapid-scan imaging of a TCNQ salt, which has an exchange-narrowed EPR line (Figure 2.33). The field-scan frequency was 333.33 Hz and the magnetic field gradient was rotated at 1 kHz. By rotating the gradients and recording the rapid-scan response, it was possible to record data much faster than that accomplished with CW EPR. Rapid-scan oscillations were not observed, so deconvolution was not required. Analogous imaging was performed with solutions of trityl radicals.

## 2.13
### Extension of the Rapid-Scan Technology to Scans That Are Not Fast Relative to Relaxation Times

EPR instrumentation evolved from direct detection of absorption spectra, to field modulated, phase-sensitive detected spectra displayed in derivative mode relatively early in the development of the field. However, difficulties of applying magnetic field modulation and distortions of lineshape by the modulation stimulated efforts to find alternatives to magnetic field modulation [93]. Many approaches toward avoiding magnetic field modulation and understanding the effect of resonator construction on signal distortion have been developed by the Hyde laboratory. The most recent approach is called NARS (nonadiabatic rapid sweep). NARS is a subset of the methodology described in this chapter, for the case in which the scan is slow relative to relaxation times.

The increased signal amplitude relative to that for a slow-scan CW spectrum acquired with conservative modulation amplitudes was used by Kittell et al. [7, 43] even without using further enhancements possible at higher scan rates and

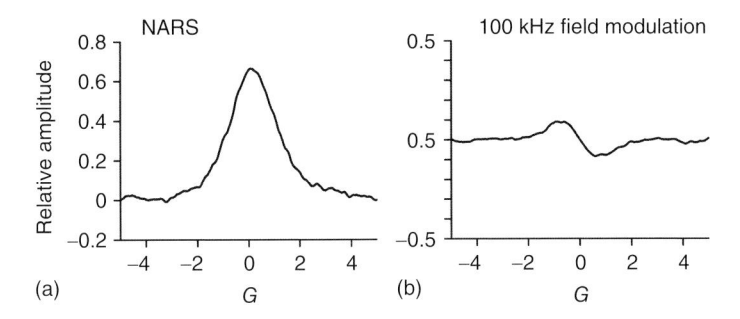

**Figure 2.34** Comparison of (a) NARS and (b) field modulation spectra of 10 μM tempol in water at L band. A factor of 5 improvement in S/N was observed with NARS detection. Data were acquired in the same acquisition time. (Source: Kittell et al., 2011 [7]. Reproduced with permission of Elsevier Limited.)

higher incident microwave power. All of the examples in this chapter from the Denver laboratory are cases in which the magnetic field scan included the entire EPR spectrum. The Hyde laboratory is developing a method of scanning spectra in small segments and combining the segments in post-acquisition processing. This latter method has been applied to nitroxides in fluid solution, and to broader spectra, such as rigid lattice nitroxide [43] and Cu(II) [94]. An example of the improvement in signal amplitude is shown in Figure 2.34.

"Rapid" magnetic field scans have been employed in other EPR experiments, but not with the same meaning as in this chapter. For example, Oikawa et al. [95] implemented "rapid" (up to 15 mT/s (150 G/s)) scan of the magnetic field of iron-core and air-core L-band EPR magnets [50, 95–98]. Hirasawa et al. [51] scanned magnetic fields rapidly (1–50 ms, 0–100 G linear scans) to observe transient radicals generated by pulse electrolysis. First-derivative EPR spectra of tetracyanoethylene anion radical in tetrahydrofuran were obtained with 100 mG modulation. Hsi et al. [71] used a trapezoidal driving voltage to produce a linear current ramp in 10 cm diameter Helmholtz coils. Sweeps of up to 75 G in times as fast as 5 ms were obtained, but not simultaneously. A 400 ms scan of the EPR spectrum of perylene in sulfuric acid was presented. Magnetic field modulation of 100 KHz was used. Zahariou et al. [72] studied tyrosine Z of photosystem II with rapid CW scans as short as 200 ms. In these examples, CW EPR spectra were recorded, which did not exhibit any of the passage effects discussed in this chapter. Since deconvolution does not alter spectra in the slow-scan regime, it can be applied to spectra that contain signals that are anywhere in the continuum from slow to rapid-scan regimes, including superpositions of such signals.

## 2.14 Summary

In rapid-scan EPR, the magnetic field or frequency is scanned and the signal is directly detected in quadrature at the resonance frequency. Signals obtained by sinusoidal or linear magnetic field or microwave frequency scans that are rapid relative to electron spin relaxation times can be deconvolved to provide the EPR absorption signal. In CW EPR, a segment of the first-derivative signal is detected by phase-sensitive detection at a modulation frequency. In rapid-scan EPR, the full amplitude of the absorption signal is obtained by direct detection, which improves $S/N$ relative to CW EPR. In addition, rapid scanning permits use of higher incident power, which further increases signal amplitude. Coherent averaging of rapid scans reduces noise, analogous to that achieved by phase-sensitive detection at a modulation frequency in CW EPR. The net result for a wide variety of samples is dramatically improved $S/N$ for rapid scan relative to CW EPR, as demonstrated by the examples in Section 2.12.

The background, theory, instrumentation, and methodology of rapid-scan EPR are described in this chapter. Rapid-scan EPR has been implemented from 250 MHz to 95 GHz EPR, and at a wide range of scan rates, from very slow to those exceeding one gigagauss per second. The advantages of rapid-scan EPR relative to CW and pulsed EPR are sample dependent. Rapid scan is particularly advantageous for samples with long spin-lattice relaxation times. For samples with long electron-spin relaxation times, rapid-scan deconvolution accounts for passage effects so that accurate lineshapes can be recovered.

Rapid-scan EPR is particularly advantageous for imaging because the amplitude of the absorption signal decreases approximately linearly with magnetic field gradient, whereas the amplitude of the first-derivative CW EPR decreases approximately quadratically. This is a substantial advantage in $S/N$ for high-gradient spectra that are needed to define spatial resolution.

The ability of rapid-scan EPR to acquire data rapidly permits higher temporal resolution for kinetics than can be achieved with CW spectroscopy. The combination of rapid scan with improvements in digital electronics provides opportunities to revolutionize the way that much EPR will be done in the future.

### Acknowledgments

The work in the Denver laboratory on rapid-scan EPR is funded by National Science Foundation grant IDBR 0753018, National Institutes of Health grant EB000557, and an NSF graduate research fellowship to DGM. We thank Martyna Elas and Janusz Koscielniak for translating parts of Ref. [30] for us.

### References

1. Poole, C.P. (1983) *Electron Spin Resonance: A Comprehensive Treatise on Experimental Techniques*, 2nd edn, John Wiley & Sons, Inc., New York.

2. Weil, J.A., Bolton, J.R., and Wertz, J.E. (1994) *Electron Paramagnetic Resonance: Elementary Theory and Practical Applications*, John Wiley & Sons, Inc., New York.

3. Keijzers, C.P., Reijerse, E.J., and Schmidt, J. (eds) (1989) *Pulsed EPR: A New Field of Applications*, North Holland Publishing Co., Amsterdam.
4. Eaton, S.S., Eaton, G.R., and Berliner, L.J. (eds) (2005) *Biomedical EPR – Part B: Methodology, Instrumentation, and Dynamics*, Kluwer Academic/Plenum Press, New York.
5. Schweiger, A. and Jeschke, G. (2001) *Principles of Pulse Electron Paramagnetic Resonance*, Oxford University Press, Oxford.
6. Dikanov, S.A. and Tsvetkov, Y.D. (1992) *Electron Spin Echo Envelope Modulation (ESEEM) Spectroscopy*, CRC Press, Boca Raton, FL.
7. Kittell, A.W., Camenisch, T.G., Ratke, J.J., Sidabras, J.W., and Hyde, J.S. (2011) *J. Magn. Reson.*, **211**, 228–233.
8. Quine, R.W., Czechowski, T., and Eaton, G.R. (2009) *Conc. Magn. Reson. B (Magn. Reson. Eng.)*, **35B**, 44–58.
9. Quine, R.W., Mitchell, D.G., Eaton, S.S., and Eaton, G.R. (2012) *Conc. Magn. Reson., Magn. Reson. Eng.*, **41B**, 95–110.
10. Mitchell, D.G., Quine, R.W., Tseitlin, M., Eaton, S.S., and Eaton, G.R. (2012) *J. Magn. Reson.*, **214**, 221–226.
11. Mitchell, D.G., Tseitlin, M., Quine, R.W., Meyer, V., Newton, M.E., Shnegg, A., George, B., Eaton, S.S., and Eaton, G.R. (2013) *Mol. Phys.*, **111**, 922–929. doi: 10.1008/00268976.00262013.00792959
12. Tseitlin, M., Dhami, A., Eaton, S.S., and Eaton, G.R. (2007) *J. Magn. Reson.*, **184**, 157–168.
13. Mitchell, D.G., Quine, R.W., Tseitlin, M., Weber, R.T., Meyer, V., Avery, A., Eaton, S.S., and Eaton, G.R. (2011) *J. Phys. Chem. B*, **115**, 7986–7990.
14. Tseitlin, M., Dhami, A., Quine, R.W., Rinard, G.A., Eaton, S.S., and Eaton, G.R. (2006) *Appl. Magn. Reson.*, **30**, 651–656.
15. Bloembergen, N., Purcell, E.M., and Pound, R.V. (1948) *Phys. Rev.*, **73**, 679–712.
16. Jacobsohn, B.A. and Wangsness, R.K. (1948) *Phys. Rev.*, **73**, 942.
17. Dadok, J. and Sprecher, R.F. (1974) *J. Magn. Reson.*, **13**, 243–248.
18. Gupta, R.K., Ferretti, J.A., and Becker, E.D. (1974) *J. Magn. Reson.*, **13**, 275–290.
19. Gupta, R.K., Ferretti, J.A., and Becker, E.D. (1974) *J. Magn. Reson.*, **16**, 505–507.
20. Beeler, R., Roux, D., Bene, G., and Extermann, R. (1955) *C. R. Acad. Sci. Fr.*, **94**, 472–473.
21. Beeler, R., Roux, D., Bene, G., and Extermann, R. (1956) *Phys. Rev.*, **102**, 296.
22. Gabillard, R. (1957) *Arch. Sci.*, 107–108.
23. Gabillard, R. and Ponchel, B. (1963) *Proc. Colloq. Ampere*, **11**, 749–757.
24. Portis, A.M. (1955) *Phys. Rev.*, **100**, 1219–1221.
25. Hyde, J.S. (1960) *Phys. Rev.*, **119**, 1483–1492.
26. Feher, G. (1956) *Phys. Rev.*, **103**, 500–501.
27. Feher, G. and Gere, E.A. (1956) *Phys. Rev.*, **103**, 501–503.
28. Feher, G. (1959) *Phys. Rev.*, **114**, 1219.
29. Weger, M. (1960) *Bell Syst. Tech. J.*, **39**, 1013–1112.
30. Czoch, R., Duchiewicz, J., Francik, A., Indyka, S., and Koscielniak, J. (1983) *Meas. Automatic Control*, **29**, 41–43.
31. Hyde, J.S. and Dalton, L. (1972) *Chem. Phys. Lett.*, **16**, 568–572.
32. Hoffman, B.M., Diemente, D.L., and Basolo, F. (1970) *J. Am. Chem. Soc.*, **92**, 61–65.
33. Seamonds, B., Blumberg, W.E., and Peisach, J. (1972) *Biochim. Biophys. Acta*, **263**, 507–514.
34. Mailer, C. and Taylor, C.P.S. (1973) *Biochim. Biophys. Acta*, **322**, 195–203.
35. Galtsev, V.E., Grinberg, O.Y., Lebedev, Y.S., and Galtseva, E.V. (1993) *Appl. Magn. Reson.*, **4**, 331–333.
36. Campbell, I.D. (1987) *J. Magn. Reson.*, **74**, 155–157.
37. Dey, S., Torgeson, D.R., and Barnes, R.G. (1986) *Appl. Phys. Lett.*, **49**, 1092–1094.
38. McGurk, J.C., Schmaltz, T.G., and Flygare, W.H. (1974) *J. Chem. Phys.*, **60**, 4181–4188.
39. Duxbury, G., Langford, N., McCulloch, M.T., and Wright, S. (2005) *Chem. Soc. Rev.*, **34**, 921–934.

40. Hyde, J.S., Strangeway, R.A., Camenisch, T.G., Ratke, J.J., and Froncisz, W. (2010) *J. Magn. Reson.*, **205**, 93–101.
41. Subramanian, S., Koscielniak, J., Devasahayam, N., Pursely, R.H., Pohida, T.J., and Krishna, M. (2007) in *EPR2007* (eds H.J. Halpern and B. Kalyanaraman), This is the proceedings for the 12th International Conference In Vivo EPR Spectroscopy and Imaging, Chicago, IL May 2007. p. O-11.
42. Subramanian, S., Koscielniak, J., Devasahayam, N., Pursely, R.H., Pohida, T.J., and Krishna, M. (2007) *J. Magn. Reson.*, **186**, 212–219.
43. Kittell, A.W., Hustedt, E.J., and Hyde, J.S. (2012) *J. Magn. Reson.*, **221**, 51–56.
44. Harbridge, J.R., Rinard, G.A., Quine, R.W., Eaton, S.S., and Eaton, G.R. (2002) *J. Magn. Reson.*, **156**, 41–51.
45. Mitchell, D.G., Quine, R.W., Tseitlin, M., Meyer, V., Eaton, S.S., and Eaton, G.R. (2011) *Radiat. Meas.*, **46**, 993–996.
46. Joshi, J.P., Eaton, G.R., and Eaton, S.S. (2005) *Appl. Magn. Reson.*, **28**, 239–249.
47. Joshi, J.P., Ballard, J.R., Rinard, G.A., Quine, R.W., Eaton, S.S., and Eaton, G.R. (2005) *J. Magn. Reson.*, **175**, 44–51.
48. Tseitlin, M., Rinard, G.A., Quine, R.W., Eaton, S.S., and Eaton, G.R. (2011) *J. Magn. Reson.*, **208**, 279–283.
49. Tseitlin, M., Mitchell, D.G., Eaton, S.S., and Eaton, G.R. (2012) *J. Magn. Res.*, **223**, 80–84.
50. Ogata, T. (1995) in *Bioradicals Detected by ESR Spectroscopy* (eds H. Ohya-Nishiguchi and L. Paker), Birkhauser Verlag, Basel, pp. 103–111.
51. Hirasawa, R., Mukaibo, T., Hasegawa, H., Kanda, Y., and Maruyama, T. (1968) *Rev. Sci. Instrum.*, **39**, 935–937.
52. Bleaney, B. (1951) *Proc. Phys. Soc. London*, **A64**, 933–935.
53. Bleaney, B. and Stevens, K.W.H. (1953) *Rep. Prog. Phys.*, **16**, 108–159.
54. Stoner, J.W., Szymanski, D., Eaton, S.S., Quine, R.W., Rinard, G.A., and Eaton, G.R. (2004) *J. Magn. Res.*, **170**, 127–135.
55. Eaton, S.S. and Eaton, G.R. (1993) *J. Magn. Reson.*, **102**, 354–356.
56. Eaton, S.S. and Eaton, G.R. (2005) in *Analytical Instrumentation Handbook*, 3rd edn (ed. J. Cazes), Marcel Dekker, New York, pp. 349–398.
57. Robinson, B.H., Mailer, C., and Reese, A.W. (1999) *J. Magn. Reson.*, **138**, 199–209.
58. Robinson, B.H., Mailer, C., and Reese, A.W. (1999) *J. Magn. Reson.*, **138**, 210–219.
59. Mailer, C., Robinson, B.H., Williams, B.B., and Halpern, H.J. (2003) *Magn. Reson. Med.*, **49**, 1175–1180.
60. Deng, Y., Pandian, R.P., Ahmad, R., Kuppusamy, P., and Zweier, J.L. (2006) *J. Magn. Reson.*, **181**, 254–261.
61. Nielsen, R.D. and Robinson, B.H. (2004) *Conc. Magn. Reson. A*, **23**, 38–48.
62. Bikineev, V.A., Zavatskii, E.A., Isaev-Ivanov, V.V., Lavrov, V.V., Lomakin, A.V., Fomichev, V.N., and Shabalin, K.A. (1995) *Tech. Phys.*, **40**, 619–625.
63. Tseitlin, M., Quine, R.W., Rinard, G.A., Eaton, S.S., and Eaton, G.R. (2011) *J. Magn. Reson.*, **213**, 119–125.
64. Quine, R.W., Rinard, G.A., Eaton, S.S., and Eaton, G.R. (2010) *J. Magn. Reson.*, **205**, 23–27.
65. Tseitlin, M., Eaton, S.S., and Eaton, G.R. (2012) *Conc. Magn. Reson.*, **40A**, 295–305.
66. Eaton, G.R., Eaton, S.S., Barr, D.P., and Weber, R.T. (2010) *Quantitative EPR*, Springer-Verlag/Wein, New York.
67. Klein, M.P. and Barton, B.W. (1963) *Rev. Sci. Instrum.*, **34**, 754–759.
68. Hyde, J.S., Jesmanowicz, A., Ratke, J.J., and Antholine, W.E. (1992) *J. Magn. Reson.*, **96**, 1–13.
69. Rinard, G.A., Quine, R.W., Eaton, S.S., Eaton, G.R., Barth, E.D., Pelizzari, C.A., and Halpern, H.J. (2002) *Magn. Reson. Eng.*, **15**, 51–58.
70. Rengan, S.K., Bhagat, V.R., Sastry, V.S.S., and Venkataraman, B. (1979) *J. Magn. Reson.*, **33**, 227–240.
71. Hsi, E.S., Fabes, L., and Bolton, J.R. (1973) *Rev. Sci. Instrum.*, **44**, 197–199.
72. Zahariou, G., Ioannidis, N., Sioros, G., and Petrouleas, V. (2007) *Biochemistry*, **46**, 14335–14341.
73. Sloop, D.J., Lin, T.-S., and Ackerman, J.J.H. (1999) *J. Magn. Reson.*, **139**, 60–66.

74. Fedin, M., Gromov, I., and Schweiger, A. (2004) *J. Magn. Reson.*, **171**, 80–89.
75. Fedin, M., Gromov, I., and Schweiger, A. (2006) *J. Magn. Reson.*, **182**, 293–297.
76. Rinard, G.A., Quine, R.A., Biller, J.R., and Eaton, G.R. (2010) *Concepts Magn. Reson. Part B: Magn. Reson. Eng.*, **37B**, 86–91.
77. Quine, R.W., Mitchell, D., and Eaton, G.R. (2011) *Concepts Magn. Reson. Part B: Magn. Reson. Eng.*, **39B**, 43–46.
78. Rinard, G.A., Quine, R.W., Eaton, G.R., and Eaton, S.S. (2002) *Magn. Reson. Eng.*, **15**, 37–46.
79. Tseitlin, M., Czechowski, T., Quine, R.W., Eaton, S.S., and Eaton, G.R. (2009) *J. Magn. Reson.*, **196**, 48–53.
80. Tseitlin, M., Quine, R.W., Rinard, G.A., Eaton, S.S., and Eaton, G.R. (2010) *J. Magn. Reson.*, **203**, 305–310.
81. Ernst, R.E. and Anderson, W.A. (1966) *Rev. Sci. Instrum.*, **37**, 93–102.
82. Eaton, S.S. and Eaton, G.R. (2000) *Biol. Magn. Reson.*, **19**, 29–154.
83. Biller, J.R., Meyer, V.M., Elajaili, H., Rosen, G.M., Eaton, S.S., and Eaton, G.R. (2012) *J. Magn. Reson.*, **225**, 52–57.
84. Lloyd, J.P. and Pake, G.E. (1954) *Phys. Rev.*, **94**, 579–591.
85. Hyodo, F., Yasukawa, K., Yamada, K., and Utsumi, H. (2006) *Magn. Reson. Med.*, **56**, 938–943.
86. Fridovich, I. (1985) in *CRC Handbook of Methods for Oxygen Radical Research* (ed. R.A. Greenwald), CRC Press, Boca Raton, FL, pp. 51–53.
87. Mitchell, D.G., Rosen, G.M., Tseitlin, M., Symmes, B., Eaton, S.S., and Eaton, G.R. (2013) *Biophys. J.*, **105**, 338–342.
88. Tsai, P., Ichikawa, K., Mailer, C., Pou, S., Halpern, H.J., Robinson, B.H., Nielsen, R.D., and Rosen, G.M. (2003) *J. Org. Chem.*, **68**, 7811–7817.
89. Talmon, Y., Shtirberg, L., Harneit, W., Rogozhnikova, O.Y., Tormyshev, V., and Blank, A. (2010) *Phys. Chem. Chem. Phys.*, **12**, 5998–6007.
90. Ernst, R.R. (1965) *Rev. Sci. Instrum.*, **36**, 1689–1695.
91. Tseitlin, M., Czechowski, T., Eaton, S.S., and Eaton, G.R. (2008) *J. Magn. Reson.*, **194**, 212–221.
92. Tikhonov, A.N. and Arsenin, V.Y. (1986) *Methods of Solving Ill-Posed Problems*. Nauka, Moscow.
93. Hyde, J.S., Sczaniecki, P.B., and Froncisz, W. (1989) *J. Chem. Soc., Faraday Trans. 1*, **85**, 3901–3912.
94. Kittell, A.W., Sidabras, J.W., Bennett, B., and Hyde, J.S. (2012) Segmental non-adiabatic rapid sweep collection of Cu(II), Rocky Mountain Conference on Analytical Chemistry, Copper Mountain, CO, Abstract 236.
95. Oikawa, K., Ogata, T., Sato, T., Kudo, R., and Kamada, H. (1995) *Anal. Sci.*, **11**, 885–888.
96. Oikawa, K., Ogata, T., Togashi, H., Yokoyama, H., Ohya-Nishiguchi, H., and Kamada, H. (1996) *Appl. Radiat. Isot.*, **47**, 1605–1609.
97. Lin, Y., Ogata, T., Watanabe, H., and Akatsuka, T. (1997) *Anal. Sci.*, **13**, 269–272.
98. Togashi, H., Matsuo, T., Shinzawa, H., Takeda, Y., Shao, L., Oikawa, K., Kamada, H., and Takahashi, T. (2000) *Magn. Reson. Imaging*, **18**, 151–156.

# 3
# Computational Modeling and Least-Squares Fitting of EPR Spectra
*Stefan Stoll*

## 3.1
## Introduction

In EPR (electron paramagnetic resonance) spectroscopy, computer simulation and least-squares fitting are essential in extracting quantitative structural and dynamic parameters from experimental spectra. Without numerical methods, this extraction would be restricted to simple systems. This chapter summarizes simulation and fitting methods that have been proposed in the literature and implemented in software. It includes an extensive, though not complete, list of references.

Emphasis is placed on methods currently implemented in the software package EasySpin [1], which covers EPR simulations in the following regimes: (i) rigid-limit continuous-wave (cw) EPR spectra for arbitrary spin systems, for both powders and single crystals, at various levels of theory including eigenfields, matrix diagonalization, and perturbation theory; (ii) dynamic EPR spectra of tumbling spin centers with one electron spin and several nuclei, implementing stochastic Liouville equation (SLE) solvers and perturbative approaches; (iii) EPR spectra in the fast-motion limit, using either a Breit–Rabi solver or perturbation theory; (iv) dynamic EPR spectra due to chemical exchange in solution, implementing a direct Liouville-space method; (v) solid-state ENDOR (electron nuclear double resonance) spectra based on either matrix diagonalization or perturbation theory; and (vi) pulse EPR spectra for general pulse sequences using the Hilbert-space density matrix formalism in the high-field limit. All these simulation regimes are reviewed in the following.

Similarly to many other programs, EasySpin also provides a range of least-squares fitting algorithms, among them Levenberg–Marquardt (LM), Nelder–Mead simplex, genetic algorithms, particle-swarm optimization, as well as simple Monte Carlo and grid searches. These algorithms, as well as the objective function choice, multicomponent fitting, and error analysis, are discussed below.

This chapter is not intended to be a complete review of all theory underlying EPR simulation methods, which would be utterly impossible. Instead, it summarizes theoretical and algorithmic aspects that are implemented in or are relevant to EasySpin. Applicability and limitations of methods are discussed as well. The chapter is not concerned with the specifics of usage of software packages. Tutorials and documentation for EasySpin can be found online at *easyspin.org*.

Many reviews have appeared over time that summarize progress in the methodology for EPR spectral simulation and fitting and that describe available simulation programs, starting with very early ones [2–4] up to more recent times [5–8]. A previous *Handbook of ESR* included a review on computer techniques [9]. A very detailed review of simulation methods and programs as of 1992 is contained in the book by Mabbs and Collison [10]. A list of software available in 1993 is published [11].

In the following, after summarizing key aspects of available simulation software packages, we discuss the basic aspects of EPR simulations and then progress to describe methods for static and dynamic cw EPR spectra, pulse EPR, ENDOR, and DEER (double electron–electron resonance) spectra. Subsequently, a section is dedicated to least-squares fitting. After a short section covering topics such as spin quantitation and data formats, we summarize in the conclusion some of the challenges that still lie ahead.

## 3.2
## Software

In this section, we describe a few details about EasySpin and other EPR simulation programs. Some of them are available online, and many others can be obtained from their authors. A few have ceased to be developed and are no longer maintained.

### 3.2.1
### EasySpin

EasySpin, developed by the author, was originally conceived as an in-house simulation program for solid-state cw EPR spectra in the laboratory of Arthur Schweiger at ETH Zurich, with a first public release in 2000. The initial work is documented in a 2003 PhD thesis [12] and, including subsequent extensions, in a 2006 article in *Journal of Magnetic Resonance* [1]. A summary of EasySpin functionality relevant to nitroxides was subsequently published [13].

Since its first publication, EasySpin has advanced on many levels. Thanks to feedback from the worldwide user community, bugs were corrected, algorithms became more robust, implementations became faster, and more regimes and experiments were added. Notably, support for pulse EPR simulations was added in 2009 [14], least-squares fitting was introduced in 2010, and chemical exchange was implemented in 2012.

The program continues to be developed, with the ultimate goal of removing the data analysis and simulation bottleneck from the EPR discovery process. Its core strengths are solid-state cw EPR spectra as well as ENDOR and ESEEM (electron spin echo envelope modulation) spectra, with growing support for slow-motion simulations and other more specialized situations.

## 3.2.2
### Other Software

EasySpin draws substantially from methods implemented in other, mostly older, EPR simulation programs. In the following, we give a partial list. The National Institute of Environmental Health Sciences (NIEHS) maintains a database of EPR simulation programs (electron spin resonance software database, ESDB) [15], including programs of limited availability and dedicated to specific problems.

Bruker ships certain spectrometers with SimFonia, a simulation program developed by Weber at Bruker in the 1990s [16]. Hanson and coworkers have developed Sophe, a widely used simulation program for solid-state EPR spectra [17–23] that has been equipped with a graphical user interface (UI) by Bruker and marketed as XSophe. A more modern UI to Sophe called *Molecular Sophe* (MoSophe) has recently been developed [24].

WinSIM is dedicated to solution spectra of spin traps and was developed at the NIEHS [25]. Hendrich [26] has developed SpinCount, a program that emphasizes spin quantitation. Slow-motion spectra of nitroxide radicals can be simulated and fitted using the suite of highly optimized SLE solvers developed by the Freed group at the ACERT center at Cornell [27–30]. Altenbach has developed a code dedicated to nitroxide labels [31]. Dipolar broadening of cw EPR spectra of nitroxides can be analyzed using DIPFIT [32]. E-SpiReS is a program for slow-motion simulation that also interfaces to quantum chemistry programs [33, 34]. At Manchester, an in-house code has been used to simulate hundreds of spectra in a book about transition metal ion EPR [10]. Weil's program EPRNMR [35] is designed for solid-state EPR and has extensive support for single-crystal spectra. DDPOW supports binuclear complexes [36]. QPOW [37] and SIMPOW6 [38] were developed at the University of Illinois. Sim is a program by Weihe that accepts arbitrary Hamiltonian matrices as input [39, 40]. SPIN, developed at the National High Magnetic Field Laboratory, is tailored toward high-spin systems. Xemr is a general-purpose EPR simulation program [41]. EPRsim32 [42] is a powder cw EPR simulation program that includes genetic fitting algorithms. Rockenbauer and Korecz [43] have developed a general simulation program that includes chemical exchange. Another still popular program for chemical exchange was created by Heinzer in the early 1970s [44, 45]. WinMOMD is a program for simulation of slow-motional nitroxide spectra using the MOMD (microscopic order, macroscopic disorder) model [46]. EWVoigt is geared toward nitroxide spectra in the fast-motion regime and utilizes convolution methods [47]. EPRSIM-C implements a variety of models for nitroxide spectra and includes evolutionary fitting algorithms [48].

Several programs were developed specifically for ENDOR and ESEEM simulations. MAGRES from Nijmegen [49, 50] was an early one. GENDOR is an ENDOR simulation program developed by Hoffman at Northwestern [51–53]. HYSCORE (hyperfine sublevel correlation) simulation programs were pioneered by Goldfarb [54] and Schweiger [55]. Astashkin's program SimBud is equipped with a UI [56]. OPTESIM [57] provides ESEEM simulations and least-squares fitting.

Many simulation programs for NMR spectra have been developed over the years and have been reviewed [58–61]. Among the many programs, SIMPSON [62], SPINEVOLUTION [63], and Spinach [64] are particularly widely used. Spinach is a very general and efficient spin dynamics code that is geared toward large NMR spin systems, but supports EPR experiments as well.

In addition to the programs mentioned, there are many excellent in-house codes developed by various research groups, but are not separately described in literature, and are either not distributed or have not seen widespread use.

## 3.3
## General Principles

### 3.3.1
### Spin Physics

The simulation of EPR spectra is based on a spin Hamiltonian that describes the interactions amongst the spins in the spin system and between the spins and the externally applied magnetic field. The following summarizes the most common terms in the spin Hamiltonian used to model EPR spectra [65]. We do not intend to outline the complete theoretical basis. Instead, the discussion is limited to some aspects that are often overlooked by users and that are important for obtaining correct simulation results. We also summarize the basic quantum dynamic equations needed to compute EPR spectra.

#### 3.3.1.1  Interactions

EPR spectra are generally simulated on the basis of a spin Hamiltonian (sH), an effective Hamiltonian that represents the subset of closely spaced and low-lying energy levels of a spin center that are accessible in EPR experiments as a spin system, a network of coupled (effective) electron spins and nuclear spins [66]. The sH model is not universally valid and becomes inadequate, for example, in the presence of very large spin–orbit coupling or in the gas phase. Essentially all common simulation programs are based on an sH. The sH (often expressed in angular frequency units) consists of a sum of interaction terms

$$H = \sum_i (H_{ez,i} + H_{zf,i}) + \sum_{i,j} H_{ss,i,j} + \sum_{i,k} H_{hf,i,k} + \sum_k (H_{ez,i} + H_{nq,i})$$

where $i$ and $j$ run over the electron spins and $k$ runs over the nuclear spins in the system. In the following, we present the conventional forms of the various terms that mostly utilize Cartesian spin vector operators, $\mathbf{S}^T = (S_x, S_y, S_z)^T$ and $\mathbf{I}^T = (I_x, I_y, I_z)^T$, where T indicates the transpose of the matrix.

The electron Zeeman interaction in angular frequency units is described by

$$H_{ez} = \frac{-\mathbf{B}^T \boldsymbol{\mu}_{el}}{\hbar} = +\left(\frac{\mu_B}{\hbar}\right) \mathbf{B}^T \mathbf{g} \mathbf{S}$$

with the 3 × 3 g-matrix g. The externally applied magnetic field $\mathbf{B}$ includes both the static and microwave fields. Another form of the term is $\left(\frac{\mu_B}{\hbar}\right) \mathbf{S}^T g \mathbf{B}$. Often, it is assumed that $g$ is symmetric, $g = g^T$, since it is not possible to determine the antisymmetric component of $g$ from conventional cw EPR spectra employing linearly polarized microwave [67]. In this case, the two forms are identical. However, g-matrices calculated by quantum chemical programs are generally asymmetric [68]. Then, the two forms are not identical [65] and one has to verify that the quantum chemistry program and the EPR simulation program assume the same form. If the EPR simulation software only supports symmetric g matrices, the effective symmetric matrix $g_{sym}$ corresponding to a given asymmetric $g$ can be obtained from $g_{sym} = (gg^T)^{\frac{1}{2}}$.

Higher order electron Zeeman terms proportional to $\mathbf{B}$ and $\mathbf{S}^3$ or $\mathbf{S}^5$ are in principle possible [66] and have been reported [69–72]. Very few programs, such as EPRNMR [73], have provisions for these terms.

The nuclear Zeeman interaction contribution to the EPR spin Hamiltonian (in angular frequency units) is given by

$$H_{nz} = \frac{-\mathbf{B}^T \boldsymbol{\mu}_{nuc}}{\hbar} = -\left(\frac{\mu_N}{\hbar}\right) \mathbf{B}^T g_n \mathbf{I}$$

where $g_n$ is the isotropic nuclear g-factor and $\mu_N$ is the nuclear Bohr magneton. Any anisotropy in $g_n$ (chemical shift anisotropy) is very small compared to $g_n$ itself and is generally neglected in EPR. The pseudo-nuclear Zeeman effect [67, 74] in high-electron-spin systems manifests itself in a significant apparent anisotropy of an effective $g_n$. It arises naturally when the full spin system is simulated, but can be taken into account explicitly in perturbational treatments with a restricted system.

The hyperfine (hf) interaction (in angular frequency units) between an electron spin $\mathbf{S}$ and a nuclear spin $\mathbf{I}$ is anisotropic and described by a general 3 × 3 coupling matrix $A$

$$H_{hf} = \mathbf{S}^T A \mathbf{I}$$

Another form of the hf interaction $\mathbf{I}^T A \mathbf{S}$ is identical to $\mathbf{S}^T A \mathbf{I}$ only if the hyperfine coupling matrix $A$ is symmetric. In general, there are three contributions to $A : A = a_{iso} + T + A_L$, where $a_{iso}$ is the isotropic Fermi contact term, $T$ is the matrix describing the magnetic dipole–dipole coupling between the electron and nucleus (axially symmetric in the limit of the point-dipole approximation of a completely localized electron spin), and $A_L$ is a generally asymmetric 3 × 3 matrix describing the orbital contribution [75]. In analogy to the g-matrix, not all programs allow the input of nonsymmetric $A$ matrices.

The interaction energy between the nuclear electric quadrupole moment of a nucleus with $I > \frac{1}{2}$ and the local electric field gradient is described in the sH by

$$H_{nq} = \mathbf{I}^T P \mathbf{I}$$

where $P$ is the traceless nuclear quadrupole tensor [75]. Although this term can have significant effects on cw EPR spectra (e.g., on Au(II) complexes [76]), it is not implemented in all EPR simulation programs.

The coupling between two electron spins is described by the general form

$$H_{ss} = S_1^T J_{12} S_2$$

with a general $3 \times 3$ coupling matrix $J_{12}$ [77]. It can contain three contributions: $J_{12} = J_{ex} + J_{dip} + J_{as}$. The first is the isotropic exchange coupling, $J_{ex}$. The associated term, called the *Heisenberg–Dirac–van Vleck Hamiltonian*, is encountered in the literature in several different forms ($J_{ex} S_1^T S_2$, $2J_{ex} S_1^T S_2$, $-2J_{ex} S_1^T S_2$), so that care has to be exercised in ensuring correct conversion between the definitions of $J_{ex}$ in the literature and in the simulation program. The second contribution is the symmetric magnetic dipole–dipole coupling, $J_{dip}$, analogous to $T$ in the hyperfine term. The third contribution, $J_{as}$, is antisymmetric and describes the Dzyaloshinskii–Moriya interaction [77]. The corresponding term can be written in vector form as $J_{as}^T (S_1 \times S_2)$.

Another term describing electron–electron coupling that is occasionally included in the spin Hamiltonian for transition metal ion dimers is biquadratic exchange [78, 79] of the form $-j(S_1^T S_2)^2$. It is supported by a few programs such as SPIN and EasySpin.

The quadratic zero-field splitting term for an electron spin $S > \frac{1}{2}$ is given by

$$H_{zf} = S^T D S$$

with the (usually made traceless) symmetric $3 \times 3$ zero-field tensor $D$, in angular frequency units [65]. Two common issues with this term are the notational ambiguity of $D$ (it indicates the full tensor as well as the scalar parameter equal to $\frac{3}{2}$ the largest principal value of the tensor) as well as the variety of axis-labeling conventions [80, 81], which determine the values and relative signs of the scalar zero-field parameters $D$ and $E$.

Beyond the most common quadratic zero-field term, a variety of higher order single-spin terms containing $S^k$ with $k > 2$ are used in the sH for high-spin transition and rare earth ions [66]. These include the two fourth-order terms with the conventional parameters $a$ and $F$ used for Fe(III) and Mn(II) ions [67, 82–85]. They have been included in early simulation programs [86]. Care has to be exercised concerning the correct axis definitions. Beyond these conventional parameters, the general form for the high-order terms is $H = \sum_{k=2}^{2S} \sum_{q=-k}^{+k} B_{kq} O_{kq}(S)$, where $B_{kq}$ represents the scalar interaction parameters and $O_{kq}(S)$ represents the tensor operator components for an electron spin $S$. There is some degree of arbitrariness in the choice of these tensor components, especially their phases. As a consequence, there are a significant number of inconsistent definitions of $O_{kq}$ in the literature. The most common ones are the extended Stevens operators [87]. The various forms and definitions of high-order terms have been extensively and critically reviewed by Rudowicz [88–90]. A review of such parameters for all 32 point groups has been compiled by Misra *et al.* [91]. The relation between $D$ and $E$ and the parameters $B_{2q}$ has also been discussed [92, 93]. Rudowicz and Chung [94] list explicit expression for operators $O_{kq}$ up to $k = 12$. Efficient methods for the computation of the matrix elements have been published recently [95, 96]. A general method for rotational transformation of these high-order tensor sets is available [97], including tesseral tensor operators. Utmost care has to be taken with these terms, as definitions and usage in the literature are often not only inconsistent, but sometimes incorrect as well [98, 99].

In the construction of the sH, the evaluation of perturbational expressions and the conversion between various energy and field units, accurate values of fundamental constants are required. The values of these fundamental constants are regularly updated every few years by CODATA based on continuous improvements in their measurements [100]. The same holds for nuclear isotope properties, which were updated last in 2011 [101]. Simulation programs should stay up-to-date in both respects.

#### 3.3.1.2 Quantum States and Spaces

In Hilbert space, a (pure) state of a spin system is described by a state vector, $|k\rangle$. In numerical representation, for an $N$-level spin system, this is an $N \times 1$ vector. The size of the Hilbert space, $N$, grows exponentially with the number of spins, $N = \prod_i (2S_i + 1) \cdot \prod_k (2I_k + 1)$. As a consequence, the simulation of large spin systems is challenging. Various basis sets are used to represent spin states, and the best choice depends on the computational problem at hand. The most common basis set for representing these vectors is the uncoupled Zeeman basis, where each basis state is a product of single-spin Zeeman states, $|S, m_S\rangle$ and $|I, m_I\rangle$, with the magnetic projection quantum numbers $m_S$ and $m_I$. Another basis in Hilbert space is the coupled basis, for example, the singlet–triplet basis for a system of two coupled spins-$\frac{1}{2}$, with the singlet state $|S\rangle$ and the triplet states $|T_{-1}\rangle$, $|T_0\rangle$, and $|T_{+1}\rangle$. Yet another basis is the eigenbasis, where the basis states are the eigenstates of the spin Hamiltonian, $H$.

State vectors can only represent pure quantum states [102] and are sufficient if only eigenstates of the spin Hamiltonian are required, such as in solid-state cw EPR simulations. For representing mixed quantum states, the density operator is required. It is a statistical state operator, $\sigma = \sum_k p_k |k\rangle\langle k|$, which can describe both pure and mixed quantum states of spin systems and spin system ensembles [102, 103]. When represented in a basis in Hilbert space, the density operator is an $N \times N$ matrix and is called the *density matrix*. It is generally advantageous to use the density matrix for computing the quantum dynamics of spin systems [75, 104, 105].

The space of all operators in Hilbert space constitutes Liouville space. The Liouville space of an $N$-level spin system is $N^2$-dimensional. In one basis choice, each Liouville state $|u\rangle|v\rangle$ corresponds to a pair of states $|u\rangle$ and $|v\rangle$, or a "transition" $|u\rangle \leftrightarrow |v\rangle$, in Hilbert space. In Liouville space, the density operator is represented as an $N^2 \times 1$ vector. Operators in Liouville space, acting on Liouville vectors such as the density matrix, are called *superoperators* and are numerically represented by $N^2 \times N^2$ matrices. The Hilbert-space Hamiltonian corresponds to the Liouville-space Hamiltonian commutation superoperator, whose matrix representation is $\widehat{\hat{H}} = H \otimes I - I \otimes H$, where $\otimes$ is the Kronecker product. Eigenvalues and eigenvectors of the Hamiltonian superoperator correspond to transition frequencies and transition state pairs. The use of Liouville space was first introduced in NMR by Banwell and Primas [106].

As in Hilbert space, there exist several basis choices in Liouville space [105]. One that is commonly used to derive analytical expressions in NMR and pulse EPR is the Cartesian Zeeman product operator basis [105]. It forms the basis of the intuitive product operator formalism [107]. A basis of irreducible spherical tensor operators (ISTOs) [64, 108–110] or linear combinations thereof [111] is less intuitive, but offers many computational advantages. In EPR, it has first been extensively utilized by Freed and Fraenkel [112]. ISTOs are related to the high-order spin operator sets used in the EPR spin Hamiltonian [72], as discussed above.

The description of systems with sets of equivalent nuclear spins can be simplified by decomposing the associated Hilbert or Liouville space into separate subspaces using the Clebsch–Gordan series of the rotation group by recursively applying $D^{j_1} \otimes D^{j_1} = D^{j_1+j_2} \otimes D^{j_1+j_2-1} \cdots \otimes D^{|j_1-j_2|}$. The sH is block-diagonal in the associated basis. Also, magnetic equivalence of $n$ nuclei means that the sH is invariant under any permutation among the equivalent spins [113–115]. Therefore, the properties of the associated permutation group $S_n$ can be leveraged to gain further insight and to reduce the size of the problem [64, 116, 117]. In EPR, internuclear couplings are generally negligible so that special considerations for the equivalence of spins with $I \geq 1$ and the effect of relaxation, as done in NMR [114, 115, 118], need not be taken fully into account.

### 3.3.1.3 Equations of Motion

There are several possible equations of motion that can be used to describe the dynamics of spin ensembles. The dynamics of a single spin-$\frac{1}{2}$ can be described classically. For such a system, the classical torque equation that describes the Larmor precession (and nutation) of its magnetic moment vector or of the macroscopic magnetization (magnetic moment per volume) in the external, possibly time-dependent, magnetic field is $\frac{d\mathbf{M}}{dt} = \gamma_e \mathbf{M} \times \mathbf{B}(t)$.

To take spin relaxation into account, Bloch [119] augmented this equation by phenomenological relaxation terms with time constants $T_1$ and $T_2$. The resulting Bloch equations in matrix form [120] are

$$\begin{pmatrix} \frac{dM_x}{dt} \\ \frac{dM_y}{dt} \\ \frac{dM_z}{dt} \end{pmatrix} = \begin{pmatrix} -T_2^{-1} & +\gamma_e B_z & -\gamma_e B_y \\ -\gamma_e B_z & -T_2^{-1} & +\gamma_e B_x \\ +\gamma_e B_y & -\gamma_e B_x & -T_1^{-1} \end{pmatrix} + \begin{pmatrix} 0 \\ 0 \\ \frac{M_z(0)}{T_1} \end{pmatrix}$$

and can be easily solved numerically.

These classical equations cannot be applied to spin systems with more than one spin-$\frac{1}{2}$. In general, quantum dynamics has to be applied. There are three forms of the quantum equation of motion for an EPR spin system: for states in Hilbert space, for density operators in Hilbert space, and for density operators (which are Liouville-space state vectors) in Liouville space.

The equation of motion for Hilbert-space state vectors $|k\rangle$ is the Schrödinger equation, $\frac{d|k(t)\rangle}{dt} = -iH|k(t)\rangle$. Its integrated form is $|k(t)\rangle = U(t, t_0)|k(t_0)\rangle$, with the time propagation operator $U$ satisfying the Schrödinger equation, $\frac{dU(t,t_0)}{dt} =$

$-iHU(t, t_0)$. If $H$ is time independent, then $U$ is a simple exponential operator $U(t, t_0) = \exp(-iH\Delta t)$ with $\Delta t = t - t_0$.

The equation of motion for the density operator $\sigma(t)$ in Hilbert space is the Liouville–von Neumann (LvN) equation, $\frac{d\sigma(t)}{dt} = -i[H, \sigma(t)]$, with the commutator $[H, \sigma] = H\sigma - \sigma H$. In integrated form, it is $\sigma(t) = U(t, t_0)\sigma(t_0)U^\dagger(t, t_0)$, with the same propagator as above. This "sandwich" time propagation product is central in Hilbert-space spin dynamics simulations. In this form, however, it is not possible to incorporate stochastic processes such as rotational diffusion or chemical exchange.

In Liouville space, the LvN equation is $\frac{d\sigma}{dt} = -i\hat{\hat{H}}\sigma$, with the Hamiltonian commutation superoperator $\hat{\hat{H}}$ and the density operator in vector form, $\sigma$ [105]. $\hat{\hat{H}}$ is often denoted $\hat{\hat{L}}$ and called the *Liouville superoperator*. In integrated form, the equation is $\sigma(t) = \hat{\hat{U}}(t, t_0)\sigma(t_0)$, with the (super)propagator $\hat{\hat{U}}$. In contrast to the sandwich product in Hilbert space, this is a simple matrix–vector product.

To include stochastic processes into the dynamic model, the Liouville-space LvN equation is extended to the SLE [121–123]. In one of its forms, the SLE is

$$\frac{d}{dt}\sigma = (-i\hat{\hat{H}} + \hat{\hat{\Gamma}} + \hat{\hat{X}})\sigma$$

with the stochastic relaxation superoperator $\hat{\hat{\Gamma}}$ and the chemical exchange superoperator $\hat{\hat{X}}$.

In Hilbert-space representation, the detected EPR signal $V$ is computed from the density operator using $V(t) = \text{trace}(D\sigma(t))$, where $D$ is the detection operator representing quadrature detection, usually one of the electron spin ladder operators $S_-$ or $S_+$. From a Hilbert-space state vector, it can be computed using the expectation value $V = \langle k|D|k \rangle$. In Liouville space, both $D$ and $\sigma$ are state vectors, and the expectation value is the scalar product $V = \langle D|\sigma \rangle$. All these equations are usually formulated in the rotating frame [75].

The two most common situations for which the above equations of motions are solved in EPR are the $\frac{\pi}{2}$ pulse experiment with FID (free induction decay) acquisition (pulse-acquire) [124] and the unsaturated steady-state limit [125]. Saturation is easily incorporated into the steady-state solution [126].

### 3.3.2
### Other Aspects

#### 3.3.2.1 Isotopologues

When magnetic nuclei are present in a spin center, the sH interactions depend on their nuclear spin quantum numbers, their $g_n$ factors (via the nuclear Zeeman and hyperfine terms), and their electric quadrupole moments. Many elements have magnetic isotopes. Several have one single dominant naturally abundant isotope (magnetic: H, F, Na, Al, P, V, Mn, Co, Rh, I, etc.; nonmagnetic: C, O, S, etc.). In these cases, there exists only one dominant isotopologue of the spin center, with other isotopologues mostly negligible. Features from naturally low-abundant $^{13}C$

are sometimes visible. On the other hand, many important elements have a mixture of two or more significantly abundant isotopes with different nuclear properties (e.g., B, K, Cl, Ti, Cr, Cu, Pd, etc.). Molecular spin centers with these elements occur as a mixture of isotopologues that differ in their properties, resulting in a series of overlapping spectra. For an accurate EPR simulation, EPR spectra of all significant isotopologues have to be simulated separately and added. For a mononuclear Cu complex with typical organic ligands, there are only two significant isotopologues ($^{63}$Cu and $^{65}$Cu). An extreme case is the cloro-borane radical anion $B_{12}Cl_{12}^{\bullet-}$ with over 16 million isotopologues [127], of which about 2800 have a relative abundance larger than 0.01. Some programs, such as EasySpin and XSophe, automatically generate and loop over all significant isotopologues. Occasionally, the same nominal material from different suppliers might have a different isotope composition. To allow for this and for isotope-enriched samples, programs such as EasySpin and XSophe provide interfaces for specifying custom isotope mixtures.

The sH parameters for different isotopes of the same element in the same molecular environment are different and must be interconverted. Hyperfine matrices $A$ for different isotopes of the same element scale with nuclear $g_n$ factors and can be converted using $A_2 = A_1 \frac{g_{n,2}}{g_{n,1}}$. For hydrogen, this conversion of the hyperfine coupling between protium ($^1$H) and deuterium ($^2$H) is not always fully accurate, as there is the possibility of structural and kinetic isotope effects [128] when substituting $^1$H for $^2$H. Nuclear electric quadrupole tensors $P$ can be converted using the nuclear electric quadrupole moments $Q_i$: $P_2 = P_1 \frac{Q_2}{Q_1}$.

### 3.3.2.2 Field Modulation for cw EPR

Essentially all cw EPR spectra are currently recorded using field modulation, producing the first harmonic of the absorption spectrum. (For a recently developed alternative, see Chapter 2.) The effect of field modulation can be easily added to a simulated absorption spectrum in a separate step. One approximate method, termed *pseudo-modulation* [129, 130], neglects sidebands and convolves the spectrum with a modulation function. This function is the Fourier transform of a Bessel function and can be represented in terms of a Chebyshev polynomial [12]. In cases where sidebands are resolved experimentally (very high modulation frequencies and/or very narrow lines), field modulation has to be modeled more completely, including the modulation frequency [131]. Various analytical expressions for field modulated lineshapes are available [132–136]. Robinson has published a series of papers on field modulation [137–139].

### 3.3.2.3 Frames and Orientations

Each second-rank tensorial interaction quantity in the sH represented by a $3 \times 3$ symmetric matrix, such as the tensors $P$ and $D$ and the symmetric parts of the matrices $g$ and $A$, possesses a frame – called the *eigenframe* or *principal-axes frame* (*PAF*) – in which it is diagonal. This frame has a fixed orientation relative to the molecular structure of the spin center. The PAFs of different tensors of the same spin center are generally not collinear. In order to build the matrix representation of the sH, all tensors have to be represented in the same frame. Commonly, an arbitrary frame fixed relative to the molecule is chosen and called the *molecular frame* (MF). It is usually chosen to be collinear with molecular symmetry axes or with the PAF of one of the tensors, for example, the one dominating the energy or the one most anisotropic.

The orientation of a tensor in the molecule can then be described by the orientation of the tensor PAF relative to the MF. The rotational transformation of the PAF to the MF can be represented in several ways: (i) by a set of three Euler angles, (ii) by a full 3×3 rotation matrix (direction cosines matrix, DCM), (iii) by a rotation axis and a rotation angle [140], and (iv) by quaternions. In publications, it is preferable to give the full DCM. Care has to be exercised when using Euler angles, as there are several possible conventions. In EPR and NMR, the prevalent one is zyz [67, 75, 105, 108, 141–144]. The three Euler angles ($\alpha, \beta, \gamma$) apply to the rotation of the tensor PAF first by angle $\alpha$ around the z axis, then by $\beta$ around the resulting y axis, and finally by $\gamma$ around the resulting z axis to bring the PAF into coincidence with the MF. This is a passive rotation (transformation of frames, change of representation) that does not rotate the tensor and must be distinguished from an active rotation that rotates the tensor [145]. Tensors of any rank can be rotated and transformed using Wigner rotation matrices [146]. In EPR, quaternions have been useful in modeling restricted anisotropic rotation diffusion [147] and in generating Brownian trajectories for nitroxide EPR simulations [148].

In a powder sample, spin centers are randomly and uniformly oriented in space relative to the spectrometer reference frame. This reference frame is called the *lab frame* (LF). Conventionally, the LF z axis is defined parallel to the static field $\boldsymbol{B}_0$, and the LF x axis is along the linearly oscillating microwave $\boldsymbol{B}_1$ field.

If the spin center is static, simulations are commonly carried out in the MF. In that frame, the only sH parameter that changes from spin center to spin center in a powder or frozen solution sample is the orientation $\boldsymbol{n}$ of the static magnetic field $\boldsymbol{B}_0^T = B_0(n_x, n_y, n_z)^T$. The sH and its matrix representation can be written as a linear combination

$$H = F + \boldsymbol{B}_0^T \boldsymbol{G} = F + B_0(n_x G_x + n_y G_y + n_z G_z)$$

where x, y, and z are the MF axes. In this form, the matrices $F$, $G_x$, $G_y$, and $G_z$ do not depend on the magnetic field orientation and can be precomputed and then reused during a powder simulation, resulting in substantial time savings. Many programs take advantage of this approach.

In some non-static cases, most importantly in the presence of rotational dynamics, it is more convenient to carry out the simulation in the LF. There, the orientation of the external fields is fixed, but all tensors reorient and are rotated from spin center to spin center. To handle tensors in the LF most efficiently, they are best represented via their ISTO components and rotated via Wigner rotation matrices. Methods for simulating EPR spectra with rotational dynamics are discussed later in the chapter.

## 3.4
## Static cw EPR Spectra

In this section, we summarize methods for the simulation of cw EPR spectra of ordered and disordered systems such as crystals, powders, glasses, and frozen solutions in the low-microwave power limit. In these types of samples, the spin centers are immobile. Therefore, the regime is called the *rigid limit*. Since dynamic processes are absent, the equations of motion are not necessary. Although they can be used [149], they are not required. Only the energy eigenstates of the sH need to be computed. We first discuss orientational properties of the sample, then discuss the various methods for computing field-swept spectra, and lastly cover frequency sweeps, inhomogeneous broadenings, and simulation artifacts.

### 3.4.1
### Crystals and Powders

Crystals and powders differ in the nature of the orientational distribution of the spin centers in the sample. While the former have an orientational distribution that consists only of a small finite set of discrete orientations, the latter have almost continuous distribution of orientations. This difference means that EPR spectra of powders are much more demanding to simulate.

#### 3.4.1.1 Crystals
Dedicated methods of analysis for single-crystal EPR data date back to the early days of EPR [150–153]. The relation between crystal symmetry and EPR spectra has been reviewed in great theoretical detail [154]. Two levels of symmetry have to be distinguished: (i) the space group of the crystal and (ii) the molecular symmetry group within the asymmetric unit (molecule, protein) of the crystal. In full powder averages, both the crystal symmetry and the molecular symmetry can be neglected.

On the basis of their point group and their translational symmetry, crystals belong to one of 219 different space groups. In spatially homogeneous static and microwave magnetic fields, EPR spectra are invariant under translation of the spin center; therefore, only the 32 crystallographic point groups underlying the space groups are relevant [154]. In addition, the EPR spectrum of a crystal is invariant under spatial inversion (substituting $B$ with $-B$ in the sH does not affect the set of its eigenvalues), so that the spectra of a crystal of a centrosymmetric point group (e.g., $D_{2h}$) and any of its non-centrosymmetric subgroups (e.g., $C_s$ or $D_2$) are identical. This is analogous to X-ray crystallography, where the diffraction pattern is inversion symmetric. Owing to this inversion symmetry, the 32 point groups fall into 11 Laue classes, each containing a centrosymmetric point group ($C_i$, $C_{2h}$, $D_{2h}$, $C_{3i}$, $D_{3d}$, $C_{4h}$, $D_{4h}$, $C_{6h}$, $D_{6h}$, $T_h$, and $O_h$) and its non-centrosymmetric subgroups. EPR can only distinguish between the 11 Laue classes. The number of asymmetric units in these classes range from 1 to 16. Only the smallest Laue class (space groups P1 and P$\bar{1}$, point groups $C_1$ and $C_i$) has a single asymmetric unit per unit cell. A crystal of any other Laue class in a general orientation (magnetic field not along any symmetry axis or in any symmetry plane) gives an EPR spectrum that is an overlap of multiple spectra from identical, but differently oriented spin centers. The terms *single-orientation* and *single-crystal* should therefore be carefully distinguished.

While the crystallographic point group determines the number and orientations of the asymmetric units in the unit cell of the crystal, each asymmetric unit (molecule, protein) might house one or more equivalent or nonequivalent spin centers. The asymmetric unit itself can have non-crystallographic symmetry such as fivefold rotational symmetry. This molecular symmetry can increase the line multiplicity and complexity of the crystal EPR spectrum.

### 3.4.1.2 Partially Ordered Samples

Between the two limiting cases of single crystals with discrete orientational order and powders or frozen solutions with complete uniform orientational disorder, there exist systems with partial orientational order. These comprise liquid crystals and solid stretched films. In these, molecules and spin centers can have an orientational distribution that is continuous but not uniform, certain orientations being more probable than others. The spin center orientation $\Omega = (\phi, \theta, \chi)$ is described relative to a frame fixed with the liquid crystal or film geometry (the director frame), and the anisotropic orientational distribution is commonly described by a function of the form $P(\Omega) = \exp\left(\frac{-U(\Omega)}{k_B T}\right)$ where $U(\Omega)$ is an ordering (pseudo)potential, $k_B$ is the Boltzmann constant, and $T$ is the temperature. $U(\Omega)$ can be an arbitrary function of orientation [155]. A very simple and common form is the axial Maier–Saupe potential based on the second-order Legendre polynomial $U(\Omega) = -k_B T \frac{\lambda(3\cos^2\theta - 1)}{2}$ with the potential coefficient $\lambda$. The axial order can be quantified by the order parameter $S_{2,0} = \left\langle \frac{(3\cos^2\theta - 1)}{2} \right\rangle$, which ranges between $-1/2$ and $+1$.

### 3.4.1.3 Disordered Systems and Spherical Grids

The orientation of a spin center in space relative to the spectrometer can be described by a set of three tilt angles $\phi$, $\theta$, and $\chi$ (Euler angles). In disordered systems such as powders, glasses, and frozen solutions, these orientations are randomly and uniformly distributed. Only $\phi$ and $\theta$ are necessary to specify the orientation of $\bm{B}_0$ in the MF. Therefore, the transition fields are independent of the third angle, $\chi$. However, $\chi$ is required to determine the orientation of $\bm{B}_1$ in the MF and thereby the transition probability.

To simulate a powder spectrum, a three-angle integration must be performed. For cw EPR spectra, the integral over the third angle $\chi$ can be performed analytically [1, 12, 156, 157]. For pulse EPR spectra, any third-angle anisotropy of the transition matrix elements that affect the pulse propagation operators has to be integrated numerically. Often, it is neglected.

The integral over the first two tilt angles $\phi$ and $\theta$ is usually approximated by a summation over a finite set of orientations $(\phi_k, \theta_k)$. These orientations can be represented as a spherical grid of points (knots) on the unit sphere. The spectrum of the sH is invariant under inversion of the magnetic field in the sense that $H(\bm{B})$ and $H(-\bm{B})$ have identical sets of eigenvalues. Therefore, the $(\phi, \theta)$ integration can be limited to four octants of the unit sphere (e.g., the upper hemisphere with $\theta \leq \frac{\pi}{2}$). Additional symmetries in the sH allow restriction of the integration range to two or one octants, or even to a quarter of a meridian, resulting in additional savings in computation time [12].

In simple cases, it is possible to derive closed-form analytical expressions for the overall powder spectrum. Several authors published and applied such expressions for axial and orthorhombic systems [158–164]. Much of this work is based on earlier results by Kneubühl [158]. More recently, analytical solutions for triplet spectra based on path integrals along field isolines in a ternary orientational diagram have been derived [165].

Since the computation of the EPR spectrum for a given orientation is time consuming, powder simulation methods try to minimize the number of orientations needed to compute the powder spectrum. Many different schemes for spherical grids have been proposed and studied over the years, with the hope of finding one that is optimal in terms of giving the fastest convergence to the correct powder spectrum as a function of the number of orientations. The performance of various integration grids [166] have been extensively compared [12, 167, 168]. Despite this effort, no grid has proven to be consistently superior. It appears that any grid that has a reasonably uniform point density over the unit sphere is about equally efficient, as long as proper weighing factors based on approximate or exact Delaunay triangles or Voronoi cells are included [1]. The differential efficiency of similar grids often depends on the particulars of the sH.

Spherical grids can be either analytical (orthogonal or non-orthogonal), randomly generated, or numerically optimized. The simplest possible grid is rectangular, where $\phi$ and $\theta$ are varied independently in fixed increments. This grid has very nonuniform density with crowding at the poles. It was very common in the early days of EPR simulation. Grids based on randomly generated points [169] are very inefficient.

One of the first analytically constructed latitude–longitude grids published steps $\theta$ in constant intervals and adjusts the number of points on the latitude circle (constant $\theta$) [37]. With increasing distance from the north pole ($\theta = 0$), an increasing number of points were placed on the latitude circles. This grid is commonly referred to as *igloo grid* [37, 170, 171], although igloos, the snow houses built by the Inuit, are usually constructed in a spiral form.

Sophe [17] and EasySpin [1] use a simple triangular latitude–longitude grid. Its grid points (over one octant) are obtained from $\theta_{k,l} = \left(\frac{k}{M}\right)\left(\frac{\pi}{2}\right)$ and $\phi_{k,l} = \left(\frac{l}{k}\right)\left(\frac{\pi}{2}\right)$ with $0 \leq k \leq M$ and $0 \leq l \leq k$. Essentially, as $\theta$ is increased in equal steps from the pole to the equator, one $\phi$ grid point is added for each $\theta$ step. This grid was originally introduced in meteorology in the 1960s [172]. It is well suited for global and local angular interpolation [17].

Several grids consisting of one or more spirals over the unit sphere have been used in EPR and other areas. One such grid employed in EPR uses numerical optimization to determine the position of the grid points on the spiral [35, 173], but it has been shown that these can be found from explicit expressions [168]. This spiral grid has also been used for DEER simulations [174]. Another spiral grid with very uniform density is the spherical Fibonacci grid [175–177], whose planar version describes the arrangement of the seeds in a sunflower head.

Other spherical grids include icosahedral [178, 179] and octahedral [180] grids, the Zaremba–Conroy–Wolfsberg scheme [60, 181], numerically optimized grids based on electrostatic repulsion between grid points [167] and similar metrics, and an iteratively generated grid [182]. In NMR powder simulations, Gaussian spherical quadrature methods have been applied [183].

Quantum chemical (QC) calculations based on density functional theory (DFT) also employ angular grids as part of a numerical integration over three-dimensional (3D) space [184]. Gaussian and ORCA, two QC software packages that are widely used in EPR, implement Lebedev grids. Lebedev grids are also extensively used in NMR simulations [185].

An iterative method was proposed that starts with a low number of orientations $N$ to approximate the powder spectrum and then doubles $N$ in each iteration [186]. With this method, convergence can be easily assessed automatically.

The problem of powder integration is still an area of active research, with new grids and integration methods continuing to appear [187–192].

One method to avoid calculating transition fields explicitly for many orientations is to leverage already computed orientations by angular interpolation. Interpolation schemes can be local [157, 193], global, or a combination thereof [12, 17]. Interpolation functions can be linear or cubic (Hermite splines, monotonic Fritsch splines). The spiral grid can be combined with one-dimensional (1D) interpolation along the spiral [35]. Two-dimensional (2D) interpolation over small triangular or rectangular patches of solid angle has been used as well [180, 193]. EasySpin uses bivariate cubic tensor product splines [1]. The SOPHE interpolation scheme combines a global cubic interpolation with an efficient local linear interpolation [17].

A method that utilizes already computed orientations maximally is the triangle projection method due to Ebert and Alderman [180, 194]. In this approach, the transition fields and intensities calculated for three close orientations are used to construct an analytical surface for the transition fields within the solid-angle triangle determined by the three orientations. This surface is then analytically projected into the field domain, yielding a triangle-shaped subspectrum. Combined with a Delaunay triangulation of the original grid, this greatly increases the convergence rate of powder simulations. For axial spectra, the projection method operates with spherical zones and is very efficient. EasySpin implements the projection method [1].

Yet another trick to speed up convergence rate as a function of the number of orientations utilizes the gradient of the transition fields with respect to orientation ($\theta$ and $\phi$) to compute an additional artificial line broadening that is applied to the lines of each orientation. For orientations where the transition fields are strongly orientation dependent (e.g., between principal axes), the resulting smoothing is strong. For orientations with vanishing gradients, that is, along principal axes of tensors, or at extra absorption directions [195], the smoothing is minimal. This gradient smoothing is implemented in EasySpin [1], sim [39, 40], and XSophe, where it is termed the *mosaic misorientation linewidth model* [22].

## 3.4.2
### Field-Swept Spectra

In field-swept cw EPR, the Zeeman terms of the sH change during the experiment. This means that spectral simulation requires more than just a single diagonalization of the sH. The main computationally intense task of any rigid-limit simulation is the determination of the transition fields and transitions intensities for each orientation. A transition field is the external magnetic field value at which two levels $|u\rangle$ and $|v\rangle$ are in resonance such that the applied microwave can induce a transition: $|E_v - E_u| = \hbar\omega_{mw}$. In the following, we summarize the main classes of methods available, starting with the most accurate and involved (eigenfields) to the least accurate and fastest (perturbation theory).

#### 3.4.2.1 Eigenfield Method

The problem of finding transition fields can be very elegantly and compactly formulated as an eigenproblem in Liouville space [196–199]. The super-Hamiltonian $\hat{\hat{H}}$ is separated into field-independent and field-dependent contributions, $\hat{\hat{H}} = \hat{\hat{F}} + B\hat{\hat{G}}$ with the zero-field superoperator $\hat{\hat{F}} = F \otimes 1 - 1 \otimes F^T$ and the Zeeman superoperator $\hat{\hat{G}} = G \otimes 1 - 1 \otimes G^T$. The Liouville-space eigenfield equation is $\hat{\hat{F}}'Z = B\hat{\hat{G}}Z$, with $\hat{\hat{F}}' = \omega_{mw}\hat{\hat{1}} - \hat{\hat{F}}$. The eigenvalues $B$ represent the transition fields, and the associated eigenvectors $Z$ contain $|u\rangle$ and $|v\rangle$, the two states involved in the transition. The associated matrix equation is a general eigenvalue problem, where the representations of $\hat{\hat{F}}'$ and $\hat{\hat{G}}$ are $N^2 \times N^2$ matrices for an $N$-level spin system.

If the microwave quantum is larger than the maximum zero-field splitting $E_N(0) - E_1(0)$, then $\hat{\hat{F}}'$ is positive-definite and can be inverted and the eigenfield equation reduces to an ordinary eigenproblem $\hat{\hat{F}}'^{-1}\hat{\hat{G}}Z = \left(\frac{1}{B}\right)Z$, whose eigenvalues are the inverse transition fields [196]. To reduce the computational effort for large systems, a perturbational treatment of the eigenfield equations was introduced [84, 198]. A method for obtaining the eigenfields via the characteristic equation [200] has been proposed as well. It has been shown that the eigenfields ansatz can be used to formulate a Hilbert-space differential equation that can be solved using the filter diagonalization method [201]. The method is, however, limited to situations with $E_N(0) - E_1(0) < \hbar\omega_{mw}$, which are easier to solve by other methods.

Transition probabilities can be calculated from the eigenvectors $Z$ in a very simple manner [196]. However, due to the large dimensionality of the superoperator space, it is often worthwhile to forgo computation of $Z$. Instead, the transition probabilities can be calculated for each transition field obtained from the eigenfield equation by solving the Hilbert-space energy eigenproblem. Another advantage of this hybrid method [200, 202] is that the frequency-to-field conversion factor [157, 203] (see below) can be obtained more easily.

Although the matrices involved in the eigenfield equation are very large, they are generally sparse. It is feasible to employ this method for systems with high spins or many coupled electron spins in case the energy level diagram gets so complicated that the Hilbert-space matrix diagonalization-based energy level diagram modeling as discussed below requires an excessive number of diagonalizations. In general, because of its mathematical and implementational simplicity, the eigenfield method can serve as a reference method for other approaches.

### 3.4.2.2 Matrix Diagonalization

All methods except the eigenfield method compute transition fields in two steps. They first determine energy levels and possibly states at one or several pre-chosen magnetic field values using a range of possible methods (matrix diagonalization, perturbation theory, or a combination thereof), and then use these energies to obtain the transition fields by interpolation or extrapolation along the field axis. Some methods combine these two steps, and others iterate between them.

To obtain the energy eigenstates and their energies for a given external field, the sH matrix can be numerically diagonalized, and transitions can be determined by comparing all energy level pair differences to the microwave quantum [86, 157, 186, 204–206]. In principle, for a field sweep, this diagonalization has to be repeated for each point along the swept field range. In this wasteful but sure-fire brute-force method, typically on the order of 1000 diagonalizations per spectrum per orientation are required. This is prohibitive for powder spectra.

The number of diagonalizations can be reduced by combining matrix diagonalization for a limited subset of field values (knots) with interpolation, extrapolation, or root-finding along the field axis. Methods based on root-finding algorithms [2, 81, 157, 186, 207–209] can locate one transition field per state pair within a narrow field range. Another method minimizes via least-squares fitting the square of the deviation between the energy difference of two levels and the microwave quantum [206, 210–214]. The simplest methods for the computation of transition fields use energies at one field only, combined with linear or quadratic extrapolation based on a Taylor-series expansion or perturbation theory [49, 193]. When the sH is diagonalized at multiple fields, instead of diagonalizing at each field value independently, eigenvalues and eigenvectors of the sH for one field can also be obtained by homotopy [19, 212, 214] or extrapolation [215] starting from the results from a nearby field.

The method implemented in EasySpin models the energy level diagram over a desired field range using an adaptive iterative bisection and interpolation algorithm [216]. Initially, $H$ is numerically diagonalized at the minimum and the maximum of the requested field range, and an approximation of the energy level diagram is constructed by Hermite cubic spline interpolation. Next, $H$ is diagonalized at the center field and the resulting eigenvalues are compared to the interpolated ones. If the difference is above a given threshold (typically a few parts per million of the microwave energy), the left and right field segments are interpolated and diagonalizations done at their centers. This procedure is repeated recursively until all segments are accurately modeled by splines. Transition fields are then determined from the spline model of the energy level diagram. The method is robust and adapts the number of diagonalizations to the complexity of the energy level diagram. EasySpin's methods were inspired by a bisective root-finding algorithm [186, 209]. Other nonadaptive interpolation methods use cubic splines [39, 217] or Chebyshev polynomials [10]. Sophe subdivides the field search range into segments, diagonalizes the Hamiltonian once per segment, and then uses second-order perturbation theory to locate transition fields [24].

Once the energy level diagram is modeled by splines, it has to be searched for transition fields. In principle, this involves searching all $\frac{N(N-1)}{2}$ unique pairs of levels for an $N$-level system. Most of these searches are in vain, as most level pairs do not give significant EPR lines because they are either off resonant over the entire field range or have negligible transition matrix elements. A simple heuristic procedure, termed *transition preselection*, can be used to narrow down the search range of level pairs [1]. It is based on the observation that the same subset of level pairs usually gives nonzero transitions for all or most orientations. It involves precomputing energies and transition intensities at the center field for one or a few orientations. From this set, the subset of level pairs with nonzero transitions is selected, and the subsequent field searches for the full powder simulation can be limited to this subset. The method, however, runs into problems for large field sweeps with significant zero-field or hyperfine interactions, where the center-field states are not representative of the entire field range. It should be avoided in systems with multiple nuclei with similar hyperfine coupling, as strongly field-dependent state mixing might occur.

No matter how they determine transition fields, all the above methods use matrix diagonalization. There are numerous diagonalization libraries available that implement very efficient algorithms for computing all eigenvalues and eigenvectors of a Hermitian matrix [218, 219]. However, algorithms differ in efficiency depending on whether the matrix is dense or sparse, on whether eigenvalues only or eigenvalues and eigenvectors are required, and on whether all or only a few of the lowest eigenvalues are requested. Efficient large-scale algorithms (Arnoldi, Lanczos, Jacobi, Davidson) are available [220].

There has been much concern about energy level crossings as a function of field magnitude and orientation and the associated ambiguity in labeling the states involved. Several methods of assigning states left and right of the crossing by tracking or Jacobi diagonalization have been proposed [35, 157, 193, 209, 214]. However, since crossings occur only at isolated points within the symmetry-unique subset of field orientations [221], these procedures are not necessary. Levels can be uniquely sorted and labeled in order of increasing or decreasing energy.

If the microwave quantum is smaller than the largest zero-field gap, $|E_N(0) - E_1(0)|$, then there can be multiple transition fields between pairs of levels that feature anticrossings [196]. For such level pairs, the number of transitions as a function of magnetic field orientation is not a constant. In a powder spectrum, the resulting looping transitions are often present only over a subset of field orientations, coalesce at orientations where the anticrossing gap matches the microwave quantum, and vanish for the rest. This makes interpolation more involved and leads to complications in the modeling of line broadenings. Accurate treatment of such looping transitions is more difficult [212, 217].

In some systems, for example, Ni(II) with $S = 1$, two-photon ("double-quantum") transitions are visible. The theoretical basis of these transitions has been discussed in several publications [10, 205, 222], but they are not routinely incorporated in simulations.

### 3.4.2.3 Perturbation Theory

When one interaction in the spin Hamiltonian dominates, perturbation theory can be used to compute the resonance field positions. Methods that treat the electron Zeeman interaction as the main interaction, the hyperfine interaction as perturbation, and neglect the nuclear Zeeman and the nuclear quadrupole interaction have been used since the early days of EPR both for solids and liquids simulations [223]. The nuclear Zeeman and quadrupole interactions can be included in a second step using sequential perturbation theory.

When the three nuclear interactions (hyperfine, Zeeman, quadrupole) are of similar strength, the treatment is more complicated. Expressions that treat hyperfine and nuclear Zeeman interaction on an equal footing have been given by Lefebvre and Maruani [224], Iwasaki [225], and others [226, 227]. When equivalent nuclei are present, transformation of the spin Hamiltonian into a coupled representation is required [112, 228]. In the presence of multiple nuclei, the inclusion of internuclear cross terms might be necessary [226]. Byfleet has developed a method based on seventh-order perturbation theory [229] applicable to a fairly general spin Hamiltonian. Generalized operator transforms based on methods developed by Bleaney and Bir can be used to derive perturbational expressions for anisotropic systems [230–232].

A specific application case for perturbation theory that has received significant attention is mononuclear $Mn^{2+}$ ($S = \frac{5}{2}$). Its sH has isotropic $g$ and $A$, but a significant anisotropic zero-field splitting. Expressions for eigenenergies and transition fields at second- and third-order levels of perturbation theory have been published many times [231, 233–237]. Perturbation theory for coupled spins using a full anisotropic spin Hamiltonian has been developed for transition metal dimers [238, 239], for transition metal–nitroxide complexes [240], and for dipolar-coupled pairs of radicals [241, 242].

From a software perspective, there are several problems with perturbation theory: (i) Most expressions are scalar and long. The resulting code is usually error prone and very difficult to debug. (ii) Scalar perturbational expressions are not general, as they almost always are limited to specific systems such as one electron spin and a certain number of nuclei, and to specific symmetries. In fact, many early programs differed mainly in the number of spins and symmetry of interaction matrices that were supported. (iii) Perturbation theory has inherently limited scope with respect to the relative strengths of the various interaction parameters. The accuracy of the simulated spectrum is a function of this, which is not a desirable software behavior from a user perspective. (iv) Lastly, very few programs check the validity of the perturbation theory approximations for the given set of input sH parameters and leave the user in the dark about the accuracy of the computed spectrum. However, with proper checks, the main advantage of perturbation methods – speed – can be harnessed for specific systems.

#### 3.4.2.4 Hybrid Models

If the interaction strengths in a spin system are such that the spins fall into two distinct groups, one with strong interactions and one with weak interactions, hybrid methods can be used. These proceed in two steps. First, they utilize matrix diagonalization for the subset of strongly coupled spins to obtain energy levels and states for the corresponding subspace. Then, they apply a perturbational theory approach to calculate the splittings resulting from the subset of weakly interacting spins. Ligand nuclei in transition metal complexes can be treated in this manner [186]. Hybrid methods have been applied to di-manganese systems [243], where the electron spins are treated exactly and the $^{55}$Mn nuclei perturbationally. Compared to full matrix diagonalization methods, hybrid methods provide a considerable saving of computer time.

### 3.4.3
### Transition Intensities

Generally, cw EPR spectra are acquired with non-saturating levels of microwave power. Then, first-order time-dependent perturbation theory and Fermi's Golden Rule are applicable. The intensity $I_{vu}$ of a transition between initial state $|u\rangle$ and final state $|v\rangle$ in a field-swept EPR spectrum is determined by three factors.

$$I_{uv} \propto |\langle v|H_1|u\rangle|^2 \cdot \left(\frac{d\Delta E}{dB}\right)^{-1} \cdot (p_u - p_v)$$

The first factor is the transition probability, the square of the transition matrix element of the microwave sH, $H_1 = \mu_B \mathbf{B}_1^T \mathbf{g} \mathbf{S}$. If the state vectors of the two states $|u\rangle$ and $|v\rangle$ are known (e.g., from matrix diagonalization or perturbation theory), then this matrix element can be evaluated numerically. Within the assumptions of first-order perturbation theory with dominant $H_{ez}$, states can be written as products of electron and nuclear substates, and the matrix element can be evaluated analytically. The expression is [225]

$$|\langle v|H_1|u\rangle|^2 = (\mu_B B_1|G|)^2 \cdot |\langle m_{S,v}|\mathbf{S}^T\mathbf{n}|m_{S,u}\rangle|^2 \cdot |\langle m_{I,v}|M_{vu}|m_{I,u}\rangle|^2$$

where $G$ is the effective g-factor along the direction of $\mathbf{B}_1$. The second factor is the spin transition moment squared. To first order, its value is $S(S+1) - m_S(m_S + 1)$ for the allowed transition $m_S \leftrightarrow m_S + 1$. The last factor is a nuclear overlap matrix element and is analogous to a Frank–Condon factor [244]. It is central to ESEEM spectroscopy (see below).

The second factor in the expression for $I_{vu}$ is the frequency-to-field conversion factor. It accounts for the fact that the spectrum is field-swept and not frequency-swept. It was originally discovered by Aasa and Vänngård [203] in 1975 for spin-$\frac{1}{2}$ systems with anisotropic g, where it is proportional to $\frac{1}{g}$. van Veen [157] gives the general expression. The presence of this factor implicitly assumes that unit-area absorption lines are employed for modeling line broadenings [157]. It was extensively discussed by Pilbrow [170, 245, 246]. Neglecting this factor can lead to errors for half-field transition intensities (used for distance measurements [247]), relative linewidths in systems with large g anisotropy, and in spin quantitation. The only systems where the factor does not affect the spectral shape are spin-$\frac{1}{2}$ systems with essentially isotropic g-matrix and only fully allowed transitions, for example, organic radicals. When this factor is used, it is crucial to use a frequency-domain (FD) line broadening model, and not a simple field-domain convolutional linewidth. If not, wrong intensities will result, for example, for half-field transitions in triplets.

The third factor in $I_{vu}$ represents the polarization of the transition. $p_u$ and $p_v$ are the populations of the initial and final state, respectively. This difference is positive and leads to absorptive lines if $p_u > p_v$, which is the case under thermal equilibrium for $E_u < E_v$. The thermal-equilibrium Boltzmann population is given by $p_u = \frac{\exp(\frac{-E_u}{k_B T})}{\sum_q \exp(\frac{-E_q}{k_B T})}$. For a spin-$\frac{1}{2}$ system with isotropic g value, the thermal polarization is $\Delta p = p_{+\frac{1}{2}} - p_{-\frac{1}{2}} = \tanh\left(\frac{\mu_B B g}{2 k_B T}\right)$. In the high-temperature limit $k_B T \gg \frac{\mu_B B g}{2}$, this simplifies to $\Delta p = \frac{\mu_B B g}{2 k_B T}$. Nonthermal equilibrium situations, for example, spin-correlated radical pairs or photoexcited high-spin states of organic molecules, can be easily accommodated.

Closed-form analytical solutions for anisotropic transition probabilities have been given for many cases: axial g tensor [248], forbidden hyperfine transitions [249], and rhombic g tensor [203, 225, 250–254]. For powder spectra, the transition intensity can be integrated analytically over the third Euler angle even for the case of a rhombic g-matrix, yielding compact expressions [225, 248, 251, 254]. A different derivation has been used by Kneubühl and Natterer [255]. With numerically obtained transition intensities, integration over the third angle can also be carried out effortlessly [156].

Most solid-state simulations assume non-saturating levels of microwave irradiation. In the case of saturation, a more complete theory based on the SLE has to be used [126]. Saturation is important in the context of saturation-transfer EPR, where a detailed theory has been developed by Dalton and Robinson [256–259].

### 3.4.4
### Isotropic Systems

For systems that are isotropic (either intrinsically, or by virtue of sufficiently fast rotational averaging), the spin Hamiltonian simplifies considerably. In each sH term, only the isotropic average survives, leaving isotropic g and A matrices, isotropic $g_n$ factors, and isotropic exchange. Zero-field splitting terms and the nuclear quadrupole term average to zero. In general, resonance fields for the isotropic spin Hamiltonian can be solved using matrix diagonalization, as discussed in Section 3.4.2. However, in the case of one electron spin $S = \frac{1}{2}$ and one nucleus with $I \geq \frac{1}{2}$, two more efficient approaches are possible: (i) Breit–Rabi solution and (ii) perturbation theory.

The Breit–Rabi formula [260–262] is an exact explicit expression for the energy levels $E(m_S, m_I)$ of a $(S, I) = \left(\frac{1}{2}, \frac{1}{2}\right)$ system. Using this, the resonance condition $E\left(+\frac{1}{2}, m_I\right) - E\left(-\frac{1}{2}, m_I\right) = \Delta E(m_I) = \hbar\omega_{mw}$ can be written in the implicit form $B = f(B)$ and solved for the transition field $B$ using a fixed-point iteration ($B_{k+1} = f(B_k)$) with the starting resonance field $B_0$ obtained from first-order perturbation theory. The solutions converge to numerical accuracy within a few iterations, even for very large hyperfine couplings. This is computationally superior to perturbation expressions and is the default method in EasySpin [1]. Fixed-point iterations can be applied to any situation where the resonance field is only known as an *implicit function of the microwave frequency* and the *spin Hamiltonian parameters* and can be formulated as $B = f(B)$. This applies to many perturbation expressions and has been applied, for example, to bisnitroxide spectra [242].

For isotropic systems, many programs implement standard perturbation theory. EasySpin provides such methods up to fifth order. The expressions are based on solving the equation $\Delta E - \hbar\omega_{mw} = 0$ with a Taylor expansion in $a_{iso}$ of the Breit–Rabi expression for $\Delta E$. For a desired perturbation order $n$, the resulting equation is multiplied by $B^{n-1}$ and truncated after the $a_{iso}^n$ term [263]. This yields a polynomial in $B$, whose roots are the resonance fields and can be determined using any root-finding algorithm, for example, the Newton–Raphson method with $B = \frac{\hbar(\omega_{mw} - m_I a_{iso})}{\mu_B g}$ as the starting value. These methods can be applied to systems with multiple nuclei as well, as long as cross terms are properly taken into account [226]. Since it is so easy to solve the Breit–Rabi equation, perturbation expressions have mostly only didactic value.

### 3.4.5
### Line Broadenings

There are two types of broadening: homogeneous and inhomogeneous. Homogeneous broadening is due to the limited lifetime of the excited states populated by microwave absorption. In solid-state EPR, this broadening is mostly negligible compared to the second type, inhomogeneous broadening.

Inhomogeneous broadening is due to site-to-site heterogeneity in the sH, which can have several different origins. The first, with the largest effect, is orientational disorder (different spin centers have different orientations) and leads to powder spectra. It needs to be treated explicitly and was discussed in Section 3.4.1.3.

The second type of inhomogeneous broadening is due to small and unresolved couplings between the electron spin(s) described in the sH and spins that are not explicitly incorporated in the sH. This includes superhyperfine (shf) interactions to nearby nuclear spins, and dipolar couplings to other electron spins. Different spin centers experience different magnetic states of spins in the nanoenvironment, which leads to a multitude of small splittings. Generally, the large number of couplings results in an overall line broadening.

The third type of inhomogeneous broadening is structural. Parameters in the sH are (sometimes) sensitive functions of the geometry of the spin center and its immediate environment. Any static structural disorder, either geometric or electronic, will lead to a distribution of magnetic parameters with a finite width. As a consequence, transition fields are shifted, and the overall spectrum broadens. Variations in geometric degrees of freedom can include variations in ligand distances and coordination geometry in a transition metal complex, variations in the length and orientation of hydrogen bonds in organic radicals, and slight misorientations in crystals.

#### 3.4.5.1 Dipolar Broadening

When a large number of unresolved couplings due to roughly homogeneously distributed magnetic moments (solvent nuclei, other spin centers) contributes to the line broadening, it can be modeled using an FD linewidth tensor that provides an orientation-dependent linewidth $\Gamma(\Omega)$ for a Gaussian lineshape. The FD linewidth can then be converted to the field domain using the frequency-to-field conversion factor discussed in Section 3.4.3. Various expressions for this linewidth tensor have been proposed and are being used [1, 23, 170], and none of the forms appears to be better or worse than the other.

If the dipolar broadening is solely due to weak dipolar coupling to a spin center at a specific distance from the spin center of interest, the cw EPR spectra can be convoluted with the dipolar Pake powder pattern. A computational analysis of this Pake broadening can be used to extract distances in dipolar coupled bisnitroxides [47, 264].

Occasionally, for accurate analysis of lineshapes in solution, it is important to explicitly compute the shf splitting pattern and use it as a convolutional line broadening function instead of a Gaussian function [265].

#### 3.4.5.2 Strains

To model geometric heterogeneity, strain models are used. Site-to-site structural variations cause corresponding variations of the sH parameters, so that each combination of sH parameter values will have a certain probability density $P(g, a, \ldots)$. Generally, Gaussian distributions of parameters are assumed. The variations of different parameters can in principle be correlated.

To include strain broadenings in the simulation, spectra have to be simulated for many points within the distribution $P$ in parameter space and then integrated. This is the only viable approach if the parameter distributions are wide. If the distributions are narrow, the energy levels and the transition fields vary mostly linearly over the narrow parameter range [170]. For a transition between two levels $|u\rangle$ and $|v\rangle$, the width of the (Gaussian) FD distribution of an sH parameter $p$ can then be directly converted to a field-domain linewidth using the derivative of the energy gap with respect to $p$ and the $\frac{1}{g}$ factor [1, 12]. The former can be obtained from the sH and the eigenstates using the Hellmann–Feynman theorem [266].

Structural heterogeneity of spin centers can lead to a distribution in g values, for example, in transition metal complexes and clusters as well as in organic radicals in frozen solution. The resulting distribution of g tensors, g strain, is often simply modeled by three uncorrelated Gaussian distribution functions of the three principal g values. This model works well in many cases, even though it can miss visible details. A more sophisticated correlated statistical model of g strain has been developed for metalloproteins by Hagen [267, 268]. Physical models for g strain based on crystal field theory have been derived and utilized as well [269, 270].

In transition metal complexes, such as in $Cu^{2+}$, variations in the principal values of g and A matrices are interrelated [271]. In frozen solution with structural heterogeneity at the spin center, this leads to correlated g and A strains [272] that have been included in simulations using a bivariate normal density distribution function [273, 274].

There has been substantial work on modeling the heterogeneity of the D tensor in solid-state samples, as the D tensor can be very sensitive to structural features. The D tensor components were related to the components of the external stresses via a spin–strain tensor [275, 276]. For general application and without extensive computational chemistry predictions, this is an overparameterized model. Simple uncorrelated distributions in the zero-field splitting parameters $D\left(=\frac{3}{2}D_z\right)$ and $E\left(=\frac{D_x-D_y}{2}\right)$ are used extensively [277–280]. In cases where the ratio $\frac{E}{D}$ (rhombicity) determines resonance line positions, such as high-spin ferric ions, the strain can be modeled with a single distribution of that ratio [281, 282]. Joint probability distributions $P(D, E)$ or $P\left(D, \frac{E}{D}\right)$ using 2D Gaussians (correlated D strain) have been used as well [38, 283, 284], for example, using complete anticorrelation between D and E [285]. Closely related to D strain is r-strain, a distribution of the inter-metal ion distance in dimers that affects the magnetic dipole coupling and consequently the linewidth [243].

Line broadening in the EPR spectra of solid crystals has been found sometimes to be due to variations in unit-cell orientations (misalignment, mosaicity, misorientation) [277, 286–288].

### 3.4.5.3 Lineshapes

For homogeneous broadenings, the Lorentzian function is used. For inhomogeneous broadenings, the Gaussian function is most common [6]. They are applied to the simulated spectrum using convolution [289]. If the broadening is anisotropic, each single-orientation spectrum is convolved separately. If it is isotropic, a single convolution of the final powder spectrum is sufficient. The convolution of a Lorentzian and a Gaussian function is called a *Voigt function*. Convolutions can be performed numerically [290], but other methods are available as well [291–293] and need to be deployed if many Voigt shapes are required. A popular approximation to the Voigt lineshape is a linear combination of Lorentzian and Gaussian functions [294], also called the *pseudo-Voigt profile*. Other shape functions for inhomogeneously broadened lines include the Holtzmark and Stoneham lineshapes [6]. Small deviations from the Lorentzian shape due to incomplete rotational averaging of anisotropies can be modeled by a two-parameter extension of the Lorentzian function [295]. The Tsallis distribution generalizes Gaussian and Lorentzian lineshapes [296].

### 3.4.6
### Frequency-Sweep Spectra

The increased use of wideband frequency-sweep EPR spectrometers up to the terahertz range has prompted the development of corresponding simulation methods [297]. Unlike in field-sweep EPR, the Zeeman terms in the sH are constant during the experiment. Therefore, the computations are more straightforward. As in the field-sweep case, matrix diagonalization, perturbation theory, or hybrid methods are applicable. For the transition intensities and linewidths, the same expressions as for the field-swept case apply, except for one important difference: The frequency-to-field conversion factor, discussed in Section 3.4.3, is not required.

### 3.4.7
### Simulation Artifacts

There are a number of common artifacts of solid-state simulations that need to be identified and remedied if present.

1) If the number of orientations in a powder simulation is small, ripples ("simulation noise" or "grass") appear in the simulated spectrum. For a given number of orientations, the severity of these ripples increases with increasing overall width of the field-swept spectrum. They can be removed by increasing the number of orientations. Simulation ripples can also be removed via low-pass filtering using Fourier transformation [267]. Another remedy is to use an iterative orientational averaging with increasing grid density until convergence is achieved [186]. Similar ripples can appear experimentally if the powder does not contain a sufficient number of microcrystallites for complete orientational averaging. Gradient smoothing [1, 22, 39] and analytical projection techniques [1, 180, 194] reduce the occurrence of ripples significantly.
2) Simulation programs differ in the implementation of the evaluation of Lorentzian and Gaussian lineshapes. Since repeated explicit evaluation of the corresponding expressions are numerically intense, look-up tables are often used. This can lead to truncation and interpolation errors. If convolutional line broadenings are applied via Fourier transformation, artifacts at the lower and upper field range limits may occur.
3) Near coalescence points, looping transitions can yield artifacts in powder spectra in the form of missing spectral intensity that leads to pairs of spurious peaks. These problems can be alleviated by substantially increasing the spherical grid density near coalescence points, or by using dedicated methods [217].
4) Some programs (such as EasySpin) apply transition preselection or screening procedures before starting a simulation. These procedures attempt to determine level pairs that will yield transition fields before starting the full powder simulation, with the goal of discarding level pairs that are never resonant. In complicated spin systems with high electron spins, high-order operators, or

large hyperfine couplings, such procedures can miss transitions, and should be used carefully.

## 3.5
## Dynamic cw EPR Spectra

Many dynamic processes can affect the shape of an EPR spectrum ([65]; see [298] for an early review). In addition to a variety of spin-relaxation processes, two types of molecular processes that are of great importance are rotational motion and chemical exchange. Both require dedicated simulation techniques and can be modeled at various levels of theory, with different scope and accuracy.

### 3.5.1
### Rotational Diffusion

The random rotational or reorientational motion (tumbling) of spin centers such as nitroxides in solution can be modeled using the SLE [27, 28, 299, 300] with a rotational diffusion superoperator. The time scale of the rotational dynamics is characterized by the rotational correlation time $\tau_c$. Based on the relation between $\tau_c$ and the rigid-limit powder spectral width $\Delta\omega$, several dynamic regimes are distinguished: (i) the fast-motion limit ($\tau_c$ essentially 0), (ii) the fast-motion regime ($\tau_c^{-1} \gg \Delta\omega$), (iii) the slow and intermediate motion regime ($\tau_c^{-1}$ similar to $\Delta\omega$ within about 2 orders of magnitude), and (iv) the rigid limit (absence of rotation, $\tau_c^{-1} = 0$).

#### 3.5.1.1  Fast-Motion Limit
Fast-motion limit spectra are isotropic [301]. All anisotropies in the spin Hamiltonian are averaged out on the time scale of the EPR experiment. Spectra can be easily simulated by the isotropic solid-state methods described in Section 3.4.4 by using a spin Hamiltonian with isotropic interaction matrices and tensors. In the isotropic limit, many of the perturbational expressions simplify considerably. Residual broadenings are due to inhomogeneous (isotropic) hyperfine couplings, Heisenberg exchange with other spin center such as dioxygen, or lifetime broadening.

The exponential growth of the number of EPR lines with the number of nuclei coupled to an electron spin makes simulations in the field or FD by summation over all possible lines very slow for large spin systems. An alternative is to perform the simulation in the inverse-field or time domain (TD), where each line can be represented by a decaying exponential, and the convolutions reduce to simple multiplications. This Fourier transform method [25, 302–305] assumes a first-order perturbational regime, that is, that the nuclear spin states do not affect the state of the electron spin. In this method, the effect of the modulation amplitude can be incorporated in the inverse domain [25, 305]. An analysis method based on Fourier transform is the cepstral analysis [306].

### 3.5.1.2 Fast-Tumbling Regime

In the fast-tumbling regime, the rotational motion can be treated using perturbational Redfield–Wangsness–Bloch relaxation theory. A detailed derivation with many references is given in Atherton's book [143]. On the basis of this theory, the linewidths can be expressed as polynomials in the nuclear projection quantum numbers, $m_I$. Kivelson has developed expressions for several cases [307, 308]. Freed has presented very general expressions [112]. A diagonalization-free implementation of Redfield relaxation theory for large spin systems has recently been developed [309].

### 3.5.1.3 Slow-Tumbling Regime

In the slow and intermediate motional regimes, the reorientational motion of the spin label is on a time scale similar to the EPR time scale, and the EPR spectrum is broadened. In these regimes, perturbation treatments have to be abandoned, and a more complete theory has to be employed.

Simulation methods for this regime differ in how the different levels of orientational dynamics are treated and can use either deterministic or stochastic models [310]. Deterministic models are based on atomistic molecular dynamics (MD) simulations and are able to treat complex local and internal dynamics. Stochastic dynamics (SD) models describe the reorientational motion as rotational diffusion of a rigid rotor and can account for simple local or global rotational dynamics.

One stochastic model, jump diffusion, assumes rotational diffusion via random jumps between multiple equivalent sites differing in the orientation of the spin center [27, 311]. This has been observed in a few cases [312]. A stochastic memory function approach that assumes random instantaneous rotational jumps by a small angle has been used to model reorientation of spin labels in supercooled water [313].

The most common stochastic model for rotational motion in solution is Brownian rotational diffusion, a random walk in 3D orientational space [314]. The anisotropy of the reorientational rate constant $\tau_c^{-1}$ due to the nonspherical shape of the spin center is described by an anisotropic diffusion tensor fixed in the MF. For a spin center freely diffusing in solution, the local environment is isotropic. On the other hand, the environment is generally anisotropic for a spin label bound to a protein or other biomolecule, resulting in preferential alignments and excluded orientations [28]. In these partially ordered samples, the free energy of the label is a function of its orientation $\Omega$ and is described by an orientational potential $U(\Omega)$. Expressions for ordering potential have been discussed above.

Simple SD models such as Brownian diffusion in a restricting potential are accommodated in the SLE by including a rotational diffusion operator [155, 314]. The SLE then describes the joint time evolution of both the quantum spin degrees of freedom and the classically treated spin center orientation $\Omega = (\alpha, \beta, \gamma)$. EPR spectra are simulated by calculating the low-power steady-state solution of the SLE, with an equilibrium orientational probability distribution of an ensemble of spin centers determined by the orientational potential. Very efficient SLE solvers were developed by Freed and coworkers [27, 299, 311, 315, 316]. A didactic review [28] summarizes the main features of the approach.

The orientational distribution of the spin center is represented in a basis of Wigner functions $\mathscr{D}^L_{KM}(\Omega)$ with $L = 0, 1, \ldots$ and $-L \leq K, M \leq +L$, which is, in principle, of infinite size. For a given spin system, the matrix size and the computational effort scale with the number of orientational basis functions. The orientational basis is usually truncated to a subset of functions with $L < L_{\max}$. (The orientational potential is also expressed as a linear combination of a few low-$L$ Wigner functions.) This basis yields manageable expressions for the matrix elements of the Liouville superoperator. For accurate simulations, the basis size needs to be increased with decreasing rotational diffusion rate and increasing complexity of the potential. In principle, the SLE approach can be used to simulate a rigid-limit spectrum. However, the basis size required for achieving converged spectra is large, so that dedicated rigid-limit methods are preferred.

In a basis truncated at $L_{\max}$, some of the basis states with $L < L_{\max}$ are negligibly populated and can be removed using heuristic pruning techniques, as introduced by Freed [27, 299]. Similar and more general state space restriction methods have recently been developed for general spin dynamics simulations [64], as discussed later.

If the rotationally diffusing spin center is attached to another entity that provides an orienting potential, not only the local dynamics of the spin center but also the global dynamics of the latter need to be included in the SLE simulation. Most commonly, this is observed for nitroxides attached to proteins. With a protein that provides an ordering potential, the slow-motion spectrum depends on the orientation of the ordering potential with respect to the external magnetic field.

Two theoretical models are available, MOMD [317] and SRLS (slowly relaxing local structure) [318]. If the protein is randomly orientationally distributed in the sample but static on the time scale of the EPR experiment, a powder average has to be computed. This is called the *MOMD model* [317]. It can be generalized to partial static order. If the protein is not static, its rotational dynamics couples to the rotational dynamics of the spin center and must be included in the simulation. In the SLE approach, the protein dynamics is also treated stochastically. A second diffusion operator is added in the SLE, and the orientational distribution of the protein is described by a second set of Wigner basis functions $\mathscr{D}^L_{KM}(\Omega_2)$. This SRLS model introduced by Freed [318–320] is a two-body coupled rotor model. It is implemented in the software available from the Freed group that includes multifrequency fitting [320, 321]. SRLS is also implemented in the software E-SpiReS [33]. The model is not only used for spin labels, but also for methyl dynamics and $^{15}$N relaxation in protein NMR [322]. A dedicated program, C++ OPPS, is available [323]. Currently, EasySpin implements the MOMD model for spin systems with one electron spin and several nuclear spins.

Mostly, SLE simulations have been limited to nitroxides with $(S, I) = \left(\frac{1}{2}, 1\right)$. However, in some cases, the SLE method has been applied to larger spin systems such as diphosphanyl radicals [324], doubly nitroxide-labeled peptide [325], nitronyl nitroxides [326], and fullerene-bisnitroxide adducts [327]. Misra [324] has derived explicit scalar matrix element expressions for the two-nuclei case. The theory of slow-motional EPR spectra of triplets is theoretically well described [328–330].

In general, the matrix elements for any spin system can be constructed in a straightforward manner from the single-spin spherical tensor operators. If the sH interactions are restricted to zero- and second-order tensors, the sH Liouville operator can be decomposed into 26 static rotational operator components [309].

As a more general alternative to solving the SLE for a stochastic model, EPR spectra can be directly computed from sets of explicit trajectories that describe the change in time of the orientation of the spin center in space. A trajectory determines the time dependence of the magnetic parameters and therefore the EPR spectrum of a moving spin center. This method is applicable to trajectories obtained from both stochastic and deterministic models.

From a trajectory, the time evolution of the magnetization following a 90° pulse is computed using density matrix or Bloch magnetization vector propagation [147, 331, 332]. From the resulting FID, the EPR spectrum is obtained by Fourier transformation. To generate a converged spectrum, FIDs of a set of trajectories generated from a range of possible initial orientations of the spin label have to be combined. The appropriate time resolution of the trajectories is determined by the spectral width via the Nyquist criterion. The length of the trajectory is determined by the required spectral resolution and has to be of the order of the transverse relaxation time $T_2$ to yield accurate EPR spectra.

The computation of SD trajectories is fast, and the number and length of trajectories required for accurate spectral simulations are easily obtained. Stochastic Brownian dynamics trajectories were introduced by Robinson *et al.* [331] and widely applied [242, 333]. On the other hand, MD trajectories [334, 335] are computationally significantly more demanding. The expense of computing long MD trajectories can be avoided. Accurate simulations can be achieved from short MD trajectories generated over the decay time of the auto-correlation function of the motion [336, 337], or by deriving an effective orientational potential from short MD trajectories and then using it to generate SD trajectories [333] or to solve the SLE [338]. A single MD trajectory can be reused several times by rotation or resampling. In general, despite being much slower than the SLE approach, trajectory methods are superior in the complexity of dynamics that can be modeled [339]. Global dynamics cannot rely on MD simulations, so stochastic models are used [148]. The methodology for simulating slow-motional EPR spectra of two coupled spin labels attached to the same macromolecule is not as established as for single labels. There exist methods based on the SLE [325] and on trajectories [242] for the simple case of a tumbling protein with two rigidly attached labels. MD trajectories have been used to simulate spectra from proteins labeled with two nitroxides [340] and to compare spin relaxation times from explicit dynamics to those obtained from Redfield–Wangsness–Bloch relaxation theory [341]. Implicit solution methods for the SLE have been investigated [342].

For multinuclear spin systems, perturbation treatments can be applied if one hyperfine coupling anisotropy dominates and the hyperfine anisotropies of the others are so small that they are in the fast-motion range. In this case, a post-convolution technique can be used [343, 344]. It has been implemented in a program for copper spectra [345] and is also available in EasySpin.

Computationally, the main challenge with SLE methods is the matrix size. It is the product of the sizes of the spin basis ($N^2$ for a full basis) and the orientational basis. The spin space dimension increases exponentially with the number of spins. In addition to pruning the orientational basis, as mentioned above, the spin space can be restricted in the high-field limit [27]. More general spin space truncation methods are applicable as well [64], but are not implemented in dedicated slow-motion EPR simulation programs.

### 3.5.2
### Chemical Exchange

Chemical exchange summarizes situations where dynamic transformations between different chemical or conformational states (called *sites*) of a spin center change the sH parameters. These changes can affect EPR spectra if the timescale of the process is not much slower than the timescale of the EPR experiment, which is equal to the inverse of the spectral shift caused by the exchange processes. Depending on this relative timescale, slow, intermediate, and fast exchange regimes are distinguished. Generally, it is assumed that the transitions between sites are "sudden," that is, negligibly short compared to the periods of the spin coherences. One prominent example of spectral effects of chemical exchange in EPR is alternating linewidths [346, 347]. They have been reviewed in great detail [298].

Two structurally distinct situations can occur [143]: intramolecular exchange, where a dynamic process transfers spin polarizations and coherences within a molecule from each transition to a unique target transition (e.g., conformational equilibria); and intermolecular exchange, where a process transfers coherences from one molecule to another and therefore from one transition to a set of transitions with different probabilities, determined by the spin state of the other molecule (e.g., intermolecular electron transfer). Mutual exchange denotes a special case of intramolecular exchange, where the molecular structures before and after the exchange are indistinguishable [348]. The spin Hamiltonians for all sites are identical except for the labeling of the nuclear spins. The fractional population of each of $N$ sites is therefore $\frac{1}{N}$, and the dynamic behavior is characterized by a single rate constant.

For a two-state kinetic system, chemical exchange effects can be modeled using modified Bloch equations [349, 350]. A more general approach is formulated in Liouville space using the SLE including an exchange superoperator or kinetic matrix that describes the transfer of coherences from one transition to another as a result of the dynamic process. This Liouville method was first developed [351–353] and then generalized and extended [125, 354] in NMR, including a simple index-permutation method [355]. It is valid for all three exchange regimes. In general, chemical exchange is modeled in the composite direct-sum space of the Liouville spaces of the various sites. In the case of mutual exchange, the Liouville matrix can be block-diagonalized, and the composite Liouville space can be reduced to the Liouville space of a single site [348]. Any kinetic network between a number of sites can be implemented, for example, for independent exchange processes occurring simultaneously such as electron transfer and internal conformational changes. Multiple superimposed intermolecular exchange processes can be modeled as well [356]. There exist efficient NMR simulation programs [357], even for nonequilibrium chemical exchange [358]. More recently, Monte Carlo approaches to the simulation of dynamic NMR spectra have been developed [359].

Freed and Fraenkel [112], in their ground-breaking work, have developed the general exchange theory for alternating linewidth effects in EPR. Intermolecular triplet transfer has been described by Hudson and McLachlan [360]. Norris has derived a simple equation for intramolecular multisite exchange under the conditions of a single average lifetime for all the sites and exchange pathways between each pair of sites [361]. The Norris equation and the Liouville method (see above paragraph) have been implemented by Grampp and Stiegler [362]. Heinzer [44] has implemented the NMR Liouville method for EPR and adapted it for least-squares fitting using analytical derivatives of the spectral and shape vectors [45]. His programs have been popular for a long time, and a more recent program has extended the functionality to biradicals [363]. Rockenbauer has combined least-squares fitting with chemical exchange simulations based on the modified Bloch equations for a two-site model that can also include fast-tumbling effects [43, 364]. EasySpin follows Heinzer's Liouville method [44, 45] and extends it to larger spin systems. Most current chemical exchange EPR simulation programs are limited to isotropic first-order spectra of $S = \frac{1}{2}$ coupled to a few nuclei. Spinach [64] supports arbitrary exchange matrices. Efficient and general programs for situations where more than a few nuclei are involved, or where the sH is anisotropic and contains nonsecular contributions, are not currently available.

As shown for the case of intramolecular multisite exchange with some equivalent sites in NMR, permutational and other symmetries in the Liouville operator can be exploited to reduce the size of the Liouville space that needs to be included in a spectral simulation [122, 365, 366]. Taking advantage of these symmetries is crucial for improving the performance of simulations for large spin systems. They have yet to be leveraged substantially in general EPR simulation programs.

Although it might appear that chemical exchange processes are not relevant in solid-state systems at low temperatures, methyl reorientations by hindered rotation (hopping) or tunneling can dynamically affect EPR and ENDOR spectra [143]. In addition, experimentally observed Jahn–Teller pseudorotation of fullerene in its photoexcited triplet state was successfully simulated using a multisite chemical exchange model [367].

## 3.6
**Pulse EPR Spectra**

To compute pulse EPR spectra, a differential equation describing the time evolution of the spins in the spin system must be solved. Depending on the particular experiment and on the algorithm, this can be done in the time or in the frequency domain. Three different levels of theory are generally employed: (i) the Bloch equations, (ii) the LvN equation in Hilbert space, and (iii) the LvN or SLE in Liouville space. All spin dynamics simulations are normally carried out in the rotating frame or a similar interaction picture [67, 75, 368].

### 3.6.1
### Bloch Equations

The classical torque equation for the magnetization vector, that is, the Bloch equation without the relaxation terms, has been used to describe spin echo phenomena since their discovery [369]. The Larmor or Bloch equations are adequate for simulating pulse EPR and NMR experiments on spin centers with a single spin with isotropic gyromagnetic ratio. They can accommodate arbitrary excitation fields. Many closed-form solutions have been derived, but it is also straightforward to solve them numerically using standard ordinary differential equation solvers. The Bloch equations form an important classical description of magnetic resonance experiments [370].

As shown by Feynman et al. [371], the state of any two-level quantum system can be described by a magnetization-like vector, and its coherent dynamics can be modeled using a Larmor- or Bloch-like equation as mentioned above. This leads to the concept of the Bloch sphere, as utilized, for example, for optical transitions.

### 3.6.2
### Hilbert space

The Hilbert-space LvN equation describing the spin dynamics in terms of the density matrix has been used for spin dynamics since the early days, for example, by Bloom [372]. Solutions of the Hilbert-space LvN equation can be derived and implemented at several levels. For simple systems and pulse sequences, scalar expressions can be obtained. For more complicated spin systems, the coherent dynamics can be modeled using numerical density matrix propagation. Longitudinal and transverse relaxation can be taken into account phenomenologically in a manner similar to the Bloch equations. In the presence of stochastic processes such as chemical exchange and rotational diffusion, the density matrix equation has to be solved in Liouville space.

#### 3.6.2.1 Scalar Equations

Scalar expressions for the simulation of two- and three-pulse ESEEM [75] traces go back to the theory developed by Mims [373–375] and others [376, 377], excellently summarized in the first chapter of a monograph by Dikanov and Tsvetkov [378]. These scalar expressions are valid only for $S = \frac{1}{2}$, isotropic g values, and $I = \frac{1}{2}$, but have been extended to $S > \frac{1}{2}$ [368, 379, 380] and anisotropic g-matrices [381]. Scalar expressions are very valuable for physical insight [382].

For $I > \frac{1}{2}$, perturbational expressions have been derived that assume that the quadrupole interaction is smaller than the hyperfine and the nuclear Zeeman interactions. For $I = 1$ and $I = \frac{3}{2}$, it is possible to incorporate the quadrupole interaction exactly, since general scalar-level expressions for the eigenfrequencies and eigenvectors of the corresponding nuclear spin Hamiltonians are known [383, 384]. A graphical method for their solution was devised [385–387].

For HYSCORE and other more advanced ESEEM experiments, explicit analytical scalar expressions are available [75, 388–392]. These types of expressions can be readily derived using algebraic methods [393–396] using programs such as Mathematica or Maple. However, for anything but the simplest pulse sequences, they are too complicated to offer much physical insight.

For the out-of-phase ESEEM observed in spin-correlated radical pairs, an analytical scalar expression based on a density matrix dynamics description has been derived and can be easily implemented [397, 398].

#### 3.6.2.2 Matrix Equations

More general methods for pulse EPR are based on the solution of the LvN equation using matrix representations of density matrix, propagator exponentials, and detection operators in Hilbert space. Mims originally introduced this description for $S = \frac{1}{2}$ in the high-field approximation [374, 375]. In that limit, hyperfine terms containing $S_x$ and $S_y$ are neglected, so that the spin Hamiltonian is block-diagonal, with two nuclear sub-Hamiltonians on the diagonal [14]. The equation can then cleanly be transformed into the rotating frame [368]. The final expressions for the TD signals contain products of elements from the unitary overlap matrix between the nuclear eigenstates in the $+\frac{1}{2}$ and $-\frac{1}{2}$ electron spin manifolds [374, 375, 378, 399, 400]. This matrix is denoted $M$ and sometimes called the *Mims matrix*. Its matrix elements are the branching factors that determine the nuclear echo envelope modulations and are analogous to the Frank–Condon factors for vibronic transitions.

In all their incarnations, Hilbert-space methods in essence generate a list of frequencies $\omega_\xi$ and complex amplitudes $Z_\xi$ that together determine the complex exponentials that constitute the final TD signal [14]. For example, for 2D data

$$V(t_1, t_2) = \sum_\xi Z_\xi \cdot e^{-i\omega_{1\xi} t_1} \cdot e^{-i\omega_{2\xi} t_2}$$

where $Z_\xi$ is a product of matrix elements from $M$, and $\omega_\xi$ are differences of eigenvalues of the nuclear sub-Hamiltonians. For a general summary of the Hilbert-space method, see [14].

Frequencies and amplitudes are used to construct either the TD directly or via the frequency domain [58, 61]. Three approaches are possible. (i) The most straightforward and most widely employed method is direct brute-force evolution in TD [55, 376, 377, 401, 402]. This is simple, but can be computationally overly expensive for situations with many peaks or large TDs. (ii) The second method utilizes the frequencies and amplitudes obtained from the density matrix calculation to construct an FD histogram, which is then converted to TD using Fourier transform. This FD "binning" method [49, 50, 54, 374, 380, 399] is very fast and advantageous for situations with many peaks, as it involves essentially no computational cost per peak. The costliest operation is the Fourier transform.

It is particularly efficient for powder simulations. However, FD binning is only approximate, as it involves rounding the frequency of each peak to the nearest discretized frequency in the FD histogram. It can lead to systematic errors, for example, incomplete phase interference in powder simulations. (iii) The third method is a variation of FD binning that addresses these shortcomings [403]. It is based on the convolution and deconvolution of a short finite impulse response filter kernel. It is much faster than the TD method and orders of magnitude more accurate than FD binning.

A bottleneck in the simulation of pulse EPR spectra is the computation of the propagator matrices $U$ via matrix exponentiation. In Hilbert space, the matrix exponential is required to compute the propagation sandwich product. Many numerical methods for computing matrix exponentials are known [404, 405], for example, Taylor-series expansion, Padé approximation, Chebyshev approximation, differential equation solvers, and matrix diagonalization.

Several ESEEM simulation programs have been developed, described, and applied in the past few decades: MAGRES, from Nijmegen, was probably the first general ESEEM simulation program based on density matrix theory [49, 50, 406]. HYSCORE simulation programs were developed and described by Goldfarb [54], Schweiger [55], and others [407]. OPTESIM, a recently described 1D ESEEM simulation program, includes a least-squares fitting algorithm [57]. SimBud [56] has a UI. Molecular Sophe [24] supports pulse EPR simulations. EasySpin supports arbitrary user-defined pulse EPR sequences [14].

ESEEM theory for high-spin systems is well developed and can be implemented in a straightforward manner. Compared to a spin-$\frac{1}{2}$ system, where only one allowed EPR transition is present, many more are present in high-spin systems. All methods compute the expectation value of the electron spin angular momentum vector $\langle \mathbf{S} \rangle_i^T = (\langle S_x \rangle_i, \langle S_y \rangle_i, \langle S_z \rangle_i)^T$ for each energy eigenstate ($i = 1, \ldots, 2S+1$) of the electron spin and then use this to construct spin Hamiltonians for the nuclear sub-manifolds:

$$H_{\text{nuc}}(i) = \sum_k \langle \mathbf{S} \rangle_i^T A_k \mathbf{I}_k - g_{n,k} \mu_N \mathbf{B}_0^T \mathbf{I}_k + \mathbf{I}_k^T P_k \mathbf{I}_k$$

Among the first high-spin ESEEM simulation examples were $Cr^{3+}$ (spin $\frac{3}{2}$) and its $Al^{3+}$ neighbors [408] in ruby, and protons in photoexcited triplet states of various organic molecules [379, 409, 410]. The theory for high-spin transition metal ions with small zero-field splittings has been detailed by Peisach [411], and graphically represented by Singel [380]. Peisach neglected nonsecular terms in the zero-field splitting, leading to $\langle S_x \rangle_i = \langle S_y \rangle_i = 0$. These terms have been shown to be important by Astashkin and Raitsimring [368]. For small zero-field splittings, $\langle S \rangle_i$ can be obtained via perturbation theory [368, 412], whereas more generally, matrix diagonalization must be used [14]. Oliete has used ESEEM simulations for fluorine ligands in an $S = 2$ $Cr^{2+}$ system [413].

When multiple nuclei are present in a spin system, the combined nuclear sub-Hamiltonians can be factored into direct products of single-spin nuclear sub-Hamiltonians. As a consequence, the overall echo modulation amplitude can be factored into a sum of products of modulations as long as pulses are nonselective. These "product rules" have been derived and published for two-pulse ESEEM [373, 374, 408], three-pulse ESEEM [378, 414, 415], and HYSCORE [416], as well as for two-pulse ESEEM on triplets [379]. A general form of the product rule was derived for and implemented in EasySpin [14].

Hilbert-space density matrix propagation methods can also be applied to multifrequency experiments. Such simulations have been performed for DEER [417] and double-quantum coherence (DQC) [300, 418] experiments.

Many software packages have been developed that provide efficient Hilbert-space and Liouville-space spin dynamics simulations for both EPR and NMR. Highly optimized packages include SMART [419], Gamma [55, 402, 420], BlochLib [421], and the more recent high-performance packages Simpson [62], SPINEVOLUTION [63], and Spinach [422]. All can be applied advantageously to EPR in some situations. There are several opportunities for parallelization in Hilbert-space methods. Once frequencies and amplitudes are calculated, they can be combined into TD traces or binned into an FD histogram in parallel. Recent work has identified ways to parallelize density matrix propagation directly using appropriate decomposition of the density matrix [423, 424].

Despite a well-developed theoretical basis, the current accuracy of pulse EPR simulation methods is not entirely satisfactory. Simulations of ESEEM and HYSCORE spectra often do not match very well with experimental data. Peak positions can often be reproduced, but peak intensities tend to be off. This discrepancy indicates that the theoretical models are too simple and should be improved. Reasons for the discrepancy are as follows: (i) Most simulations assume ideal rectangular pulses, whereas in practice pulses are neither infinitely short nor perfectly rectangular. (ii) Because of the product rules, simulations of ESEEM spectra of one nucleus can only be accurate if all other nuclei are included. This is currently not done. (iii) The position and width of the detection integration window relative to the echo transient affect the modulation intensities (observer blind spots [389]). For example, an integration window with nonzero width acts as a low-pass filter. Most simulations assume a single-point detection at the simple $\tau$ point, which need not even be the echo maximum. (iv) Site-to-site heterogeneity and correlated hyperfine strains between different nuclei on the same spin center can affect line positions and intensities, but are never taken into account.

In NMR, simulation methods for time-dependent problems such as magic angle spinning (MAS) have been developed [61, 425, 426]. Two recent reviews summarize the application of Floquet theory to such problems in solid-state NMR simulations [427, 428]. For EPR, these techniques can be used to describe experiments with simultaneous irradiation at multiple frequencies such as rf-driven (radiofrequency) ESEEM [429], cw multi-quantum EPR [430], double-modulation EPR [431], and standard cw EPR [131].

### 3.6.3
### Liouville Space

The most general equation for the simulation of spin dynamics in pulse magnetic resonance experiments is the SLE. In NMR, it is widely applied [64, 432]. In EPR, apart from its central importance in the simulation of slow-motion, saturation, and chemical exchange cw EPR spectra, the SLE has been used only in a few pulse EPR experiments: 1D and 2D Fourier transform EPR experiments in ordered and viscous fluids [316, 433–435], 2D DQC experiments [436], EXSCY [437], and a few others [75]. In general, the SLE has not seen widespread use in pulse EPR, since most current pulse applications are on solids and do not require the incorporation of stochastic processes.

Since the matrices associated with the SLE are very large, simulations tend to be slow. Two approaches are possible to improve performance: (i) more efficient algorithms and (ii) reduction of space dimensionality.

Much effort has been spent in finding more efficient algorithms. Sparse matrix methods can be advantageous, and methods based on Lanczos methods [27] are routinely utilized. An improved Lanczos-based method for matrix reduction to tridiagonal form has recently been presented [438]. Finite-element methods have also been proposed [439]. In Liouville space, in contrast to Hilbert space, the costly computation of the matrix exponential can be avoided and replaced by faster methods that directly apply the propagator to a density matrix (in vector form) as a matrix–vector multiplication. Methods that reduce the dimensionality of Liouville space by pruning the basis set to exclude insignificant dimensions were originally introduced in slow-motion EPR simulations that employed the Lanczos algorithm [440, 441], as discussed in Section 3.5.1.3. More recently, Kuprov has developed state space restriction methods for NMR that reduce the size of the Liouville space and the size of the associated density vector and Liouville matrix, leading to large gains in performance compared to a brute-force approach [64, 422, 442–444]. NMR simulation methods for large spin systems is an active area of research [445, 446].

## 3.7
### Pulse and cw ENDOR Spectra

ENDOR spectra are acquired in the presence of a constant external magnetic field, so that the sH does not change during an experiment. Therefore, ENDOR spectra in the rigid and fast-motion limits can be simulated much more easily than cw EPR spectra in the same regimes. No field-dependent Zeeman energy level diagram needs to be constructed. As will be described below, the main difficulty is the efficient and accurate computation of line intensities.

## 3.7.1
### Transition Frequencies

The computation of the transition frequencies for ENDOR is identical in complexity to that for frequency-swept EPR spectra. For this step, two levels of theory, matrix diagonalization and perturbation theory, are used. Using matrix diagonalization, the full sH matrix with the chosen external static field is diagonalized once to determine all energy eigenvalues and eigenstates. The differences of the eigenvalues give the transition frequencies. This approach has been implemented in several programs [49, 208] including EasySpin and works well for small spin systems. However, the size of the Hamiltonian matrix scales exponentially with the number of ENDOR nuclei.

Therefore, in the case of a large number of nuclei (e.g., when modeling matrix lines), analytical first- and second-order perturbation theory expressions [225, 227, 447–449] are used to obtain approximations to the eigenvalues. For multiple nuclei, cross terms between hyperfine couplings within pairs of nuclei are present [226]. They cannot be neglected if one of the hyperfine couplings is substantial. A relevant example is $^1$H ENDOR in Cu complexes with large Cu hyperfine couplings. Perturbative treatments can become unacceptably inaccurate for systems with multiple large hyperfine couplings such as $^{55}$Mn ENDOR on oligonuclear Mn clusters. As in the case of cw EPR spectra, the validity of perturbational approximations should be checked carefully. Unfortunately, few programs do this consistently, leaving the user uninformed on whether the chosen level of theory is accurate enough. Hybrid methods, as used for cw EPR and ESEEM, are applicable to ENDOR as well.

## 3.7.2
### Intensities

While the computation of ENDOR line positions is straightforward, the accurate and general calculation of line intensities is more challenging. To the best of our knowledge, no generally available program is currently able to do this. Unlike cw EPR, where Fermi's Golden Rule gives accurate approximations to the line intensities in the experimentally relevant non-saturating limit, ENDOR involves either saturating irradiation (cw ENDOR) or bandwidth-limited and spectrally selective excitation (pulse ENDOR). For cw EPR, the full steady-state LvN equation needs to be solved, including relaxation rate constants. This often leads to overparameterization. For pulse ENDOR, at least the Hilbert-space LvN equation for the density matrix is required.

In a simplified picture, the intensity of an ENDOR transition between two levels $|u\rangle$ and $|v\rangle$ can be written as a product of several factors

$$I_{uv}^{\mathrm{ENDOR}} \propto t_{uv}(\boldsymbol{B}_2) \cdot \alpha_{uv}(\omega_{\mathrm{mw}}, \tau, t_{\mathrm{p}}) \cdot \Delta p_{uv}(T) \cdot f(\omega_{\mathrm{rf}} - \omega_{uv})$$

where $t_{uv}$ is the transition moment, $\alpha_{uv}$ is a selectivity factor that depends on experimental settings, $\Delta p_{uv}$ is the Boltzmann polarization of the transition, and $f$ is the excitation profile of the rf excitation [450].

One aspect of the transition moment that affects both cw and pulse ENDOR spectral intensities, and distinguishes ENDOR from NMR, is hyperfine enhancement [451–453], the fact that the presence of the hyperfine-coupled electron spin amplifies the driving rf field strength at the nucleus. Theoretically, this is accounted for by including the electron Zeeman Hamiltonian in addition to the nuclear Zeeman operator in the transition operator [1]. In cw ENDOR, this results in increasing intensity with increasing ENDOR frequency. In pulse ENDOR, this results in increased nutation frequencies and effective flip angles during rf pulses with increasing ENDOR frequency, affecting spectral intensities in a nonlinear way. In principle, these distortions can be modeled using a complete Hilbert-space density matrix treatment. One approximation uses a product of NMR and EPR transition moments, where the latter are summed over all EPR transitions that share one level with the given ENDOR transition [49]. In general, ENDOR simulations tend to be most quantitative for single-isotope pulse ENDOR spectra over a relatively narrow frequency range, for example, ENDOR of weakly coupled $^1$H at Q-band, where transition moments do not vary much across the spectrum.

The second aspect that distinguishes ENDOR from NMR is the selectivity $\alpha$ imposed by the narrow-bandwidth microwave excitation. The intensity of an ENDOR transition not only depends on matrix elements of the Zeeman operator, but also on the microwave frequency. Only a nuclear transition for which one of the nuclear levels is part of an EPR transition that is resonant with the microwave frequency will yield significant ENDOR intensity. This leads to strong orientation selectivity in anisotropic systems and transition selectivity in systems with hyperfine couplings that are larger than the microwave excitation bandwidth. For powder simulations, this selectivity is a significant computational burden, since often only a small fraction of computed orientations of the spherical grid exhibit nonvanishing ENDOR intensity. The majority of evaluated orientations are not significant to the final spectrum. There seems to be no general remedy against these superfluous computations. In some cases, a viable work-around is an orientation preselection procedure: the selectivity is computed for all orientations for a reduced spin system consisting only of spins with anisotropic interactions larger than the electron excitation bandwidth. In a second step, the ENDOR spectrum of the full system is computed only for those orientations where the selectivity factor is above a chosen threshold [1, 14]. Alternatively, the simulation in this second step can be restricted to a spin system containing only the nuclei of interest, using an effective hyperfine field [14, 55, 454]. All pulse EPR experiments are orientation selective if the spectral width exceeds the excitation bandwidth of the pulses, as is most commonly the case.

In pulse ENDOR, there exist additional selectivity effects. Intensities in Mims ENDOR spectra are affected by $\tau$-dependent blind spots. In Davies ENDOR spectra, intensities are attenuated by the nonselectivity of the first inversion pulse (self-suppression) of length $t_p$. Again, these effects arise automatically if full density matrix simulations are performed. However, an intermediate approach that circumvents density matrix dynamics solvers, but still reproduces the selectivity amplitude features of specific pulse ENDOR experiments, uses analytical approximate excitation and detection envelopes [450] that can model effects from limited excitation bandwidth and $\tau$-dependent blind spots. The simulation of blind spots and matrix suppression in Mims and Davies ENDOR have been investigated in great detail [455–457].

High-spin pulse ENDOR simulation methods were developed for Mn systems [368, 458, 459], and iron(III) [460] and diiron(II) centers [461]. They follow along the lines of high-spin ESEEM simulations, as discussed in Section 3.6.2.2.

Full-density matrix methods have been used to simulate pulse ENDOR spectra [462] and can include both coherence and relaxation [463]. These models tend to suffer from over-parameterization, but can be trimmed down. For instance, most pulse ENDOR experiments are based on polarization transfers [75]. For such experiments, a simplified density matrix dynamics approach is often used [464]. Coherences are neglected, and the vector of populations, containing only the diagonal elements from the density matrix arranged in a column, is propagated in time. This is equivalent to restricting the Liouville-space spin basis for the density operator to longitudinal terms ($S_z$, $I_z$, $S_z I_z$, and identity operator). $T_1$ relaxation and saturation effects are easily included. Multi-sequence pulse ENDOR experiments [465, 466] have been modeled and analyzed using this approach.

Some of the difficulty of quantitatively simulating ENDOR intensities is due to instrumental imperfections. In contrast to EPR and NMR, ENDOR involves a broadband rf transmitter. Over the broad rf ranges that are swept in ENDOR, the frequency characteristics of the transmitter (source power, amplifier gain, coil impedance) are rarely completely flat. Any non-flatness directly affects the power delivered to the sample and therefore the ENDOR intensities. This could in principle be taken into account by convolving the spectrum with the transmitter characteristics, but would require extensive rf characterization of the instrumentation. Spurious resonances in the rf transmitter can lead to spurious features in the ENDOR spectrum.

### 3.7.3
**Broadenings**

Line broadenings in ENDOR spectra are usually modeled by simple convolution with a combination of Gaussian and/or Lorentzian lineshapes. Anisotropies in these broadenings are rarely modeled. No physical models are used. Often, $A$ strain is thought to contribute to the broadening. Broadening effects due to the pulse length of the ENDOR pulse can be included using convolution with a sinc function.

## 3.8
## Pulse DEER Spectra

Although pulse DEER methods can be simulated using the density matrix methods outlined in Section 3.6 [300, 417, 418], the main need for computational methods for DEER data is in data analysis. Fitting of experimental DEER data requires intricate data analysis methods. Therefore, we mention some of these recent methods. Methods for the extraction of distance distributions from dipolar time traces based on Tikhonov regularization [467, 468] and maximum entropy [469] have been developed. DeerAnalysis [470] is a widely used program that provides a large number of analysis and fitting methods for DEER data. DEFit [471, 472] provides multi-Gaussian distance distribution models. MMM [473] can use protein structures to derive a distributional model for nitroxide pair distances and orientations, and compute the DEER trace from that.

For molecules with two flexible spin labels, the computation of the solid-state DEER spectrum involves summation over all possible relative orientations and distances between the two spin labels [174, 474]. For a powder simulation, this requires integration over a total of six degrees of freedom: two Euler angles that describe the orientation of the molecule with respect to the external magnetic field, three additional Euler angles that describe the relative orientation between the two spin labels, and the interlabel distance. The combined conformer/distance distribution is generally described by a multidimensional probability distribution function. As a consequence, the corresponding simulations doing this brute-force integration are painfully slow. However, parallelization is trivial and can result in tremendous speed-ups. Computational aspects of the dynamics and conformational distributions of spin labels have been reviewed very recently [475]. PRONOX is an algorithm for rapid computation of distance distributions based on conformer distributions [474]. MtsslWizard is a plugin to PyMOL that allows *in silico* spin labeling and generation of distance distributions [476].

## 3.9
## Least-Squares Fitting

Methods for "automatic" least-squares fitting of simulated spectra based on an sH model to experimental spectra have been developed from the very early days of EPR [477, 478]. In these methods, a set of sH parameters is varied in consecutive simulations until an optimal match between the simulated spectrum and the experimental spectrum is achieved. The match is generally quantified by an objective function (goodness-of-fit function, error function, target function) that depends on some measure of the difference between the simulated and experimental spectra.

Least-squares fitting methods have also been used in other aspects of EPR data analysis, for example, to computationally determine the center of symmetry of a spectrum [479]. These applications lie beyond the scope of this overview. Least-squares fitting methods are used in conjunction with locating transition fields [211] and have been mentioned in the corresponding section above.

In this section, we review the various choices available for the objective function and for the fitting algorithm, as well as their application to multicomponent and multispectral problems. Unless otherwise stated, cw EPR is assumed. In addition, we discuss error analysis. The early literature on fitting methods in EPR have been reviewed in the book by Mabbs and Collison [10]. For a review of direct search methods that do not rely on derivatives, see [480].

### 3.9.1
### Objective Function

To assess the quality of the fit between a simulated and an experimental spectrum, the sum of squared deviations is most commonly used:

$$\chi^2 \propto \sum_i \left( \frac{y_{\exp,i} - a y_{\sin,i} - b_i}{w_i} \right)^2$$

where $y_{\exp,i}$ are the data points of the experimental spectrum, $y_{\sin,i}$ are the corresponding ones from the simulated spectrum, $a$ is a scaling factor, and $b_i$ are the data points from a baseline correction function. Each difference in $\chi^2$ is additionally weighted by $w_i$, for example, the error in the measurement. Most often, $w_i = 1$.

Conventionally, for cw EPR, the $y$ values in the above expression are taken directly from the first-harmonic spectrum. However, this is not the only choice. The spectral derivative, the integral [47], or even the double integral [481, 482] can be used as basis for $\chi^2$. Both the integral and double integral as target functions have the advantage that search algorithms are less likely to get trapped in local minima in the case of EPR spectra with many resolved lines. Integration also reduces the effect of noise. A multistep procedure has been proposed where the fit is initially based on the double-integrated spectrum, in a middle stage on the integrated spectrum, and finally on the spectrum as recorded [181, 182].

For spectra of organic radicals with many hyperfine lines, it might be beneficial to fit the Fourier transform of the spectrum [304], since this dramatically smooths the error function by removing many local minima. Another objective function that can be used for spectra with distinct maxima and minima is the aggregate mismatch in line positions between experiment and simulation. This has recently been used in linear combination with the conventional $\chi^2$ function [483]. An objective function that subdivides the spectral deviation into segments has been proposed [484].

Despite the added flexibility with a choice of objective function, there are still situations where almost all fitting algorithms can get stuck. A prototypical example is a two-component mixture of two spin-$\frac{1}{2}$ systems with orthorhombic g-matrices, with two maximum g values separated from the others and resolved among themselves. This leads to two minima of almost identical depth in any objective function. If the two g values are assigned to the wrong components in the fitting, it is almost impossible to get out of this minimum by change of objective function alone. Only permutation of the values, followed by further optimization, will improve the fit. To prevent these types of failures, additional physical constraints on the ranges and correlations of the fitting parameters need to be taken into account. Ideally, additional experimental data are included to resolve this ambiguity.

### 3.9.2
### Search Range and Starting Point

In addition to a smooth error function, two other keys to success in least-squares fitting with iterative methods are (i) a choice of the starting parameter values close to their values at the expected global minimum and (ii) the choice of an adequate search range. The choice of the fitting algorithm is secondary to these aspects.

A poor choice of starting point will often lead to non-convergence or to the convergence to a physically nonsensical solution. The search range should be restricted to a parameter subspace large enough to contain the expected solution, but small enough to be searchable in a reasonable amount of time. The set of parameters can also be transformed from one basis to another to deal with correlations between parameters. For example, instead of searching the space of two g-factors $(g_\perp, g_\parallel)$, the search can be done with the transformed coordinates $(\Delta g, \bar{g})$ with $\Delta g = g_\parallel - g_\perp$ and $\bar{g} = \frac{(g_\parallel + 2g_\perp)}{3}$.

### 3.9.3
### Fitting Algorithms

The choice of the algorithm for least-squares fitting has profound influence on the convergence rate and on the robustness of the search, that is, on the ability to find the global minimum despite local minima and noise. The dependence of EPR spectra on magnetic parameters is nonlinear, so that nonlinear least-squares methods are used [485]. Broadly, they fall into two groups: local and global. Local methods are generally fast, but are only able to locate a minimum in the objective function close to a starting parameter set. Global methods search the parameter space more widely and are able to locate a global minimum in the objective function, although mostly at the cost of a significantly slower convergence rate. EasySpin provides the user with a selection of algorithms as do XSophe [23] and other programs [486].

#### 3.9.3.1 Local Methods
Local methods involving derivatives based on analytical expressions [215] and Feynman's theorem [487] have been employed. The first instances of local least-squares fitting of EPR spectra used the Gauss–Newton or gradient descent methods with analytical derivatives [45, 477, 478, 488, 489]. These algorithms are still used occasionally. Another simple algorithm, the Newton method, is based on a local quadratic approximation of the error function. Misra [211, 213, 214, 490, 491]

has used it together with explicit first and second derivatives of $\chi^2$ in a general least-squares fitting program.

The *de facto* standard method of local nonlinear least-squares search is the Levenberg-Marquardt (LM) algorithm. The LM method adaptively varies the step size and direction. Far from a minimum of $\chi^2$, it acts similarly to a gradient descent method, stepping in the direction of steepest descent of $\chi^2$. Close to a minimum, it acts similarly to the Gauss–Newton method that assumes that $\chi^2$ is locally quadratic and steps accordingly. Typically, this method converges rapidly toward a minimum. The LM algorithm has been deployed many times in EPR, both with analytical [290, 492, 493] and numerical gradients. For the fitting of slow-motional nitroxide spectra, the LM algorithm [494] as well as a trust-region modification [433] were implemented. These two approaches and the simplex methods were recently compared [46].

The popular Nelder–Mead simplex method is simple and relatively robust. For the search of an $N$-dimensional parameter space, it sets up a set of $N + 1$ parameter values that geometrically constitute the vertices of a simplex in parameter space (e.g., a triangle for $N = 2$, a tetrahedron for $N = 3$). On the basis of the $\chi^2$ values at each of the vertices of the simplex, new vertices are chosen and evaluated. The simplex "walks" through parameter space to a nearby local minimum. The method is robust, but not very fast. It is used extensively in EPR [295, 484, 495–500].

Other local-search methods used in EPR simulations include the Hooke and Jeeves pattern search [20, 22] and Powell's conjugate gradient method [42, 501]. Multidimensional fits can be performed by consecutive 1D minimizations [43].

### 3.9.3.2 Global Methods

Global methods are able to search the parameter space more completely, increasing the likelihood of locating the global minimum. Most of them rely on an element of randomness and often involve large sets of parameter sets.

Many forms of random Monte Carlo search methods have been proposed and implemented. They range from simple random-step downhill walk [350, 502] to simulated annealing [20, 22, 23, 242, 484, 503–506], an adaptation of the Metropolis algorithm.

In recent years, several variations of the genetic (evolutionary) algorithm have been applied to EPR [498]. In these methods, the goodness-of-fit ("fitness") for a group of $M$ candidate parameter sets are computed. Each parameter set is treated as an individual, and a process akin to natural selection based on the fitness values is applied to the $M$ sets by crossover and mutation to form the next generation of $M$ parameter sets. This process is repeated until the population converges to a minimum. The main appeal of these methods lies in the fact that they are random, but directed. Genetic methods have been combined with local-search algorithms such as Powell or simplex [42, 501, 507] to accelerate convergence once close to a minimum.

Artificial neural network models have been used to extract rotational correlation times from motional spectra [508]. Other nature-inspired algorithms such as particle-swarm optimization and bacterial foraging have not seen visible usage in EPR.

Systematic grid searches [86, 509] form another set of methods that can locate the global minimum. In contrast to the methods above, they are nonrandom and systematically scan the entire parameter space. For large parameter spaces, they are very slow. For example, for searching a six-dimensional parameter space with 10 points along each dimension for each parameter, $10^6$ simulations are necessary. However, they can be expanded into weighted tree searches and combined with local-search methods to improve efficiency. Their main appeal lies in the fact that they return the global minimum. They are well suited for massively parallel computer architectures, as all the simulations are independent of each other.

### 3.9.4
### Multicomponent and Multispectral Fits

In spectra consisting of multiple components, the fitting parameter space includes not only a set of magnetic parameters for each component but also the relative weights of each component. In this case, the spectrum is a linear combination of nonlinear functions of the spin Hamiltonian parameters. These types of problems can be efficiently solved using separable nonlinear least-squares methods [510]. A special case of multicomponent spectra is a spectrum that contains baseline distortions. These can be treated as an additional component and included in the fit, usually as a linear or quadratic function (see $\chi^2$ definition above).

Multicomponent spectra are in general very challenging to analyze. If no good starting guess for the component parameters of a multicomponent spectrum is available, an attempt at decomposition into single components can be made using principal component analysis (PCA) [511] or maximum-likelihood common-factor analysis [512]. A fitting program for multicomponent nitroxide spectra has been implemented [31]. EasySpin supports general multicomponent fits.

The simultaneous fit of multiple cw EPR spectra acquired at different microwave frequencies (e.g., S, X, Q, and W band) can help constrain the parameter space and reduce the number of local minima for a multidimensional search space. One simple way to implement such multispectral fits is to concatenate all experimental spectra into a 1D array with appropriate weights and then use standard methods on this concatenated spectrum. A set of ENDOR or ESEEM spectra acquired at different magnetic fields is another common multispectral fitting problem. Simultaneous fits of spectra acquired at different temperatures can constrain static and dynamic parameters together, for example, in chemical exchange problems [43, 513].

### 3.9.5
### Limits of Automatic Fitting

Most fitting algorithms depend on a series of parameters, such as default step sizes and damping factors. The values of these parameters can affect the convergence rate and determine whether fitting will be successful or not. Therefore, there are several levels of user choice in least-squares fitting: (i) the dimensionality and representation of the parameter space, (ii) the starting point(s), (iii) the search range, (iv) the objective function, (v) the fitting algorithm, and (vi) the configuration of the fitting algorithm. These choices depend strongly on the problem at hand and on the expertise of the user. This renders a general "automatic" least-squares fitting procedure applicable to all types of EPR spectra essentially impossible. Despite tremendous effort, the goal of fully automated fitting of EPR spectra is as distant as ever. However, the host of developed algorithms provides a rich toolbox that can greatly assist the search for a good fit. EasySpin provides an interface that lets users make and change settings at all levels of choice.

### 3.9.6
### Error Analysis

Once converged, least-squares fitting algorithms return a set of supposedly optimal parameters. Several questions need to be answered before these values should be taken as a final result: (i) Is the fit close enough? Even if the fit returns the global minimum, the $\chi^2$ error might be too large. In this case, the physical model needs to be modified. (ii) Are the optimal parameters physically meaningful? If not, other minima with similar $\chi^2$ values need to be examined, or the model needs to be modified. (iii) How accurate are the obtained parameters? To answer this question, a statistical error analysis that establishes estimates for the parameter variances needs to be performed, for example, based on the covariance matrix or on Monte Carlo simulations of synthetic parameter sets [514]. Misra was the first to present a method based on the curvature matrix (matrix of second derivatives of $\chi^2$) [515]. Others have provided similar approaches [57, 214, 433, 486, 516]. However, in the experimental literature, this crucial aspect of least-squares fitting is very often neglected.

## 3.10
## Various Topics

### 3.10.1
### Spin Quantitation

Accurate quantitation of spin centers in an EPR sample is very desirable in many applications [517]. Two different principles of quantitation can be used: comparison of double integrals and quantitation by simulation. Among software programs, SpinCount has special provisions for spin quantitation [26]. EasySpin returns simulated cw EPR spectra with calibrated intensities for all systems and regimes, so that quantitation by simulation is possible.

The experimental method of double integration and comparison to a separate concentration or quantitation standard is feasible for certain classes of spin centers, for example, organic radicals and other species with narrow spectra. This method usually relies on a set of assumptions that can all introduce systematic errors if not valid: (i) sample geometry, placement, fill factors, and Q factors of the analyte and standard sample are identical; (ii) the transition moments of all lines in the spectra of both samples are identical; (iii) the full EPR spectra of both samples are acquired; and (iv) neither spectrum is saturated. Progress has been made recently in addressing some of these potential biases. The effect of sample geometry and spatial $B_1$ distribution within EPR resonators has been incorporated in commercial software [517].

The second method uses computer simulation to quantitatively simulate the EPR spectra of both the analyte and the standard sample [282]. From the scaling factors, relative concentrations can be derived. Compared to the first approach, this method does not make assumptions about intensities and spectral extent. It correctly includes differences and anisotropies in transition moments, and it can easily deal with spectra that are partially out of range. However, it still has to be ascertained that sample geometry and placement are controlled, and that neither sample is saturated.

### 3.10.2
**Smoothing and Filtering**

Many filtering and resolution-enhancement techniques have been proposed, but very few are regularly utilized, probably because of the reluctance of spectroscopists to tinker with experimental raw data. Among digital techniques, moving-average and Savitzky–Golay filtering [478] and smoothing as well as lineshape deconvolutions are the most common. Analog filtering is based on a resistor–capacitor filter as implemented in hardware in most EPR spectrometers, but is clearly limited compared to the myriad digital filtering tools available for spectral post-processing, and is now generally discouraged. An adaptive digital filtering technique can be used to increase the signal-to-noise ratio of cw EPR spectra [518].

### 3.10.3
**Data Formats**

The two most common data formats are Bruker's old ESP format (file extensions par and spc) and Bruker's current BES$^3$T format (file extensions DTA and DSC). Both store the data in binary form in one file, and experimental parameters, plus other metainformation such as details on the data storage format, in a separate second file. Both formats are open source and documented, although the older ESP format exists in a somewhat confusing variety of versions. JEOL spectrometers store data in a proprietary single-file binary format. Custom-built spectrometers generally employ simple text files to store data. Reading and storing these formats is straightforward. There exists an EPR version of the JCAMP-DX data format standard developed by the International Union of Pure and Applied Chemistry (IUPAC) [519]. This format has yet to gain traction, although the universal adoption of a standard format would have clear benefits [520].

## 3.11
## Outlook

Currently, EPR spectra can be modeled reasonably well on the basis of currently available methods described in this chapter. A large range of systems and experiment types are supported, and simulations are fast enough in many cases to allow interactive fitting for smaller spin systems. However, both the scope and the speed of current methods can be significantly increased. Extending and automating approximation methods should enable a more flexible choice of theory level free of user interference. This will benefit the simulation of larger spin systems. Parallel computing, either multi-CPU or GPU-based, will significantly enhance the performance of simulation methods, as many algorithms, from matrix diagonalization to powder averaging and least-squares fitting, can be trivially parallelized.

A major challenge for simulation methods is the increasing size of spin systems that are of current interest, such as oligometallic clusters and molecular magnets in EPR and entire proteins in NMR. Efficient simulation of these systems requires the development of dedicated and highly optimized large-scale methods.

Additional work is necessary to develop more general and usable methods and software for multispectral fits, which would allow the simultaneous analysis of multifrequency data (X-band and high-field) or multi-method data (EPR, ENDOR, ESEEM). It remains to be seen whether multiple types of spectra can be fitted with a single underlying model without an inordinate increase in the dimensionality of the parameter space.

Except for very simple spin systems and pulse sequences, pulse EPR simulation methods are still slow and quantitatively not entirely reliable. As a consequence, they lag far behind cw EPR simulation methodology and hinder the development of the field. More work is needed to implement faster and more accurate pulse EPR simulation methods and to calibrate these against experimental data, for example, for HYSCORE. This is especially important given the recent emergence of optimal control pulses [521, 522] that will open up a wide array of new possibilities in pulse EPR.

A general trend is observed where EPR simulation methods are increasingly combined with other computational methods, to provide complete end-to-end solutions for certain areas of research. This includes integration with computational chemistry methods such as DFT [33] and direct fitting of molecular structures to EPR spectra [523]. MD methods are of increasing importance in spin-label studies, as discussed in the section on slow-motion simulations. Advances in these areas will be significant [475].

As an ultimate goal, all types of EPR spectra, no matter whether from solids or liquids, or including any number of relaxation or other dynamic effects, should be in principle analyzable on a time scale that is short relative to the effort of sample preparation and spectral acquisition. This would eliminate the current bottleneck of spectral simulation and analysis. I hope that this goal is reached in the near future.

## References

1. Stoll, S. and Schweiger, A. (2006) EasySpin, a comprehensive software package for spectral simulation and analysis in EPR. *J. Magn. Reson.*, **178**, 42–55.
2. Swalen, J.D. and Gladney, H.M. (1964) Computer analysis of electron paramagnetic resonance spectra. *IBM J. Res. Dev.*, **8** (5), 515–526.
3. Taylor, P.C., Baugher, J.F., and Kritz, H.M. (1975) Magnetic resonance spectra in polycrystalline solids. *Chem. Rev.*, **75** (2), 203–240.
4. Vancamp, H.L. and Heiss, A.H. (1981) Computer applications in electron paramagnetic resonance. *Magn. Reson. Rev.*, **7**, 1–40.
5. Morse, P.D. and Smirnov, A.I. (1996) in *Electron Spin Resonance* (eds B.C. Gilbert, N.M. Atherton, and M.J. Davies), Royal Society of Chemistry, pp. 244–267.
6. Pilbrow, J.R. (1996) Principles of computer simulation in EPR. *Appl. Magn. Reson.*, **10**, 45–53.
7. Weil, J.A. (1999) The simulation of EPR spectra: a mini-review. *Mol. Phys. Rep.*, **26**, 11–24.
8. Lund, A. and Liu, W. (2013) Continuous wave EPR of radicals in solids, in *EPR of Free Radicals in Solids I* (eds A. Lund and M. Shiotani), Springer.
9. Kirste, B. (1994) in *Handbook of Electron Spin Resonance* (eds C.P. Poole and H.A. Farach), AIP Press, New York, pp. 27–50.
10. Mabbs, F.E. and Collison, D. (1992) *Electron Paramagnetic Resonance of d Transition Metal Compounds*, Elsevier, Amsterdam.
11. Morse, P.D. and Madden, K.P. (1993) The computer corner. *EPR Newsl.*, **5**, 4–9.
12. Stoll, S. (2003) *Spectral Simulations in Solid-State Electron Paramagnetic Resonance*, Eidgenössische Technische Hochschule Zürich.
13. Stoll, S. and Schweiger, A. (2007) EasySpin: simulating cw ESR spectra. *Biol. Magn. Reson.*, **27**, 299–321.
14. Stoll, S. and Britt, R.D. (2009) General and efficient simulation of pulse EPR spectra. *Phys. Chem. Chem. Phys.*, **11**, 6614–6625.
15. Cammack, R., Fann, Y.C., and Mason, R.P. Electron Spin Resonance Software Database, http://tools.niehs.nih.gov/stdb/index.cfm/spintrap/epr_home (accessed 22 August 2013).
16. Weber, R.T. (2011) EPR simulation at bruker. *EPR Newsl.*, **20**, 26–28.
17. Wang, D. and Hanson, G.R. (1995) A new method for simulating randomly oriented powder spectra in magnetic resonance: the Sydney Opera House (SOPHE) method. *J. Magn. Reson., Ser. A*, **117**, 1–8.
18. Wang, D. and Hanson, G.R. (1996) New methodologies for computer simulation of paramagnetic resonance spectra. *Appl. Magn. Reson.*, **11**, 401–415.
19. Gates, K.E. et al. (1998) Computer simulation of magnetic resonance spectra employing homotopy. *J. Magn. Reson.*, **135**, 104–112.
20. Griffin, M. et al. (1999) XSophe, a computer simulation software suite for the analysis of electron paramagnetic resonance spectra. *Mol. Phys. Rep.*, **26**, 60–84.
21. Heichel, M. et al. (2000) Xsophe-Sophe-XeprView Bruker's professional CW-EPR simulation suite. *Bruker Rep.*, **2000** (148), 6–9.
22. Hanson, G.R. et al. (2003) in *EPR of Free Radicals in Solids I. Trends in Methods and Applications* (eds A. Lund and M. Shiotani), Springer, New York, pp. 197–237.
23. Hanson, G.R. et al. (2004) XSophe-Sophe-XeprView: a computer simulation software suite (v. 1.1.3) for the analysis of continuous wave EPR spectra. *J. Inorg. Biochem.*, **98**, 903–916.
24. Hanson, G.R., Noble, C.J., and Benson, S. (2009) Molecular Sophe, an integrated approach to the structural characterization of metalloproteins. The next generation of computer simulation software. *Biol. Magn. Reson.*, **28**, 105–174.

25. Duling, D.R. (1994) Simulation of multiple isotropic spin-trap EPR spectra. *J. Magn. Reson. B*, **104**, 105–110.
26. Hendrich, M.P. SpinCount, *http://www.chem.cmu.edu/groups/hendrich/facilities/index.html* (accessed 22 August 2013).
27. Schneider, D.J. and Freed, J.H. (1989) Calculating slow motional magnetic resonance spectra. *Biol. Magn. Reson.*, **8**, 1–76.
28. Earle, K.A. and Budil, D.E. (2006) in *Advanced ESR Methods in Polymer Research* (ed. S. Schlick), John Wiley & Sons, Inc., Hoboken, NJ, pp. 53–83.
29. Chiang, Y.W., Liang, Z., and Freed, J.H. (2007) Software available from ACERT website. *EPR Newsl.*, **16** (4), 19–20.
30. Freed, J.H. (2007) ACERT software: simulation and analysis of ESR spectra. *Biol. Magn. Reson.*, **27**, 283–285.
31. Altenbach, C. LabVIEW Programs for the Analysis of EPR Data, *https://sites.google.com/site/altenbach/* (accessed 22 August 2013).
32. Steinhoff, H.J. *et al.* (1997) Determination of interspin distances between spin labels attached to insulin: comparison of electron paramagnetic resonance data with the X–ray structure. *Biophys. J.*, **73**, 3287–3298.
33. Zerbetto, M., Polimeno, A., and Barone, V. (2009) Simulation of electron spin resonance spectroscopy in diverse environments: an integrated approach. *Comput. Phys. Commun.*, **180**, 2680–2697.
34. Zerbetto, M. *et al.* (2013) Computational tools for the interpretation of electron spin resonance spectra in solution. *Mol. Phys.*, **111**, 2746–2756.
35. Mombourquette, M.J. and Weil, J.A. (1992) Simulation of magnetic resonance powder spectra. *J. Magn. Reson.*, **99**, 37–44.
36. Li, L. *et al.* (2006) Targeted guanine oxidation by a dinuclear copper(II) complex at single strand/double stranded DNA junctions. *Inorg. Chem.*, **45**, 7144–7159.
37. Nilges, M.J. (1979) Electron paramagnetic resonance studies of low symmetry nickel(I) and molybdenum(III) complexes. PhD thesis. University of Illinois, Urbana, IL.
38. Weisser, J.T. *et al.* (2006) EPR investigation and spectral simulations of iron-catecholate complexes and iron-peptide models of marine adhesive cross-links. *Inorg. Chem.*, **45**, 7736–7747.
39. Glerup, J. and Weihe, H. (1991) Magnetic susceptivility and EPR spectra of μ-cyano-bis[pentaaminechromium(III)] perchlorate. *Acta Chem. Scand.*, **45**, 444–448.
40. Jacobsen, C.J.H. *et al.* (1993) ESR characterization of trans-$V^{II}(py)_4X_2$ and trans-$Mn^{II}(py)_4X_2$ (X = NCS, Cl, Br, I; py = pyridine). *Inorg. Chem.*, **32**, 1216–1221.
41. Eloranta, J. (1999) Xemr – A general purpose electron magnetic resonance software system. *EPR Newsl.*, **10** (4), 3.
42. Spałek, T., Pietrzyk, P., and Sojka, Z. (2005) Application of the genetic algorithm joint with the powell method to nonlinear least-squares fitting of powder EPR spectra. *J. Chem. Inf. Model.*, **45**, 18–29.
43. Rockenbauer, A. and Korecz, L. (1996) Automatic computer simulations of ESR spectra. *Appl. Magn. Reson.*, **10**, 29–43.
44. Heinzer, J. (1971) Fast computation of exchange-broadened isotropic E.S.R. spectra. *Mol. Phys.*, **22** (1), 167–177.
45. Heinzer, J. (1974) Least-squares analysis of exchange–broadened ESR spectra. *J. Magn. Reson.*, **13**, 124–136.
46. Khairy, K., Budil, D.E., and Fajer, P.G. (2006) Nonlinear-least-squares analysis of slow motional regime EPR spectra. *J. Magn. Reson.*, **183**, 152–159.
47. Smirnov, A.I. (2007) EWVOIGT and EWVOIGTN: inhomogeneous line shape simulation and fitting programs. *Biol. Magn. Reson.*, **27**, 289–297.
48. Štrancar, J. (2007) EPRSIM-C: a spectral analysis package. *Biol. Magn. Reson.*, **27**, 323–341.
49. Keijzers, C.P. *et al.* (1987) MAGRES: a general program for electron spin resonance, ENDOR and ESEEM. *J. Chem. Soc., Faraday Trans. 1*, **83** (12), 3493–3503.

50. Reijerse, E.J. and Keijzers, C.P. (1987) Model calculations of frequency-domain ESEEM spectra of disordered systems. *J. Magn. Reson.*, **71**, 83–96.
51. Hoffman, B.M., Martinsen, J., and Venters, R.A. (1984) General theory of polycrystalline ENDOR pattens. g and hyperfine tensors of arbitrary symmetry and relative orientation. *J. Magn. Reson.*, **59**, 110–123.
52. Hoffman, B.M., Venters, R.A., and Martinsen, J. (1985) General theory of polycrystalline ENDOR patterns. Effects of finite EPR and ENDOR component linewidths. *J. Magn. Reson.*, **62**, 537–542.
53. Hoffman, B.M. *et al.* (1989) Electron nuclear double resonance (ENDOR) of metalloenzymes, in *Advanced EPR: Applications in Biology and Biochemistry* (ed. A.J. Hoff), Elsevier.
54. Szosenfogel, R. and Goldfarb, D. (1998) Simulations of HYSCORE spectra obtained with ideal and non-ideal pulses. *Mol. Phys.*, **95** (6), 1295–1308.
55. Mádi, Z.L., Van Doorslaer, S., and Schweiger, A. (2002) Numerical simulation of One- and Two-dimensional ESEEM experiments. *J. Magn. Reson.*, **154**, 181–191.
56. Astashkin, A.V. SimBud, http://www.cbc.arizona.edu/facilities/eprfacility'software (accessed 22 August 2013).
57. Sun, L., Hernandez-Guzman, J., and Warncke, K. (2009) OPTESIM, a versatile toolbox for numerical simulation of electron spin echo envelope modulation (ESEEM) that features hybrid optimization and statistical assessment of parameters. *J. Magn. Reson.*, **200**, 21–28.
58. Hodgkinson, P. and Emsley, L. (2000) Numerical simulation of solid-state NMR experiments. *Prog. Nucl. Magn. Reson. Spectrosc.*, **36**, 201–239.
59. Edén, M. (2003) Computer simulations in solid-state NMR. I. Spin dynamics theory. *Concepts Magn. Reson. Part A*, **17**, 117–154.
60. Edén, M. (2003) Computer simulation in solid-state NMR. III. Powder averaging. *Concepts Magn. Reson. Part A*, **18**, 24–55.
61. Edén, M. (2003) Computer simulations in solid-state NMR. II. Implementations for static and rotating samples. *Concepts Magn. Reson. Part A*, **18**, 1–23.
62. Bak, M., Rasmussen, J.T., and Nielsen, N.C. (2000) SIMPSON: a general simulation program for solid-state NMR spectroscopy. *J. Magn. Reson.*, **147**, 296–330.
63. Veshtort, M. and Griffin, R.G. (2006) SPINEVOLUTION: a powerful tool for the simulation of solid and liquid state NMR experiments. *J. Magn. Reson.*, **178**, 248–282.
64. Hogben, H.J. *et al.* (2011) Spinach – A software library for simulation of spin dynamics in large spin systems. *J. Magn. Reson.*, **208**, 179–194.
65. Weil, J.A. and Bolton, J.R. (2007) *Electron Paramagnetic Resonance. Elementary Theory and Practical Applications*, John Wiley & Sons Inc., Hoboken, NJ.
66. Pake, G.E. and Estle, T.L. (1973) *The Physical Principles of Electron Paramagnetic Resonance*, Benjamin, London.
67. Abragam, A. and Bleaney, B. (1986) *Electron Paramagnetic Resonance of Transition Ions*, Dover, New York.
68. Neese, F. (2001) Prediction of electron paramagnetic resonance g values using coupled perturbed Hartree-Fock and Kohn-Sham theory. *J. Chem. Phys.*, **115**, 11080–11096.
69. McGavin, D.G., Tennant, W.C., and Weil, J.A. (1990) High-spin Zeeman terms in the spin Hamiltonian. *J. Magn. Reson.*, **87**, 92–109.
70. Claridge, R.F.C., Tennant, W.C., and McGavin, D.G. (1997) X-band EPR of $Fe^{3+}/CaWO_4$ at 10K: evidence for large magnitude high spin Zeeman interactions. *J. Phys. Chem. Solids*, **58**, 813–820.
71. Chen, N. *et al.* (2002) Electron paramagnetic resonance spectroscopic study of synthetic fluorapatite: part II. $Gd^{3+}$ at the Ca1 site, with a neighboring Ca2 vacancy. *Am. Mineral.*, **87**, 47–55.
72. McGavin, D.G. and Tennant, W.C. (2009) Higher-order Zeeman and spin terms in the electron paramagnetic resonance spin Hamiltonian; their description in irreducible form using Cartesian, tesseral spherical

tensor and Stevens' operator expressions. *J. Phys. Condens. Matter*, **2**, 245501/1–245501/14.
73. Weil, J.A. EPRNMR, http://www.chem.queensu.ca/eprnmr/ (accessed 22 August 2013).
74. Sottini, S. and Groenen, E.J.J. (2012) A comment of the pseudo-nuclear Zeeman effect. *J. Magn. Reson.*, **218**, 11–15.
75. Schweiger, A. and Jeschke, G. (2001) *Principles of Pulse Electron Paramagnetic Resonance*, Oxford University Press, Oxford.
76. Shaw, J.L. et al. (2006) Redox non-innocence of thioether macrocycles: elucidation of the electronic structures of mononuclear complexes of gold(II) and silver(II). *J. Am. Chem. Soc.*, **128**, 13827–13839.
77. Bencini, A. and Gatteschi, D. (1990) *Electron Paramagnetic Resonance of Exchange Coupled Systems*, Springer, Berlin.
78. Huang, N.L. and Orbach, R. (1964) Biquadratic superexchange. *Phys. Rev. Lett.*, **12**, 275–276.
79. Semenaka, V.V. et al. (2010) $Cr^{III}$-$Cr^{III}$ interactions in two alkoxo-bridged heterometallic $Zn_2Cr_2$ complexes self-assembled from zinc oxide, Reinecke's salt, and diethanolamine. *Inorg. Chem.*, **49**, 5460–5471.
80. Poole, C.P., Farrach, H.A., and Jackson, W.K. (1974) Standardization of convention for zero field splitting parameters. *J. Chem. Phys.*, **61**, 2220–2221.
81. Gaffney, B.J. and Silverstone, H.J. (1993) Simulation of the EMR spectra of high-spin iron in proteins. *Biol. Magn. Reson.*, **13**, 1–57.
82. Bleaney, B. and Trenam, R.S. (1954) Paramagnetic resonance spectra of some ferric alums, and the nuclear magnetic moment of $^{57}$Fe. *Proc. R. Soc. London, Ser. A*, **223**, 1–14.
83. Doetschman, D.C. and McCool, B.J. (1975) Electron paramagnetic resonance studies of transition metal oxalates and their photochemistry in single crystals. *Chem. Phys.*, **8**, 1–16.
84. Scullane, M.I., White, L.K., and Chasteen, N.D. (1982) An efficient approach to computer simulation of EPR spectra of high-spin Fe(III) in rhombic ligand fields. *J. Magn. Reson.*, **47**, 383–397.
85. Jain, V.K. and Lehmann, G. (1990) Electron paramagnetic resonance of $Mn^{2+}$ in orthorhombic and higher symmetry crystals. *Phys. Status Solidi B*, **159**, 495–544.
86. Buckmaster, H.A. et al. (1971) Computer analysis of EPR data. *J. Magn. Reson.*, **4**, 113–122.
87. Stevens, K.W.H. (1952) Matrix elements and operator equivalents connected with magnetic properties of rare earth ions. *Proc. Phys. Soc.*, **65**, 209–215.
88. Roitsin, A.B. (1981) Generalized spin-Hamiltonian and low-symmetry effects in paramagnetic resonance. *Phys. Status Solidi B*, **104**, 11–35.
89. Rudowicz, C. (1987) Concept of spin Hamiltonian, forms of zero field splitting and electronic Zeeman Hamiltonians and relations between parameters used in EPR. A critical review. *Magn. Reson. Rev.*, **13**, 1–89.
90. Rudowicz, C. and Misra, S.K. (2001) Spin-Hamiltonian formalisms in electron magnetic resonance (EMR) and related spectroscopies. *Appl. Spectrosc. Rev.*, **36**, 11–63.
91. Misra, S.K., Poole, C.P., and Farrach, H.A. (1996) A review of spin Hamiltonian forms for various point-group site symmetries. *Appl. Magn. Reson.*, **11**, 29–46.
92. Rudowicz, C. and Madhu, S.B. (1999) Orthorhombic standardization of spin-Hamiltonian parameters for transition-metal centres in various crystals. *J. Phys. Condens. Matter*, **11**, 273–287.
93. Rudowicz, C. (2000) On the relations between the zero-field splitting parameters in the extended Stevens operator notation and the conventional ones used in EMR for orthorhombic and lower symmetry. *J. Phys. Condens. Matter*, **12**, L417–L423.
94. Rudowicz, C. and Chung, C.Y. (2004) The generalization of the extended Stevens operators to higher ranks and spins, and a systematic review of the

95. Ryabov, I.D. (1999) On the generation of operator equivalents and the calculation of their matrix elements. *J. Magn. Reson.*, **140**, 141–145.
96. Waldmann, O. and Güdel, H.U. (2005) Many-spin effects in inelastic neutron scattering and electron paramagnetic resonance of molecular nanomagnets. *Phys. Rev. B*, **72**, 094422.
97. Tennant, W.C. et al. (2000) Rotation matrix elements and further decomposition functions of two-vector tesseral spherical tensor operators; their uses in electron paramagnetic resonance spectroscopy. *J. Phys. Condens. Matter*, **12**, 9481–9495.
98. Yang, Z.Y. and Wei, Q. (2005) On the relations between the crystal field parameter notation in the "wybourne" notation and the conventional ones for $3d^N$ ions in axial symmetric crystal field. *Physica B*, **370**, 137–145.
99. Golding, R.M. (2007) Interpreting electron-nuclear-magnetic field interactions from a Hamiltonian expressed in tensorial notation. *J. Magn. Reson.*, **187**, 52–56.
100. Mohr, P.J., Taylor, B.N., and Newell, D.B. (2012) CODATA recommended values of the fundamental physical constants: 2010. *Rev. Mod. Phys.*, **84**, 1527–1605.
101. Stone, N.J. (2011) *Table of Nuclear Magnetic Dipole and Electric Quadrupole Moments*, Oxford Physics, Clarendon Laboratory, Oxford.
102. Sakurai, J.J. (1994) *Modern Quantum Mechanics*, Addison-Wesley, New York.
103. Blum, K. (1996) *Density Matrix Theory and Applications*, Plenum Press, New York.
104. Fano, U. (1957) Description of states in quantum mechanics by density matrix and operator techniques. *Rev. Mod. Phys.*, **29**, 74–93.
105. Ernst, R.R., Bodenhausen, G., and Wokaun, A. (1991) *Principles of Nuclear Magnetic Resonance in One and Two Dimensions*, Clarendon Press, Oxford.
106. Banwell, C.N. and Primas, H. (1963) On the analysis of high-resolution nuclear magnetic resonance spectra. I. Methods of calculating N.M.R. spectra. *Mol. Phys.*, **6** (3), 225–256.
107. Sørensen, O.W. et al. (1983) Product operator formalism for the description of NMR pulse experiments. *Prog. Nucl. Magn. Reson. Spectrosc.*, **16**, 163–192.
108. Mehring, M. (1983) *Principles of High-Resolution NMR in Solids*, Springer, Berlin.
109. van Beek, J.D. et al. (2005) Spherical tensor analysis of nuclear magnetic resonance signals. *J. Chem. Phys.*, **122**, 244510/1–244510/12.
110. Mueller, L.J. (2011) Tensors and rotations in NMR. *Concepts Magn. Reson. Part A*, **38**, 221–235.
111. Allard, P. and Härd, T. (2001) A complete Hermitian operator basis set for any spin quantum number. *J. Magn. Reson.*, **153**, 15–21.
112. Freed, J.H. and Fraenkel, G.K. (1963) Theory of linewidths in electron spin resonance spectra. *J. Chem. Phys.*, **39**, 326–348.
113. Musher, J.I. (1967) Equivalence of nuclear spins. *J. Chem. Phys.*, **46**, 1537–1538.
114. Saupe, A. and Nehring, J. (1967) Magnetic equivalence of nuclear spins in oriented molecules. *J. Chem. Phys.*, **47**, 5459–5460.
115. Musher, J.I. (1967) Magnetic equivalence of nuclear spins. *J. Chem. Phys.*, **47**, 5460–5461.
116. Nokhrin, S.M., Weil, J.A., and Howarth, D.F. (2005) Magnetic resonance in systems with equivalent spin-1/2 nuclides. Part 1. *J. Magn. Reson.*, **174**, 209–218.
117. Nokhrin, S.M., Howarth, D.F., and Weil, J.A. (2008) Magnetic resonance in systems with equivalent spin-1/2 nuclides. Part 2: energy values and spin states. *J. Magn. Reson.*, **193**, 1–9.
118. Szymański, S. (1997) Magnetic equivalence between nuclei of spin greater than 1/2 in presence of relaxation. *J. Magn. Reson.*, **127**, 199–205.
119. Bloch, F. (1946) Nuclear induction. *Phys. Rev.*, **70**, 460–474.
120. Jaynes, E.T. (1955) Matrix treatment of nuclear induction. *Phys. Rev.*, **98**, 1099–1105.

121. Kubo, R. and Tomita, K. (1954) A general theory of magnetic resonance absorption. *J. Phys. Soc. Jpn.*, **9**, 888–919.
122. Gamliel, D. and Levanon, H. (1995) *Stochastic Processes in Magnetic Resonance*, World Scientific, Singapore.
123. Tanimura, Y. (2006) Stochastic Liouville, Langevin, Fokker-Planck, and master equation approaches to quantum dissipative systems. *J. Phys. Soc. Jpn.*, **75**, 082001/1–082001/39.
124. Bain, A.D. and Duns, G.J. (1996) A unified approach to dynamic NMR based on a physical interpretation of the transition probability. *Can. J. Chem.*, **74**, 819–824.
125. Binsch, G. (1969) A unified theory of exchange effects on nuclear magnetic resonance line shapes. *J. Am. Chem. Soc.*, **91**, 1304–1309.
126. Freed, J.H., Bruno, G.V., and Polnaszek, C.F. (1971) Electron spin resonance lines shapes and saturation in the slow motional regime. *J. Phys. Chem.*, **75**, 3385–3399.
127. Boere, R.T. et al. (2011) Oxidation of closo-$[B_{12}Cl_{12}]^{2-}$ to the radical anion $[B_{12}Cl_{12}]^{-}$ and to neutral $B_{12}Cl_{12}$. *Angew. Chem. Int. Ed.*, **50**, 549–552.
128. Weber, S. et al. (2005) Probing the N(5)-H bond of the isoalloxazine moiety of flavin radicals by X- and W-band pulsed electron-nuclear double resonance. *ChemPhysChem*, **6**, 292–299.
129. Hyde, J.S. et al. (1990) Pseudo field modulation in EPR spectroscopy. *Appl. Magn. Reson.*, **1**, 483–496.
130. Hyde, J.S. et al. (1992) Pseudomodulation: a computer-based stragegy for resolution enhancement. *J. Magn. Reson.*, **96**, 1–13.
131. Kälin, M., Gromov, I., and Schweiger, A. (2003) The continuous wave electron paramagnetic resonance experiment revisited. *J. Magn. Reson.*, **160**, 166–182.
132. Anderson, W.A. (1960) *Magnetic Field Modulation for High Resolution NMR*, Pergamon, New York, pp. 180–184.
133. Berger, P.A. and Günthart, H.H. (1962) The distortion of electron spin resonance signal shapes by finite modulation amplitudes. *Z. Angew. Math. Phys.*, **13**, 310–323.
134. Wilson, G.V.H. (1963) Modulation broadening of NMR and ESR line shapes. *J. Appl. Phys.*, **34**, 3276–3285.
135. Haworth, O. and Richards, R.E. (1966) The use of modulation in magnetic resonance. *Prog. Nucl. Magn. Reson. Spectrosc.*, **1**, 1–14.
136. Dulčić, A. and Ravkin, B. (1983) Frequency versus field modulation in magnetic resonance. *J. Magn. Reson.*, **52**, 323–325.
137. Mailer, C. et al. (2003) Spectral fitting: the extraction of crucial information from a spectrum and a spectral image. *Magn. Reson. Chem.*, **49**, 1175–1180.
138. Nielsen, R.D. et al. (2004) Formulation of Zeeman modulation as a signal filter. *J. Magn. Reson.*, **170**, 345–371.
139. Nielsen, R.D. and Robinson, B.H. (2004) The effect of field modulation on a simple resonance line shape. *Concepts Magn. Reson. Part A*, **23**, 38–48.
140. Siemens, M., Hancock, J., and Siminovitch, D. (2007) Beyond Euler angles: exploiting the angle-axis parametparameter in a multipole expansion of the rotation operator. *Solid State Nucl. Magn. Reson.*, **31**, 35–54.
141. Rose, M.E. (1957) *Elementary Theory of Angular Momentum*, John Wiley & Sons Inc., New York.
142. Edmonds, A.R. (1957) *Angular Momentum in Quantum Mechanics*, Princeton University Press, Princeton, NJ.
143. Atherton, N.M. (1993) *Princples of Electron Spin Resonance*, Ellis Horwood.
144. Schmidt-Rohr, K. and Spiess, H.W. (1994) *Multidimensional Solid-State NMR and Polymers*, Academic Press, New York.
145. Millot, Y. and Man, P.P. (2012) Active and passive rotations with Euler angles in NMR. *Concepts Magn. Reson. Part A*, **40**, 215–252.
146. Brink, D.M. and Satcher, G.R. (1993) *Angular Momentum*, Oxford University Press.
147. Sezer, D., Freed, J.H., and Roux, B. (2008) Simulating electron spin resonance spectra of nitroxide spin labels

148. DeSensi, S.C. et al. (2008) Simulation of nitroxide electron paramagnetic resonance spectra from Brownian trajectories and molecular dynamics simulations. *Biophys. J.*, **94**, 3798–3809.
149. Teki, Y. (2008) General simulation of transient ESR and continuous-wave ESR spectra for high-spin states using a density matrix formalism. *ChemPhysChem*, **9**, 393–396.
150. Weil, J.A. and Anderson, J.H. (1958) Determination of the g tensor in magnetic resonance. *J. Chem. Phys.*, **28**, 864–866.
151. Schonland, D.S. (1959) On the determination of the principal g-values in electron spin resonance. *Proc. Phys. Soc.*, **73**, 788–792.
152. Waller, W.G. and Rogers, M.T. (1973) A generalization of methods for determining the g tensor. *J. Magn. Reson.*, **9**, 92–107.
153. Morton, J.R. and Preston, K.F. (1983) EPR spectroscopy of single crystals using a two-circle goniometer. *J. Magn. Reson.*, **52**, 457–474.
154. Weil, J.A., Buch, T., and Clapp, J.E. (1973) Crystal point group symmetry and microscopic tensor properties in magnetic resonance spectroscopy. *Adv. Magn. Reson.*, **6**, 183–257.
155. Freed, J.H. (1976) Theory of slow tumbling ESR spectra for nitroxides, in *Spin Labeling: Theory and Applications* (ed. L.J. Berliner), Academic Press, New York.
156. Wasserman, E., Snyder, L.C., and Yager, W.A. (1964) ESR of the triplet state of randomly oriented molecules. *J. Chem. Phys.*, **41** (6), 1763–1772.
157. van Veen, G. (1978) Simulation and analysis of EPR spectra of paramagnetic ions in powders. *J. Magn. Reson.*, **30**, 91–109.
158. Kneubühl, F.K. (1960) Line shapes of electron paramagnetic resonance signals produced by powders, glasses, and viscous liquids. *J. Chem. Phys.*, **33** (4), 1074–1078.
159. Siderer, Y. and Luz, Z. (1980) Analytical expressions for magnetic resonance lineshapes of powder samples. *J. Magn. Reson.*, **37** (3), 449–463.
160. Beltrán-López, V. and Castro-Tello, J. (1982) Powder pattern of systems with axially anisotropic g and A tensors. The EPR spectrum of copper phthalocyanine. *J. Magn. Reson.*, **47**, 19–27.
161. Varner, S.J., Vold, R.L., and Hoatson, G.L. (1996) An efficient method for calculating powder patterns. *J. Magn. Reson., Ser. A*, **123**, 72–80.
162. Beltrán-López, V., Mile, B., and Rowlands, C.C. (1996) Exact analytical solution for the powder pattern of orthorhombic-g systems. *J. Chem. Soc., Faraday Trans.*, **92** (12), 2203–2210.
163. Bikchantaev, I.G. (1988) Rapid analysis of the anisotropic ESR spectra of polycrystalline and amorphous systems formed by species with orthorhombic symmetry of the g value. *J. Struct. Chem.*, **29** (3), 388–391.
164. Beltrán-López, V. (1999) Closed form solutions in EPR computer simulations. *Mol. Phys. Rep.*, **26**, 25–38.
165. Mi, Q., Ratner, M.A., and Wasielewski, M.R. (2010) Accurate and general solutions to three-dimensional anisotropies: applications to EPR spectra of triplets involving dipole-dipole, spin-orbit interactions and liquid crystals. *J. Phys. Chem. C*, **114**, 13853–13860.
166. Saff, E.B. and Kuijlaars, A.B.J. (1997) Distributing many points on a sphere. *Math. Intell.*, **19**, 5–11.
167. Bak, M. and Nielsen, N.C. (1997) REPULSION, a novel approach to efficient powder averaging in solid-state NMR. *J. Magn. Reson.*, **125**, 132–139.
168. Ponti, A. (1999) Simulation of magnetic resonance static powder lineshapes: a quantitative assessment of spherical codes. *J. Magn. Reson.*, **138**, 288–297.
169. Galindo, S. and Gonzáles-Tovany, L. (1981) Monte Carlo simulation of EPR spectra of polycrystalline samples. *J. Magn. Reson.*, **44**, 250–254.
170. Pilbrow, J.R. (1990) *Transition Ion Electron Paramagnetic Resonance*, Clarendon Press, Oxford.
171. Crittenden, R.G. and Turok, N.G. (2008) Exactly Azimuthal Pixelations of

the Sky. arXiv: astro-ph/9806374v1, pp. 1–17.
172. Kurihara, Y. (1965) Numerical integration of the primitive equations on a spherical grid. *Mon. Weather Rev.*, **93**, 399–415.
173. Wong, S.T.S. and Roos, M.S. (1994) A strategy for sampling on the sphere applied to 3D selective RF pulse design. *Magn. Reson. Med.*, **32**, 778–784.
174. Hustedt, E.J. et al. (2006) Dipolar coupling between nitroxide spin labels: the development and application of a tether-in-a-cone model. *Biophys. J.*, **90**, 340–356.
175. Hannay, J.H. and Nye, J.F. (2004) Fibonacci numerical integration on a sphere. *J. Phys. A: Math. Gen.*, **37**, 11591–11601.
176. Swinbank, R. and Purser, R.J. (2006) Fibonacci grids: a novel approach to global modelling. *Q. J. Roy. Meteorol. Soc.*, **132**, 1769–1793.
177. Gonzáles, Á. (2010) Measurment of areas on a sphere using fibonacci and latitude-longitude grids. *Math. Geosci.*, **42**, 49–64.
178. Baumgardner, J.R. and Frederickson, P.O. (1985) Icosahedral discretization of the two-sphere. *SIAM J. Numer. Anal.*, **22**, 1107–1115.
179. Randall, D.A. et al. (2002) Climate modeling with spherical geodesic grids. *Comput. Sci. Eng.*, **4**, 32–41.
180. Alderman, D.W., Solum, M.S., and Grant, D.M. (1986) Methods for analyzing spectroscopic line shapes. NMR solid powder patterns. *J. Chem. Phys.*, **84**, 3717–3725.
181. Koons, J.M. et al. (1995) Extracting multitensor solid-state NMR parameters from lineshapes. *J. Magn. Reson., Ser. A*, **114**, 12–23.
182. Purser, R.J. and Rančić, M. (1998) Smooth quasi-homogeneous gridding of the sphere. *Q. J. R. Meteorol. Soc.*, **124**, 637–647.
183. Edén, M. and Levitt, M.H. (1998) Computation of orientational averages in solid-state NMR by Gaussian spherical quadrature. *J. Magn. Reson.*, **132**, 220–239.
184. Treutler, O. and Ahlrichs, R. (1994) Efficient molecular numerical integration schemes. *J. Chem. Phys.*, **102**, 346–354.
185. Stevensson, B. and Edén, M. (2006) Efficient orientational averaging by the extension of Lebedev grids via regularized octahedral symmetry expansion. *J. Magn. Reson.*, **181**, 162–176.
186. Nettar, D. and Villafranca, J.J. (1985) A program for EPR powder spectrum simulation. *J. Magn. Reson.*, **64**, 61–65.
187. Gorski, K.M. et al. (2005) HEALPix: a framework for high-resolution discretization and fast analysis of data distributed on the sphere. *Astrophys. J.*, **622**, 759–771.
188. Hüttig, C. and Stemmer, K. (2008) The spiral grid: a new approach to discretize the sphere and its application to mantle convection. *Geochem. Geophys. Geosyst.*, **9**, Q02018.
189. Craciun, C. (2010) Application of the SCVT orientation grid to the simulation of CW EPR powder spectra. *Appl. Magn. Reson.*, **38**, 279–293.
190. Roşca, D. (2010) New uniform grids on the sphere. *Astron. Astrophys.*, **520**, A63.
191. Stevensson, B. and Edén, M. (2011) Interpolation by fast Wigner transform for rapid calculations of magnetic resonance spectra from powders. *J. Chem. Phys.*, **134**, 124104.
192. Koay, C.G. (2011) Analytically exact spiral scheme for generating uniformly distributed points on the unit sphere. *J. Comput. Sci.*, **2**, 88–91.
193. Gribnau, M.C.M., van Tits, J.L.C., and Reijerse, E.J. (1990) An efficient general algorithm for the simulation of magnetic resonance spectra of orientationally disordered solids. *J. Magn. Reson.*, **90**, 474–485.
194. Ebert, H., Abart, J., and Voitländer, J. (1983) Simulation of quadrupole disturbed NMR field spectra by using perturbation theory and the triangle integration method. *J. Chem. Phys.*, **79**, 1719–1723.
195. Ovchinnikov, I.V. and Konstantinov, V.N. (1978) Extra absorption peaks in EPR spectra of systems with anisotropic g-tensors and hyperfine

196. Belford, G.G., Belford, R.L., and Burkhalter, J.F. (1973) Eigenfields: a practical direct calculation of resonance fields and intensities for field-swept fixed–frequency spectrometers. *J. Magn. Reson.*, **11**, 251–265.
197. Belford, R.L. and Belford, G.G. (1973) Eigenfield expansion technique for efficient computation of field-swept fixed-frequency spectra from relaxation master equations. *J. Chem. Phys.*, **59**, 853–854.
198. Belford, R.L. et al. (1974) Computation of field-swept EPR spectra for systems with large interelectronic interactions. *Adv. Chem. Ser.*, **5**, 40–50.
199. McGregor, K.T., Scaringe, R.P., and Hatfield, W.E. (1975) E.P.R. calculations by the eigenfield method. *Mol. Phys.*, **30**, 1925–1933.
200. Čugunov, L., Mednis, A., and Kliava, J. (1994) Computer simulation of the EPR spectra of ions with $S > 1/2$ by the Eigenfield and related methods. *J. Magn. Reson., Ser. A*, **106**, 153–158.
201. Magon, C.J. et al. (2007) The harmonic inversion of the field-swept fixed-frequency resonance spectrum. *J. Magn. Reson.*, **184**, 176–183.
202. Matsuoka, H. et al. (2003) Importance of fourth-order zero-field splitting terms in random-orientation EPR spectra of Eu(II)-doped strontium aluminate. *J. Phys. Chem. A*, **107**, 11539–11546.
203. Aasa, R. and Vänngård, T. (1975) EPR signal intensity and powder shapes: a reexamination. *J. Magn. Reson.*, **19**, 308–315.
204. Tynan, E.C. and Yen, T.F. (1970) General purpose computer program for exact ESR spectrum calculations with applications to vanadium chelates. *J. Magn. Reson.*, **3**, 327–335.
205. Mabbs, F.E. and Collison, D. (1999) The use of matrix diagonalisation in the simulation of the EPR powder spectra of d-transition metal compounds. *Mol. Phys. Rep.*, **26**, 39–59.
206. Misra, S.K. (1999) Angular variation of electron paramagnetic resonance spectrum: simulation of a polycrystalline EPR spectrum. *J. Magn. Reson.*, **137**, 83–92.
207. Mackey, J.H. et al. (1969) in *Electron Spin Resonance of Metal Complexes* (ed. T.F. Yen), Plenum Press, pp. 33–57.
208. Kreiter, A. and Hüttermann, J. (1991) Simultaneous EPR and ENDOR powder-spectra synthesis by direct matrix diagonalization. *J. Magn. Reson.*, **93**, 12–26.
209. Morin, G. and Bonnin, D. (1999) Modeling EPR powder spectra using numerical diagonalization of the spin Hamiltonian. *J. Magn. Reson.*, **136**, 176–199.
210. Uhlin, J. (1973) A computational method of determination of constants of spin Hamiltonian of fine structure ESR spectra. *Czech. J. Phys. B*, **23**, 551–557.
211. Misra, S.K. (1976) Evaluation of spin Hamiltonian parameters from EPR data by the method of least-squares fitting. *J. Magn. Reson.*, **23** (3), 403–410.
212. Misra, S.K. and Vasilopoulos, P. (1980) Angular variation of electron paramagnetic resonance spectra. *J. Phys. C: Solid State Phys.*, **13**, 1083–1092.
213. Misra, S.K. (1983) Evaluation of spin Hamiltonian parameters of electron-nuclear spin-coupled systems from EPR data by the method of least-squares fitting. *Physica B*, **121**, 193–201.
214. Misra, S.K. (1999) A rigorous evaluation of spin-Hamiltonian parameters and linewidths from a polycrystalline EPR spectrum. *J. Magn. Reson.*, **140**, 179–188.
215. Lund, A. (2004) Applications of automatic fittings to powder EPR spectra of free radicals, $S > 1/2$, and coupled systems. *Appl. Magn. Reson.*, **26**, 365–385.
216. Stoll, S. and Schweiger, A. (2003) An adaptive method for computing resonance fields for continuous-wave EPR spectra. *Chem. Phys. Lett.*, **380**, 464–470.
217. Gaffney, B.J. and Silverstone, H.J. (1998) Simulation methods for looping transitions. *J. Magn. Reson.*, **134**, 57–66.

218. Parlett, B.N. (1987) *The Symmetric Eigenvalue Problem*, SIAM, Philadelphia, PA.
219. Golub, G.H. and Loan, C.F.V. (2012) *Matrix Computations*, Johns Hopkins University Press.
220. Saad, Y. (2011) *Numerical Methods for Large Eigenvalue Problems*, SIAM.
221. Kato, T. (1982) *A Short Introduction to Perturbation Theopy for Linear Operators*, Springer, NewYork.
222. van Dam, P.J. et al. (1998) Application of high frequency EPR to integer spin systems: unusual behavior of the double-quantum line. *J. Magn. Reson.*, **130**, 140–144.
223. Taylor, P.C. and Bray, P.J. (1970) Computer simulations of magnetic resonance spectra observed in polycrystalline and glassy samples. *J. Magn. Reson.*, **2**, 305–331.
224. Lefebvre, R. and Maruani, J. (1965) Use of computer programs in the interpretation of electron paramagnetic resonance spectra of dilute radicals in amorphous solid samples. I. High-field treatment. X-band spectra of $-electron unconjugated hydrocarbon radicals. *J. Chem. Phys.*, **42**, 1480–1496.
225. Iwasaki, M. (1974) Second-order perturbation treatment of the general spin Hamiltonian in an arbitrary coordinate system. *J. Magn. Reson.*, **16**, 417–423.
226. Weil, J.A. (1975) Comments on second-order spin-Hamiltonian energies. *J. Magn. Reson.*, **18**, 113–116.
227. Lund, A. and Erickson, R. (1998) EPR and ENDOR simulations for disordered systems: the balance between efficiency and accuracy. *Acta Chem. Scand.*, **52**, 261–274.
228. Schoemaker, D. (1968) Spin Hamiltonian of two equivalent nuclei: application to the $I_2^-$ centers. *Phys. Rev.*, **174**, 1060–1068.
229. Byfleet, C.R. et al. (1970) Calculation of EPR transition fields and transition probabilities for a general spin Hamiltonian. *J. Magn. Reson.*, **2**, 69–78.
230. Bleaney, B. (1951) Hyperfine structure in paramagnetic salts and nuclear alignment. *Philos. Mag.*, **42**, 441–458.
231. Upreti, G.C. (1974) Study of the intensities and positions of allowed and forbidden hyperfine transitions in the EPR of $Mn^{2+}$ doped in single crystals of $Cd(CH_3COO)_2 \cdot H_2O$. *J. Magn. Reson.*, **13**, 336–347.
232. Schweiger, A. et al. (1976) Theory and applications of generalized operator transforms for diagonalization of spin Hamiltonians. *Chem. Phys.*, **17**, 155–185.
233. Piper, W.W. and Prener, J.S. (1972) Electron-paramagnetic-resonance study of $Mn^{2+}$ in calcium chlorophosphate. *Phys. Rev. B*, **6**, 2547–2554.
234. Markham, G.D., Rao, B.D.N., and Reed, G.H. (1979) Analysis of EPR powder pattern lineshapes for Mn(II) including third-order perturbation corrections. Applications to Mn(II) complexes in enzymes. *J. Magn. Reson.*, **33**, 595–602.
235. Hagston, W.E. and Holmes, B.J. (1980) Matrix methods for spin Hamiltonians of low symmetry. *J. Phys. B: At. Mol. Phys.*, **13**, 3505–3519.
236. Misra, S.K. (1994) Estimation of the $Mn^{2+}$ zero-field splitting parameter from a polycrystaline EPR spectrum. *Physica B*, **302**, 193–200.
237. Garribba, E. and Micera, G. (2006) Determination of the hyperfine coupling constant and zero-field splitting in the ESR spectrum of $Mn^{2+}$ in calcite. *Magn. Reson. Chem.*, **44**, 11–19.
238. Smith, T.D. and Pilbrow, J.R. (1974) The determination of structural properties of dimeric transition metal ion complexes from EPR spectra. *Coord. Chem. Rev.*, **13**, 173–278.
239. Boas, J.F. et al. (1978) Interpretation of electron spin resonance spectra due to some $B_{12}$-dependent enzyme reactions. *J. Chem. Soc., Faraday Trans. 2*, **74**, 417–431.
240. Eaton, S.S. et al. (1983) Metal-nitroxyl interactions. 29. EPR studies of spin-labeled copper complexes in frozen solution. *J. Magn. Reson.*, **52**, 435–449.
241. Kirste, B., Krüger, A., and Kurreck, H. (1982) ESR and ENDOR investigations of spin exchange in mixed galvinoxyl/nitroxide biradicals. Syntheses. *J. Am. Chem. Soc.*, **104**, 3850–3858.
242. Hustedt, E.J. et al. (1997) Molecular distances from dipolar coupled

242. spin-labels: the global analysis of multifrequency continuous wave electron paramagnetic resonance data. *Biophys. J.*, **74**, 1861–1877.
243. Golombek, A.P. and Hendrich, M.P. (2003) Quantitative analysis of dinuclear manganese(II) EPR spectra. *J. Magn. Reson.*, **165**, 33–48.
244. Harris, D.C. and Bertolucci, M.D. (1989) *Symmetry and Spectroscopy: An Introduction to Vibrational and Electronic Spectroscopy*, Dover, New York.
245. Pilbrow, J.R. et al. (1983) Asymmetric lines in field-swept EPR: $Cr^{3+}$ looping transitions in ruby. *J. Magn. Reson.*, **52**, 386–399.
246. Pilbrow, J.R. (1984) Lineshapes in frequency-swept and field-swept EPR for spin 1/2. *J. Magn. Reson.*, **58**, 186–203.
247. Eaton, S.S. et al. (1983) Use of the EPR half-field transition to determine the interspin distance and the orientation of the interspin vector in systems with two unpaired electrons. *J. Am. Chem. Soc.*, **105**, 6560–6567.
248. Bleaney, B. (1960) Electron spin resonance intensity in anisotropic substances. *Proc. Phys. Soc.*, **75**, 621–623.
249. Bleaney, B. (1961) Explanation of some forbidden transitions in paramagnetic resonance. *Proc. Phys. Soc.*, **77**, 103–112.
250. Holuj, F. (1966) The spin Hamiltonian and intensities of the ESR spectra originating from large zero-field effects on $^6S$ states. *Can. J. Phys.*, **44**, 503–508.
251. Isomoto, A., Watari, H., and Kotani, M. (1970) Dependence of EPR transition probability on magnetic field. *J. Phys. Soc. Jpn.*, **29**, 1571–1577.
252. Pilbrow, J.R. (1969) Anisotropic transition probability factor in E.S.R. *Mol. Phys.*, **16**, 307–309.
253. Weil, J.A. (1987) in *Electronic Magnetic Resonance of the Solid State* (ed. J.A. Weil), Canadian Society of Chemistry, pp. 1–19.
254. Lund, A. et al. (2008) Automatic fitting procedures for EPR spectra of disordered systems: matrix diagonalization and perturbation methods applied to fluorocarbon radicals. *Spectrochim. Acta, Part A*, **69**, 1294–1300.
255. Kneubühl, F.K. and Natterer, B. (1961) Paramagnetic resonance intensity of anisotropic substances and its influence on line shapes. *Helv. Phys. Acta*, **34**, 710–717.
256. Dalton, L.R. et al. (1976) in *Advances in Magnetic Resonance VIII* (ed. J.S. Waugh), Academic Press, pp. 149–259.
257. Robinson, B.H. and Dalton, L.R. (1979) EPR and saturation transfer EPR spectra at high microwave field intensities. *Chem. Phys.*, **36**, 207–237.
258. Robinson, B.H. et al. (1986) *EPR and Advanced EPR Studies of Biological Systems*, CRC Press, Boca Raton, FL.
259. Howard, E.C. et al. (1993) Simulation of saturation transfer electron paramagnetic resonance spectra for rotational motion with restricted angular amplitude. *Biophys. J.*, **64**, 581–593.
260. Breit, G. and Rabi, I.I. (1931) Measurement of nuclear spin. *Phys. Rev.*, **38**, 2082–2083.
261. Weil, J.A. (1971) The analysis of large hyperfine splitting in paramagnetic resonance spectroscopy. *J. Magn. Reson.*, **4**, 394–399.
262. Dickson, R.S. and Weil, J.A. (1991) Breit-Rabi Zeeman states of atomic hydrogen. *Am. J. Phys.*, **59** (2), 125–129.
263. Fessenden, R.W. and Schuler, R.H. (1965) ESR spectra and structure of the fluorinated methyl radicals. *J. Chem. Phys.*, **43** (8), 2704–2712.
264. Rabenstein, M.D. and Shin, Y.K. (1995) Determination of the distance between two spin labels attached to a macromolecule. *Proc. Natl. Acad. Sci. U.S.A.*, **92**, 8239–8243.
265. Robinson, B.H., Mailer, C., and Reese, A.W. (1999) Linewidth analysis of spin labels in liquids. *J. Magn. Reson.*, **138**, 199–209.
266. Feynman, R.P. (1939) Forces in molecules. *Phys. Rev.*, **56**, 340–343.
267. Hagen, W.R. et al. (1985) Quantitative numerical analysis of g strain in the EPR of distributed systems and its importance for multicenter metalloproteins. *J. Magn. Reson.*, **61**, 233–244.

268. Hagen, W.R. et al. (1985) A statistical theory for powder EPR in distributed systems. *J. Magn. Reson.*, **61**, 220–232.
269. More, C., Bertrand, P., and Gayda, J.P. (1987) Simulation of the EPR spectra of metalloproteins based on a physical description of the "g-strain" effect. *J. Magn. Reson.*, **73**, 13–22.
270. More, C., Gayda, J.P., and Bertrand, P. (1990) Simulations of the g-strain broadening of low-spin hemoprotein EPR spectra based on the $t_{2g}$ model. *J. Magn. Reson.*, **90**, 486–499.
271. Kivelson, D. and Neiman, R. (1961) ESR studies on the bonding in copper complexes. *J. Chem. Phys.*, **35**, 149–155.
272. Froncisz, W. and Hyde, J.S. (1980) Broadening by strains of lines in the g-parallel region of $Cu^{2+}$ EPR spectra. *J. Chem. Phys.*, **73**, 3123–3131.
273. Giugliarelli, G. and Cannistraro, S. (1985) Simulation of EPR spectra of $Cu^{2+}$ complexes with statistical distribution of the g-factor and hyperfine splitting. *Chem. Phys.*, **98**, 115–122.
274. Aqualino, A. et al. (1991) Correlated distributions in g and A tensors at a biologically active low-symmetric cupric site. *Phys. Rev. A*, **44**, 5257–5271.
275. Feher, E.R. (1964) Effect of uniaxial stress on the paramagnetic spectra of $Mn^{3+}$ and $Fe^{3+}$ in MgO. *Phys. Rev. A*, **136**, 145–157.
276. Clare, J.F. and Devine, S.D. (1984) The determination of instrinsic strain in a crystal from EPR linewidths and the spin-strain coupling tensor. *J. Phys. C: Solid State Phys.*, **17**, 2801–2812.
277. Wenzel, R.F. and Kim, Y.W. (1965) Linewidth of the electron paramagnetic resonance of $(Al_2O_3)_{1-x}(Cr_2O_3)_x$. *Phys. Rev.*, **5A**, 1592–1598.
278. Coffman, R.E. (1975) Inhomogeneously broadened line shapes and information content of calculated paramagnetic resonance spectra of biological molecules containing high-spin iron(III). *J. Phys. Chem.*, **79** (11), 1129–1136.
279. Scholz, G. et al. (2001) Modeling of multifrequency EPR spectra of $Fe^{3+}$ ions in crystalline and amorphous materials: a simplified approach to determine statistical distributions of spin-spin coupling parameters. *Appl. Magn. Reson.*, **21**, 105–123.
280. Yahiaoui, E.M. et al. (1994) Electron paramagnetic resonance of $Fe^{3+}$ ions in borate glass: computer simulations. *J. Phys. Condens. Matter*, **6**, 9415–9428.
281. Brill, A.S. et al. (1985) Density of low-energy vibrational states in a protein solution. *Phys. Rev. Lett.*, **54**, 1864–1867.
282. Yang, A.S. and Gaffney, B.J. (1987) Determination of relative spin concentration in some high-spin ferric proteins using E/D-distribution in electron paramagnetic resonance simulations. *Biophys. J.*, **51**, 55–67.
283. Peterson, G.E., Kurkjian, C.R., and Carnevale, A. (1974) Random structure models and spin resonance in glass. *Phys. Chem. Glasses*, **15**, 52–58.
284. Kliava, J. (1986) EPR of impurity ions in disordered solids. *Phys. Status Solidi B*, **134**, 411–455.
285. Hagen, W.R. (2007) Wide zero field interaction distributions in the high-field EPR of metalloproteins. *Mol. Phys.*, **105**, 2031–2039.
286. Eisenberger, P. and Pershan, P.S. (1967) Magnetic resonance studies of MetMyoglobin and myoglobin azide. *J. Chem. Phys.*, **47**, 3327–3333.
287. Mailer, C. and Taylor, C.P.S. (1972) Electron paramagnetic resonance study of single crystals of horse heart ferricytochrome c at 4.2K. *Can. J. Biochem.*, **50**, 1048–1055.
288. Klitgaard, S.K., Galsbøl, F., and Weihe, H. (2006) Angular variation of linewidths in single-crystal EPR spectra. *Spectrochim. Acta, Part A*, **63**, 836–839.
289. Poole, C.P. and Farach, H.A. (1979) Line shapes in electron spin resonance. *Bull. Magn. Reson.*, **1**, 162–194.
290. Smirnov, A.I. and Belford, R.L. (1995) Rapid quantitation from inhomogeneously broadened EPR spectra by a fast convolution algorithm. *J. Magn. Reson., Ser. A*, **113**, 65–73.
291. Grivet, J.P. (1997) Accurate numerical approximation to the Gauss-Lorentz lineshape. *J. Magn. Reson.*, **125**, 102–106.

292. Higinbotham, J. and Marshall, I. (2001) NMR lineshapes and lineshape fitting procedures. *Annu. Rep. NMR Spectrosc.*, **43**, 59–120.

293. Abrarov, S.M., Quine, B.M., and Jagpal, R.K. (2010) Rapidly convergent series for high-accuracy calculation of the Voigt function. *J. Quant. Spectrosc. Radiat. Transfer*, **111**, 372–375.

294. Bruce, S.D. et al. (2000) An analytical derivation of a popular approximation of the Voigt function for quantification of NMR spectra. *J. Magn. Reson.*, **142**, 57–63.

295. Tränkle, E. and Lendzian, F. (1989) Computer analysis of spectra with strongly overlapping lines. Application to TRIPLE resonance spectra of the chlorophyll a cation radical. *J. Magn. Reson.*, **84**, 537–547.

296. Howarth, D.F., Weil, J.A., and Zimpel, Z. (2003) Generalization of the lineshape useful in magnetic resonance spectroscopy. *J. Magn. Reson.*, **161**, 215–221.

297. Kirchner, N., van Slageren, J., and Dressel, M. (2007) Simulation of frequency domain magnetic resonance spectra of molecular magnets. *Inorg. Chim. Acta*, **360**, 3813–3819.

298. Hudson, A. and Luckhurst, G.R. (1969) The electron resonance line shapes of radicals in solution. *Chem. Rev.*, **69**, 191–225.

299. Schneider, D.J. and Freed, J.H. (1989) Spin relaxation and motional dyanmics. *Adv. Chem. Phys.*, **73**, 387–527.

300. Misra, S.K. (ed.) (2011) *Multifrequency Electron Paramagnetic Resonance*, Wiley-VCH Verlag GmbH.

301. Gerson, F. and Huber, W. (2001) *Electron Spin Resonance Spectroscopy of Organic Radicals*, Wiley-VCH Verlag GmbH.

302. Silsbee, R.H. (1966) Fourier transform analysis of hyperfine structure in ESR. *J. Chem. Phys.*, **45**, 1710–1714.

303. Lebedev, Y.S. and Dobryakov, S.N. (1967) Analysis of the EPR spectra of free radicals. *J. Struct. Chem.*, **8**, 757–769.

304. Evans, J.C., Morgan, P.H., and Renaud, R.H. (1978) Simulation of electron spin resonance spectra by fast Fourier transform. *Anal. Chim. Acta*, **103**, 175–187.

305. Evans, J.C. and Morgan, P.H. (1983) Simulation of electron spin resonance spectra by fast Fourier transform. *J. Magn. Reson.*, **52**, 529–531.

306. Das, R. et al. (2007) Simplifcation of complex EPR spectra by cepstral analysis. *J. Phys. Chem. A*, **111**, 4650–4657.

307. Kivelson, D. (1960) Theory of ESR linewidths of free radicals. *J. Chem. Phys.*, **33**, 1094–1106.

308. Wilson, R. and Kivelson, D. (1966) ESR linewidths in solution. I. Experiments on anisotropic and spin-rotational effects. *J. Chem. Phys.*, **44**, 154–168.

309. Kuprov, I. (2011) Diagonalization-free implementation of spin relaxation theory for large spin systems. *J. Magn. Reson.*, **209**, 31–38.

310. Stoll, S. (2012) in *Encyclopedia of Biophysics* (ed. G. Roberts), Springer, pp. 2316–2319.

311. Meirovitch, E. et al. (1982) Electron-spin relaxation and ordering in smectic and supercooled nematic liquid crystals. *J. Chem. Phys.*, **77**, 3915–3938.

312. Andreozzi, L. et al. (1996) Jump reorientation of a molecular probe in the glass transition region of o-terphenyl. *J. Phys. Condens. Matter*, **8**, 3795–3809.

313. Banerjee, D. et al. (2009) ESR evidence for 2 coexisting liquid phases in deeply supercooled bulk water. *Proc. Natl. Acad. Sci. U.S.A.*, **106**, 11448–11453.

314. Favro, L.D. (1965) *Flutuation Phenomena in Solids*, Academic Press, pp. 70–101.

315. Earle, K.A. (1993) 250-GHz EPR of nitroxides in the slow-motional regime: models of rotational diffusion. *J. Phys. Chem.*, **97**, 13289–13297.

316. Lee, S., Budil, D.E., and Freed, J.H. (1994) Theory of two-dimensional Fourier transform electron spin resonance for ordered and viscous fluids. *J. Chem. Phys.*, **101**, 5529–5558.

317. Meirovitch, E., Nayeem, A., and Freed, J.H. (1984) Analysis of protein-lipid interactions based on model simulations of electron spin resonance spectra. *J. Phys. Chem.*, **88**, 3454–3465.

318. Polimeno, A. and Freed, J.H. (1993) A many-body stochastic approach to rotational motions in liquids. *Adv. Chem. Phys.*, **83**, 89–206.
319. Polimeno, A. and Freed, J.H. (1995) Slow motional ESR in complex fluids: the slowly relaxing local structure model of solvent cage effects. *J. Phys. Chem.*, **99**, 10995–11006.
320. Liang, Z. and Freed, J.H. (1999) An assessment of the applicability of multifrequency ESR to study the complex dyanmics of biomolecules. *J. Phys. Chem. B*, **103**, 6384–6396.
321. Liang, Z. et al. (2004) A multifrequency electron spin resonance study of T4 lysozyme dynamics using the slowly relaxing local structure model. *J. Phys. Chem. B*, **108**, 17649–17659.
322. Zerbetto, M., Polimeno, A., and Meirovitch, E. (2009) General theoretical/computational tool for interpreting NMR spin relaxation in proteins. *J. Phys. Chem. B*, **113**, 13613–13625.
323. Zerbetto, M., Polimeno, A., and Meirovitch, E. (2010) C++ OPPS, a new software for the interpretation of protein dynamics from nuclear magnetic resonance measurements. *Int. J. Quantum Chem.*, **110**, 387–405.
324. Misra, S.K. (2007) Simulation of slow-motion CW EPR spectra using stochastic Liouville equation for an electron spin coupled to two nuclei with arbitrary spins: matrix elements of the Liouville superoperator. *J. Magn. Reson.*, **189**, 59–77.
325. Zerbetto, M. et al. (2007) Ab initio modeling of CW ESR spectra of the double spin labeled peptide Fmoc-(Aib-Aib-TOAC)$_2$-Aib-OMe in acetonitrile. *J. Phys. Chem. B*, **111**, 2668–2674.
326. Collauto, A. et al. (2012) Interpretation of cw-ESR spectra of p-methyl-thio-phenyl-nitronyl nitroxide in a nematic liquid crystalline phase. *Phys. Chem. Chem. Phys.*, **14**, 3200–3207.
327. Polimeno, A. et al. (2006) Stochastic modeling of CW-ESR spcetroscopy of [60]fulleropyrrolidine bisadducts with nitroxide probes. *J. Am. Chem. Soc.*, **128**, 4734–4741.
328. Freed, J.H., Bruno, G.V., and Polnaszek, C. (1971) ESR line shapes for triplets undergoing slow rotational reorientation. *J. Chem. Phys.*, **55**, 5270–5281.
329. Gamliel, D. and Levanon, H. (1992) Electron paramagnetic resonance lines shapes of photoexcited triplets with rotational diffusion. *J. Chem. Phys.*, **97**, 7140–7159.
330. Blank, A. and Levanon, H. (2005) Triplet line shape simulation in continuous wave electron paramagnetic resonance experiments. *Concepts Magn. Reson. Part A*, **25**, 18–39.
331. Robinson, B.H., Slutsky, L.J., and Auteri, F.P. (1992) Direct simulation of continuous wave electron paramagnetic resonance spectra from Brownian dynamics trajectories. *J. Chem. Phys.*, **96**, 2609–2616.
332. Sezer, D., Freed, J.H., and Roux, B. (2008) Using Markov models to simulate electron spin resonance spectra from molecular dynamics trajectories. *J. Phys. Chem. B*, **112**, 11014–11027.
333. Steinhoff, H.J. and Hubbell, W.L. (1996) Calculation of electron paramagnetic resonance spectra from Brownian dynamics trajectories: application to nitroxide side chains in proteins. *Biophys. J.*, **71**, 2201–2212.
334. Hakansson, P. et al. (2001) A direct simulation of EPR slow-motion spectra of spin-labelled phospholipids in liquid crystalline bilayers based on a molecular dynamics simulation of the lipid dynamics. *Phys. Chem. Chem. Phys.*, **3**, 5311–5319.
335. Beier, C. and Steinhoff, H.J. (2006) A structure-based simulation approach for electron paramagnetic resonance spectra using molecular and stochastic dynamics simulations. *Biophys. J.*, **91**, 2647–2664.
336. Oganesyan, V.S. (2007) A novel approach to the simulation of nitroxide spin label EPR spectra from a single truncated dynamical trajectory. *J. Magn. Reson.*, **188**, 196–205.
337. Oganesyan, V.S. (2011) A general approach for prediction of motional EPR spectra from Molecular Dynamics (MD) simulations: application to spin labelled protein. *Phys. Chem. Chem. Phys.*, **13**, 4724–4737.

338. Budil, D.E. et al. (2006) Calculating slow-motional electron paramagnetic resonance spectra from molecular dynamics using a diffusion operator approach. *J. Phys. Chem. A*, **110**, 3703–3713.

339. Sezer, D., Freed, J.H., and Roux, B. (2009) Multifrequency electron spin resonance spectra of a spin-labeled protein calculated from molecular dynamics simulations. *J. Am. Chem. Soc.*, **131**, 2597–2605.

340. Sezer, D. and Sigurdsson, S.T. (2011) Simulating electron spin resonance spectra of macromolecules labeled with two dipolar-coupled nitroxide spin labels from trajectories. *Phys. Chem. Chem. Phys.*, **13**, 12785–12797.

341. Rangel, D.P., Baveye, P.C., and Robinson, B.H. (2012) Direct simulation of magnetic resonance relaxation rates and line shapes from molecular trajectories. *J. Phys. Chem. B*, **116**, 6233–6249.

342. Hakansson, P. and Nair, P.B. (2011) Implicit numerical schemes for the stochastic Liouville equation in Langevin form. *Phys. Chem. Chem. Phys.*, **13**, 9578–9589.

343. Della Lunga, G., Pogni, R., and Basosi, R. (1994) Computer simulation of ESR spectra in the slow-motion region for copper complexes with nitrogen ligands. *J. Phys. Chem.*, **98**, 3937–3942.

344. Pasenkiewicz-Gierula, M., Subczynski, W.K., and Antholine, W.E. (1997) Rotational motion of square planar copper complexes in solution and phospholipid bilayer membranes. *J. Phys. Chem. B*, **101**, 5596–5606.

345. Della Lunga, G. et al. (2003) A new program based on stochastic Liouville equation for the analysis of superhyperfine interaction in CW-ESR spectroscopy. *J. Magn. Reson.*, **164**, 71–77.

346. Bolton, J.R. and Carrington, A. (1962) Line width alternation in the electron spin resonance spectrum of the durosemiquinone radical. *Mol. Phys.*, **5**, 161–167.

347. Freed, J.H. and Fraenkel, G.K. (1962) Anomalous alternating linewidths in ESR spectra. *J. Chem. Phys.*, **37** (5), 1156–1157.

348. Bain, A.D. (2003) Chemical exchange in NMR. *Prog. Nucl. Magn. Reson. Spectrosc.*, **43**, 63–103.

349. McConnell, H.M. (1958) Reaction rates by nuclear magnetic resonance. *J. Chem. Phys.*, **28**, 430–431.

350. Kirste, B. (1987) Least-squares fitting of EPR spectra by Monte Carlo methods. *J. Magn. Reson.*, **73**, 213–224.

351. Gordon, R.G. and McGinnis, R.P. (1968) Line shapes in molecular spectra. *J. Chem. Phys.*, **49**, 2455–2456.

352. Alexander, S. (1962) Exchange of interacting nuclear spins in nuclear magnetic resonance. I. Intramolecular exchange. *J. Chem. Phys.*, **37**, 967–974.

353. Alexander, S. (1962) Exchange of interacting nuclear spins in nuclear magnetic resonance. II. Chemical exchange. *J. Chem. Phys.*, **37**, 974–980.

354. Binsch, G. (1968) The direct method for calculating high-resolution nuclear magnetic resonance spectra. *Mol. Phys.*, **15**, 469–478.

355. Kaplan, J.I. and Fraenkel, G. (1972) Effect of molecular reorganization on nuclear magnetic resonance lineshapes. Permutation of indices method. *J. Am. Chem. Soc.*, **94**, 2907–2912.

356. Limbach, H.H. (1979) NMR lineshape theory of superimposed intermolecular spin exchange reactions and its application to the system acetic acid/methanol/tetrahydrofuran-$d_8$. *J. Magn. Reson.*, **36**, 287–300.

357. Cuperlovic, M. et al. (2000) Spin relaxation and chemical exchange in NMR simulations. *J. Magn. Reson.*, **142**, 11–23.

358. Helgstrand, M., Härd, T., and Allard, P. (2000) Simulations of NMR pulse sequences during equilibrium and non-equilibrium chemical exchange. *J. Biomol. NMR*, **18**, 9–63.

359. Szalay, Z. and Rohonczy, J. (2011) Kinetic Monte Carlo simulation of DNMR spectra. *Annu. Rep. NMR Spectrosc.*, **73**, 175–215.

360. Hudson, A. and McLachlan, A.D. (1965) Line shapes of triplet ESR spectra: the effects of intemolecular exciton transfer. *J. Chem. Phys.*, **43**, 1518–1524.

361. Norris, J.R. (1967) Rapid computation of magnetic resonance line shapes for exchange among many sites. *Chem. Phys. Lett.*, **1**, 333–334.
362. Grampp, G. and Stiegler, G. (1986) Application of the density-matrix formalism to the simulation of kinetic ESR spectra if intermolecular electron-transfer reactions. *J. Magn. Reson.*, **70**, 1–10.
363. Sankarapandi, S. et al. (1993) Fast computation of dynamic EPR spectra of biradicals. *J. Magn. Reson., Ser. A*, **103**, 163–170.
364. Rockenbauer, A. (1999) Determination of chemical exchange parameters in ESR. *Mol. Phys. Rep.*, **26**, 117–127.
365. Gamliel, D., Luz, Z., and Vega, S. (1986) Complex dynamic NMR spectra in the fast exchange limit. *J. Chem. Phys.*, **85**, 2516–2527.
366. Levitt, M.H. and Beshah, K. (1987) NMR in chemically exchanging systems. Is the number of sites equal to the number of frequencies? *J. Magn. Reson.*, **75**, 222–228.
367. Bennati, M., Grupp, A., and Mehring, M. (1997) Pulsed-EPR on the photoexcited triplet state of $C_{60}$. *Synth. Met.*, **86**, 2321–2324.
368. Astashkin, A.V. and Raitsimring, A.M. (2002) Electron spin echo envelope modulation theory for high electron spin systems in weak crystal field. *J. Chem. Phys.*, **117**, 6121–6132.
369. Hahn, E.L. (1950) Spin echoes. *Phys. Rev.*, **80**, 580–594.
370. Hanson, L.G. (2008) Is quantum mechanics necessary for understanding magnetic resonance? *Concepts Magn. Reson. Part A*, **32**, 329–340.
371. Feynman, R.P., Vernon, F.L., and Hellwarth, R.W. (1957) Geometrical representation of the schrödinger equation for solving maser problems. *J. Appl. Phys.*, **28**, 49–52.
372. Bloom, A.L. (1955) Nuclear induction in inhomogeneous fields. *Phys. Rev.*, **98**, 1105–1111.
373. Rowan, L.G., Hahn, E.L., and Mims, W.B. (1965) Electron-spin-echo envelope modulation. *Phys. Rev. A*, **137**, 61–71.
374. Mims, W.B. (1972) Envelope modulation in spin-echo experiments. *Phys. Rev. B*, **5**, 2409–2419.
375. Mims, W.B. (1972) Amplitudes of superhyperfine frequencies displayed in the electron-spin-echo envelope. *Phys. Rev. B*, **6**, 3543–3545.
376. Zhidomirov, G.M. and Salikhov, K.M. (1968) Modulation effects in free-radical spin-echo signals. *Teor. Eksp. Khim.*, **4**, 514–519.
377. Zhidomirov, G.M. and Salikhov, K.M. (1971) Modulation effects in free-radical spin-echo signals. *Theor. Exp. Chem.*, **4**, 332–334.
378. Dikanov, S.A. and Tsvetkov, Y.D. (1992) *Electron Spin Echo Envelope Modulation (ESEEM) Spectroscopy*, CRC Press, Boca Raton, FL.
379. Sloop, D.J. et al. (1981) Electron spin echoes of a photoexcited triplet: pentacene in p-terphenyl crystals. *J. Chem. Phys.*, **75**, 3746–3757.
380. Larsen, R.G., Halkides, C.J., and Singel, D.J. (1993) A geometric representation of nuclear modulation effects: the effects of high electron spin multiplicity on the electron spin echo envelope modulation spectra of $Mn^{2+}$ complexes of N-ras p21. *J. Chem. Phys.*, **98**, 6704–6721.
381. Maryasov, A.G. and Bowman, M.K. (2012) Spin dynamics of paramagnetic centers with anisotropic g tensor and spin 1/2. *J. Magn. Reson.*, **221**, 69–75.
382. Reijerse, E.J. and Dikanov, S.A. (1991) Electron spin echo envelope modulationspectroscopyu on orientationally disordered systems: line shape singularities in S = 1/2, I = 1/2 spin systems. *J. Chem. Phys.*, **95**, 836–845.
383. Muha, G.M. (1980) Exact solution of the NQR I = 1 eigenvalue problem for an arbitrary asymmetry parameter and Zeeman field strength and orientation. *J. Chem. Phys.*, **73**, 4139–4140.
384. Muha, G.M. (1983) Exact solution of the eigenvalue problem for a spin 3/2 system in the presence of a magnetic field. *J. Magn. Reson.*, **53**, 85–102.
385. Kottis, P. and Lefebvre, R. (1963) Calculation of the electron spin resonance line shape of randomly oriented molecules in a triplet state. I. The

$\Delta m = 2$ transition with a constant linewidth. *J. Chem. Phys.*, **39**, 393–403.

386. Astashkin, A.V., Dikanov, S.A., and Tsvetkov, Y.D. (1984) Modulation effects from N-14 and N-15 nitrogen nuclei in the electron-spin echo of imidazole nitroxyl radicals containing the 2-oximinoalkyl group. *J. Struct. Chem.*, **25**, 45–55.

387. Flanagan, H.L. and Singel, D.J. (1987) Analysis of $^{14}$N ESEEM patterns of randomly oriented solids. *J. Chem. Phys.*, **87**, 5606–5616.

388. Gemperle, C. et al. (1990) Phase cycling in pulse EPR. *J. Magn. Reson.*, **88**, 241–256.

389. Gemperle, C., Schweiger, A., and Ernst, R.R. (1991) Novel analytical treatments of electron spin-echo envelope modulation with short and extended pulses. *J. Magn. Reson.*, **91**, 273–288.

390. Gemperle, C., Schweiger, A., and Ernst, R.R. (1991) Electron-spin-echo envelope modulation with improved modulation depth. *Chem. Phys. Lett.*, **178**, 565–572.

391. Jeschke, G. (1996) New concepts in solid-state pulse electron spin resonance.PhD thesis. ETH Zurich.

392. Kasumaj, B. and Stoll, S. (2008) 5- and 6-pulse electron spin echo envelope modulation (ESEEM) of multi-nuclear spin systems. *J. Magn. Reson.*, **190**, 233–247.

393. Jerschow, A. (2005) MathNMR: spin and spatial tensor manipulations in mathematica. *J. Magn. Reson.*, **176**, 7–14.

394. Güntert, P. (2006) Symbolic NMR product operator calculations. *Int. J. Quantum Chem.*, **106**, 344–350.

395. Filip, X. and Filip, C. (2010) SD-CAS: spin dynamics by computer algebra systems. *J. Magn. Reson.*, **207**, 95–113.

396. Levitt, M.H., Rantaharju, J., and Brinkmann, A. SpinDynamica, http://www.spindynamica.soton.ac.uk (accessed 22 August 2013).

397. Salikhov, K.M., Kandrashkin, Y.E., and Salikhov, A.K. (1992) Peculiarities of free induction and primary electron spin echo signals for spin-correlated radical pairs. *Appl. Magn. Reson.*, **3**, 199–261.

398. Tang, J., Thurnauer, M.C., and Norris, J.R. (1994) Electron spin echo envelope modulation due to exchange and dipolar interactions in a spin-correlated radical pair. *Chem. Phys. Lett.*, **219**, 283–290.

399. Mims, W.B., Peisach, J., and Davies, J.L. (1977) Nuclear modulation of the electron spin echo envelope in glassy materials. *J. Chem. Phys.*, **66**, 5536–5550.

400. Reijerse, E.J. et al. (1991) One- and two-dimensional ESEEM on disordered systems; applications to nitrogen coordinated oxo-vanadium complexes, in *Electron Paramagnetic Resonance of Disordered Systems*, World Scientific.

401. Shane, J.J. (1993) Electron spin echo envelope modulation spectroscopy of disordered solids. PhD thesis. University of Nijmegen.

402. Shane, J.J., Liesum, L.P., and Schweiger, A. (1998) Efficient simulation of ESEEM spectra using gamma. *J. Magn. Reson.*, **134**, 72–75.

403. Stoll, S. and Schweiger, A. (2003) Rapid construction of solid-state magnetic resonance powder spectra from frequencies and amplitudes as applied to ESEEM. *J. Magn. Reson.*, **163**, 248–256.

404. Sidje, R.B. (1998) ExpoKit: a software package for computing matrix exponentials. *ACM Trans. Math. Softw.*, **24**, 130–156.

405. Moler, C. and Loan, C.V. (2003) Nineteen dubious ways to compute the exponential of a matrix, twenty-five years later. *SIAM Rev.*, **45**, 3–49.

406. Reijerse, E.J. et al. (1986) Comparison of ESEEM, ESE-ENDOR, and CW-ENDOR on $^{14}$N in a powder. *J. Magn. Reson.*, **67**, 114–124.

407. Benetis, N.P. and Sørnes, A.R. (2000) Automatic spin-Hamiltonian diagonalization for electronic doublet coupled to anisotropic nuclear spins applied in one- and two-dimensional electron spin-echo experiments. *Concepts Magn. Reson.*, **12**, 410–433.

408. Grischkowsky, D. and Hartmann, S.R. (1970) Behavior of electron-spin echoes

and photon echoes in high field. *Phys. Rev. B*, **2**, 60–74.
409. Lin, T.S. (1984) Electron spin echo spectroscopy of organic triplets. *Chem. Rev.*, **84**, 1–15.
410. Singel, D.J. *et al.* (1984) Complete determination of $^{14}$N hyperfine and quadrupole interactions in the metastable triplet state of free-base porphin via electron spin echo envelope modulation. *J. Chem. Phys.*, **81**, 5453–5461.
411. Coffino, A. and Peisach, J. (1992) Nuclear modulation effects in high-spin electron systems with small zero-field splittings. *J. Chem. Phys.*, **97**, 3072–3091.
412. Benetis, N.P., Dave, P.C., and Goldfarb, D. (2002) Characteristics of ESEEM and HYSCORE spectra of $S > 1/2$ centers in orientationally disordered systems. *J. Magn. Reson.*, **158**, 126–142.
413. Oliete, P.B., Orera, V.M., and Alonso, P.J. (1996) Structure of the Jahn-Teller distorted $Cr^{2+}$ defect in $SrF_2$:Cr by electron-spin-echo envelope modulation. *Phys. Rev. B*, **54**, 12099–12108.
414. Dikanov, S.A., Yudanov, V.F., and Tsvetkov, Y.D. (1979) Electron spin-echo studies of weak hyperfine interactions with ligands in some $VO^{2+}$ complexes in frozen glassy solution. *J. Magn. Reson.*, **34**, 631–645.
415. Dikanov, S.A., Shubin, A.A., and Parmon, V.N. (1981) Modulation effects in the electron spin echo resulting from hyperfine interaction with a nucleus of arbitrary spin. *J. Magn. Reson.*, **42**, 474–487.
416. Tyryshkin, A.M., Dikanov, S.A., and Goldfarb, D. (1993) Sum combination harmonics in four-pulse ESEEM spectra. Study of the ligand geometry in aqua-vanadyl complexes in polycrystalline and glass matrices. *J. Magn. Reson., Ser. A*, **105**, 271–283.
417. Yulikov, M. *et al.* (2012) Distance measurements in Au nanoparticles functionalized with nitroxide radicals and $Gd^{3+}$-DTPA chelate complexes. *Phys. Chem. Chem. Phys.*, **14**, 10732–10746.
418. Misra, S.K., Borbat, P.P., and Freed, J.H. (2009) Calculation of double-quantum-coherence two-dimensional spectra: distance measurements and orientational correlations. *Appl. Magn. Reson.*, **36**, 237–258.
419. Studer, W. (1988) SMART, a general purpose pulse experiment simulation program using numerical density matrix calculations. *J. Magn. Reson.*, **77**, 424–438.
420. Smith, S.A. *et al.* (1994) Computer simulations in magnetic resonance. An object-oriented programming approach. *J. Magn. Reson., Ser. A*, **106**, 75–105.
421. Blanton, W.B. (2003) BlochLib: a fast NMR C++ toolkit. *J. Magn. Reson.*, **162**, 269–283.
422. Hogben, H.J., Hore, P.J., and Kuprov, I. (2010) Strategies for state space restriction in densely coupled spin systems with applications to spin chemistry. *J. Chem. Phys.*, **132**, 174101, 1–10.
423. Skinner, T.E. and Glaser, S.J. (2002) Representation of a quantum ensemble as a minimal set of pure states. *Phys. Rev. A*, **66**, 032112.
424. Edwards, L.J. and Kuprov, I. (2012) Parallel density matrix propagation in spin dynamics simulations. *J. Chem. Phys.*, **136**, 044108.
425. Edén, M., Lee, Y.K., and Levitt, M.H. (1996) Efficient simulation of periodic problems in NMR. Application to decoupling and rotational resonance. *J. Magn. Reson., Ser. A*, **120**, 56–71.
426. Charpentier, T., Fermon, C., and Virlet, J. (1998) Efficient time propagation techniques for MAS NMR simulation: application to quadrupolar nuclei. *J. Magn. Reson.*, **132**, 181–190.
427. Leskes, M., Madhu, P.K., and Vega, S. (2010) Floquet theory in solid-state nuclear magnetic resonance. *Prog. Nucl. Magn. Reson. Spectrosc.*, **57**, 345–380.
428. Scholz, I., van Beek, J.D., and Ernst, M. (2010) Operator-based Floquet theory in solid-state NMR. *Solid State Nucl. Magn. Reson.*, **37**, 39–59.
429. Kälin, M. and Schweiger, A. (2001) Radio-frequency-driven electron spin echo envelope modulation spectroscopy

on spin systems with isotropic hyperfine interactions. *J. Chem. Phys.*, **115**, 10863–10875.

430. Sczaniecki, P.B., Hyde, J.S., and Froncisz, W. (1990) Continous wave multi-quantum electron-paramagnetic resonance spectroscopy. *J. Chem. Phys.*, **93**, 3891–3898.
431. Giordano, M. et al. (1988) Double-modulation electron-spin-resonance spectroscopy – experimental observations and theoretical comprehensive interpretation. *Phys. Rev. A*, **38**, 1931–1942.
432. Bain, A.D. and Berno, B. (2011) Liouvillians in NMR: the direct method revisited. *Prog. Nucl. Magn. Reson. Spectrosc.*, **59**, 223–244.
433. Budil, D.E. et al. (1996) Nonlinear-least-squares analysis of slow-motion EPR spectra in one and two dimensions using a modified Levenberg-Marquardt algorithm. *J. Magn. Reson., Ser. A*, **120**, 155–189.
434. Liang, Z., Crepeau, R.H., and Freed, J.H. (2005) Effects of finite pulse width on two-dimensional Fourier transform electron spin resonance. *J. Magn. Reson.*, **177**, 247–260.
435. Chiang, Y.W., Costa-Filho, A., and Freed, J.H. (2007) 2D-ELDOR using full $S_{c-}$ fitting and absorption lineshapes. *J. Magn. Reson.*, **188**, 231–245.
436. Saxena, S. and Freed, J.H. (1997) Theory of double-quantum two-dimensional electron spin resonance with application to distance measurements. *J. Chem. Phys.*, **107**, 1317–1340.
437. Plüschau, M. and Dinse, K.P. (1994) 2D EPR study of a photoinduced proton abstraction in the system anthraquinone and 4-methyl-2,6-di-tert-butylphenol in 2-propanol. *J. Magn. Reson., Ser. A*, **109**, 181–191.
438. Chiang, Y.W. and Freed, J.H. (2011) A new Lanczos-based algorithm for simulating high-frequency two-dimensional electron spin resonance spectra. *J. Chem. Phys.*, **134**, 034112.
439. Zientara, G.P. and Freed, J.H. (1979) The variational method and the stochastic-Liouville equation. I. A finite element solution to the CIDN(E)P problem. *J. Chem. Phys.*, **70**, 2587–2598.
440. Moro, G. and Freed, J.H. (1981) Calculation of ESR spectra and related Fokker-Planck forms by the use of the Lanczos algorithm. *J. Chem. Phys.*, **74**, 3757–3773.
441. Vasavada, K.V., Schneider, D.J., and Freed, J.H. (1987) Calculation of ESR spectra and related Fokker-Planck forms by the use of the Lanczos algorithm. II. Criteria for truncation of basis sets and recursive steps utilizing conjugate gradients. *J. Chem. Phys.*, **86**, 647–661.
442. Kuprov, I., Wagner-Rundell, N., and Hore, P.J. (2007) Polynomially scaling spin dynamics simulation algorithm based on adaptive state-space restriction. *J. Magn. Reson.*, **189**, 241–250.
443. Kuprov, I. (2008) Polynomially scaling spin dynamics II: further state-space compression using Krylov subspace techniques and zero track elimination. *J. Magn. Reson.*, **195**, 45–51.
444. Krzystyniak, M., Edwards, L.J., and Kuprov, I. (2011) Destination state screening of active spaces in spin dynamics simulations. *J. Magn. Reson.*, **210**, 228–232.
445. Dumez, J.N., Butler, M.C., and Emsley, L. (2010) Numerical simulation of free evolution in solid-state magnetic resonance using low-order correlations in Liouville space. *J. Chem. Phys.*, **133**, 224501.
446. Castillo, A.M., Patiny, L., and Wist, J. (2011) Fast and accurate algorithm for the simulation of NMR spectra of large spin systems. *J. Magn. Reson.*, **209**, 123–130.
447. Dalton, L.R. and Kwiram, A.L. (1972) ENDOR studies in molecular crystals. II. Computer analysis of the polycrystalline ENDOR spectra of Low symmetry materials. *J. Chem. Phys.*, **57**, 1132–1145.
448. Toriyama, K., Nunome, K., and Iwasaki, M. (1976) ENDOR studies of methyl radicals in irradiated single crystals of $CH_3COOLi \cdot 2H_2O$. *J. Chem. Phys.*, **64**, 2020–2026.
449. Erickson, R. (1996) Simulation of ENDOR spectra of radicals with

anisotropic hyperfine and nuclear quadrupole interactions in disordered solids. *Chem. Phys.*, **202**, 263–275.

450. Thomann, H. and Bernardo, M. (1993) Pulsed electron nuclear multiple resonance spectroscopic methods for metalloproteins and metalloenzymes. *Methods Enzymol.*, **227**, 118–189.

451. Whiffen, D.H. (1966) ENDOR transition moments. *Mol. Phys.*, **10**, 595–596.

452. Schweiger, A. (1982) Electron nuclear double resonance of transition metal complexes with organic ligands. *Struct. Bond.*, **51**, 1–119.

453. Schweiger, A. and Günthart, H.H. (1982) Transition probabilities in electron-nuclear double- and multiple-resonance spectroscopy with non-coherent and coherent radio-frequency fields. *Chem. Phys.*, **70**, 1–22.

454. Hutchison, C.A. and McKay, D.B. (1977) The determination of hydrogen coordination in lanthanum nicotinate dihydate crystals by $Nd^{3+}$-proton double resonance. *J. Chem. Phys.*, **66**, 3311–3330.

455. Fan, C. et al. (1992) Quantitative studies of Davies pulsed ENDOR. *J. Magn. Reson.*, **98**, 62–72.

456. Astashkin, A.V. and Kawamori, A. (1998) Matrix line in pulsed electron-nuclear double resonance spectra. *J. Magn. Reson.*, **135**, 406–417.

457. Doan, P.E. et al. (2010) Simulating suppression effects in pulsed ENDOR, and the 'Hole in the Middle' of Mims and Davies ENDOR spectra. *Appl. Magn. Reson.*, **37**, 763–779.

458. Tan, X. et al. (1993) Pulsed and continuous wave electron nuclear double resonance patterns of aquo protons coordinated in frozen solution to high spin $Mn^{2+}$. *J. Chem. Phys.*, **98**, 5147–5157.

459. Sturgeon, B.E. et al. (1994) $^{55}$Mn electron spin echo ENDOR of $Mn^{2+}$ complexes. *J. Phys. Chem.*, **98**, 12871–12883.

460. Vardi, R. et al. (1997) X-band pulsed ENDOR study of $^{57}$Fe-substituted sodalite – The effect of the zero-field splitting. *J. Magn. Reson.*, **126**, 229–241.

461. Hoffman, B.M. (1994) ENDOR and ESEEM of a non-Kramers doublet in an integer-spin system. *J. Phys. Chem.*, **98**, 11657–11665.

462. Liao, P.F. and Hartmann, S.R. (1973) Determination of Cr-Al hyperfine and electric quadrupole interaction parameters in ruby using spin-echo electron-nuclear double resonance. *Phys. Rev. B*, **8**, 69–80.

463. Stillman, A.E. and Schwartz, R.N. (1978) ENDOR spin-echo spectroscopy. *Mol. Phys.*, **35**, 301–313.

464. Epel, B. et al. (2001) The effect of spin relaxation on ENDOR spectra recorded at high magnetic fields and low temperatures. *J. Magn. Reson.*, **148**, 388–397.

465. Morton, J.J.L. et al. (2008) Nuclear relaxation effects in Davies ENDOR variants. *J. Magn. Reson.*, **191**, 315–321.

466. Doan, P.E. (2011) Combining steady-state and dynamic methods for determining absolute signs of hyperfine interactions: Pulsed ENDOR Saturation and Recovery (PESTRE). *J. Magn. Reson.*, **208**, 76–86.

467. Jeschke, G. et al. (2004) Data analysis procedures for pulse ELDOR measurements of broad distance distributions. *Appl. Magn. Reson.*, **26**, 223–244.

468. Chiang, Y.W., Borbat, P.P., and Freed, J.H. (2005) The determination of pair distance distributions by pulsed ESR using Tikhonov reguarlization. *J. Magn. Reson.*, **172**, 279–295.

469. Chiang, Y.W., Borbat, P.P., and Freed, J.H. (2005) Maximum entropy: a complement to Tikhonov regularization for determination of pair distance distributions by pulsed ESR. *J. Magn. Reson.*, **177**, 184–196.

470. Jeschke, G. et al. (2006) DeerAnalysis2006—a comprehensive software package for analyzing pulsed ELDOR data. *Appl. Magn. Reson.*, **30**, 473–498.

471. Sen, K.I., Logan, T.M., and Fajer, P.G. (2007) Protein dynamics and monomer-monomer interactions in AntR activation by electron paramagnetic resonance and double electron-electron resonance. *Biochemistry*, **46**, 11639–11649.

472. Sen, K.I. and Fajer, P.G. (2009) Analysis of DEER signals with DEFit. *EPR Newsl.*, **19** (1-2), 26–28.
473. Polyhach, Y., Bordignon, E., and Jeschke, G. (2011) Rotamer libraries of spin labelled cysteines for protein studies. *Phys. Chem. Chem. Phys.*, **13**, 2356–2366.
474. Hatmal, M.M.M. *et al.* (2011) Computer modeling of nitroxide spin labels on proteins. *Biopolymers*, **97**, 35–44.
475. Jeschke, G. (2013) Conformational dynamics and distribution of nitroxide spin labels. *Prog. Nucl. Magn. Reson. Spectrosc.*, **72**, 42–60.
476. Hagelueken, G. *et al.* (2012) MtsslWizard: in silico spin-labeling and generation of distance distributions in PyMOL. *Appl. Magn. Reson.*, **42**, 377–391.
477. Johnston, T.S. and Hecht, H.G. (1965) An automatic fitting procedure for the determination of anisotropic g-tensors from EPR studies of powder samples. *J. Mol. Spectrosc.*, **17**, 98–107.
478. Bauder, A. and Myers, R.J. (1968) Least squars curve fitting of EPR spectra. *J. Mol. Spectrosc.*, **27**, 110–116.
479. Dračka, O. (1985) Computer-assisted analysis of linear isotropic EPR spectra. *J. Magn. Reson.*, **65**, 187–205.
480. Kolda, T.G., Lewis, R.M., and Torczon, V. (2003) Optimization by direct search: new perspectives on some classical and modern methods. *SIAM Rev.*, **45**, 385–482.
481. Rakitin, Y.V., Larin, G.M., and Minin, V.V. (1993) *Interpretation of EPR Spectra of Coordination Compounds*, Nauka, Moscow.
482. Mor, H.H., Weihe, H., and Bendix, J. (2010) Fitting of EPR spectra: the importance of a flexible bandwidth. *J. Magn. Reson.*, **207**, 283–286.
483. Żurek, S.G. (2011) Genetic algorithm with peaks adaptive objective function used to fit the EPR powder spectrum. *Appl. Soft Comput.*, **11**, 1000–1007.
484. Štrancar, J., Šentjurc, M., and Schara, M. (2000) Fast and accurate characterization of biological membranes by EPR spectral simulations of nitroxides. *J. Magn. Reson.*, **142**, 254–265.
485. Björck, A. (1996) *Numerical Methods for Least-Squares Problems*, SIAM, Philadelphia, PA.
486. Barzaghi, M. and Simonetta, M. (1983) Iterative computer analysis of complex dynamic EPR bandshapes. Fast motional regime. *J. Magn. Reson.*, **51**, 175–204.
487. Lund, A. *et al.* (2006) Automatic fitting to 'powder' EPR spectra of coupled paramagnetic species employing Feynman's theorem. *Spectrochim. Acta, Part A*, **63**, 830–835.
488. Brumby, S. (1980) Numerical analysis of ESR spectra. 3. Iterative least–squares analysis of significance plots. *J. Magn. Reson.*, **39**, 1–9.
489. Hrabański, R. and Lech, J. (1990) Optimization of spin-Hamiltonian parameters by the method of nonlinear least-squares fitting. *Phys. Status Solidi B*, **162**, 275–280.
490. Misra, S.K. (1984) Evaluation of anisotropic non-coincident g and A tensors from EPR and ENDOR data by the method of least-squares fitting. *Physica B*, **124**, 53–61.
491. Misra, S.K. (1986) Evaluation of spin Hamiltonian parameters from ESR data of single crystals. *Magn. Reson. Rev.*, **10**, 285–331.
492. Chachaty, C. and Soulié, E.J. (1995) Determination of electron spin resonance static and dynamic parameters by automated fitting of the spectra. *J. Phys. III France*, **5**, 1927–1952.
493. Soulié, E.J. and Berclaz, T. (2005) Electron paramagnetic resonance: nonlinear least-squares fitting of the Hamiltonian parameters from powder spectra with the Levenberg-Marquardt algorithm. *Appl. Magn. Reson.*, **29**, 401–416.
494. Freed, J.H. (1990) Modern techniques in electron paramagnetic resonance spectroscopy. *J. Chem. Soc., Faraday Trans.*, **86**, 3173–3180.
495. Beckwith, A.L.J. and Brumby, S. (1987) Numerical analysis of EPR spectra. 7. The simplex algorithm. *J. Magn. Reson.*, **72**, 252–259.
496. Fajer, P.G. *et al.* (1990) General method for multparameter fitting of high-resolution EPR spectra using a

simplex algorithm. *J. Magn. Reson.*, **88**, 111–125.
497. Brumby, S. (1992) ESR spectrum simulation: the simplex algorithm with quadratic convergence and error estimation. *Appl. Spectrosc.*, **46**, 176–178.
498. Filipič, B. and Štrancar, J. (2001) Tuning EPR spectral parameters with a genetic algorithm. *Appl. Soft Comput.*, **1**, 83–90.
499. Carl, P.J., Isley, S.L., and Larsen, S.C. (2001) Combining theory and experiment to interpret the EPR spectra of $VO^{2+}$-exchanged zeolites. *J. Phys. Chem. A*, **105**, 4563–4573.
500. Nilges, M.J., Matteson, K., and Belford, R.L. (2006) SIMPOW6: a software package for the simulation of ESR powder-type spectra. *Biol. Magn. Reson.*, **27**, 261–281.
501. Spałek, T., Pietrzyk, P., and Sojka, Z. (2005) Application of genetic algorithm for extraction of the parameters from powder EPR spectra. *Acta Phys. Pol.*, **108**, 95–102.
502. Kirste, B. (1992) Methods for automated analysis and simulation of electron paramagnetic resonance spectra. *Anal. Chim. Acta*, **265**, 191–200.
503. Puma, M. *et al.* (1988) Computer analysis of electron paramagnetic resonance data using the Monte Carlo method. *J. Phys. C: Solid State Phys.*, **21**, 5555–5564.
504. Della Lunga, G., Pogni, R., and Basosi, R. (1998) Global versus local minimization procedures for the determination of spin Hamiltonian parameters from electron spin resonance spectra. *Mol. Phys.*, **95**, 1275–1281.
505. Calvo, R. *et al.* (2000) EPR study of the molecular and electronic structure of the semiquinone biradical $Q_A^-$ $Q_B^-$ in photosynthetic reaction centers from rhodobacter sphaeroides. *J. Am. Chem. Soc.*, **122**, 7327–7341.
506. Basosi, R., Lunga, G.D., and Pogni, R. (2005) Copper biomolecules in solution. *Biol. Magn. Reson.*, **23**, 385–416.
507. Kavalenka, A.A. *et al.* (2005) Speeding up a genetic algorithm for EPR-based spin label characterization of biosystem complexity. *J. Chem. Inf. Model.*, **45**, 1628–1635.
508. Martinez, G.V. and Millhauser, G.L. (1998) A neural network approach to the rapid computation of rotational correlation times from slow motional ESR spectra. *J. Magn. Reson.*, **134**, 124–130.
509. Misra, S.K. (1976) Analysis of EPR data characterized by spin Hamiltonian with large off-diagonal elements. *J. Magn. Reson.*, **23**, 191–198.
510. Golub, G. and Pereyra, V. (2003) Separable nonlinear least squares: the variable projection method and its applications. *Inverse Prob.*, **19**, R1–R26.
511. Steinbock, O. *et al.* (1997) A demonstration of principal component analysis for EPR spectroscopy: identifying pure component spectra from complex spectra. *Anal. Chem.*, **69**, 3708–3713.
512. Moens, P. *et al.* (1993) Maximum-likelihood common-factor analysis as a powerful tool in decomposing multicomponent EPR powder spectra. *J. Magn. Reson., Ser. A*, **101**, 1–15.
513. Zalibera, M. *et al.* (2013) Monotrimethylene-bridged Bis-p-phenylenediamine radical cations and dications: spin states, conformations, and dynamics. *J. Phys. Chem. A*, **117**, 1439–1448.
514. Press, W.H. *et al.* (1992) *Numerical Recipes in C*, Cambridge University Press.
515. Misra, S.K. and Subramanian, S. (1982) Calculation of parameter errors in the analysis of electron paramagnetic resonance data. *J. Phys. C: Solid State Phys.*, **15**, 7199–7207.
516. Fursman, C.E. and Hore, P.J. (1999) Distance determination in spin-correlated radical pairs in photosynthetic reaction centers by electron spin echo envelope modulation. *Chem. Phys. Lett.*, **303**, 593–600.
517. Eaton, G.R. *et al.* (2010) *Quantitative EPR*, Springer.
518. Cochrane, C.J. and Lenahan, P.M. (2008) Real time exponential weighted recursive least squares adaptive signal averaging for enhancing the sensitivity of continuous wave magnetic resonance. *J. Magn. Reson.*, **195**, 17–22.

519. Cammack, R. *et al.* (2006) JCAMP-DX for electron magnetic resonance (EMR). *Pure Appl. Chem.*, **78**, 613–631.
520. Cammack, R. (2010) EPR spectra of transition-metal proteins: the benefits of data deposition in standard formats. *Appl. Magn. Reson.*, **37**, 257–266.
521. Spindler, P.E. *et al.* (2012) Shaped optimal control pulses for increased excitation bandwidth in EPR. *J. Magn. Reson.*, **218**, 49–58.
522. Doll, A. *et al.* (2013) Adiabatic and fast passage ultra-wideband inversion in pulsed EPR. *J. Magn. Reson.*, **230**, 27–39.
523. Charnock, G.T.P., Krzystyniak, M., and Kuprov, I. (2012) Molecular structure refinement by direct fitting of atomic coordinates to experimental ESR spectra. *J. Magn. Reson.*, **216**, 62–68.

# 4
# Multifrequency Transition Ion Data Tabulation

*Sushil K. Misra, Sean Moncrieff, and Stefan Diehl*

## 4.1
## Introduction

The published spin-Hamiltonian parameters (SHPs) for transition metal ions in various hosts, which maybe single crystals, powders, and glasses, are listed in the table included in this chapter, covering the period 1993–2012. The SHPs reported in the period from the 1960s to 1992s were published in the *Handbook of Electron Spin Resonance*, Volume 2 (Eds. C. P. Poole, Jr. and H. A. Farach, AIP Press, Springer Verlag, New York, 1999; Chapter 9: Transition ion data tabulation by S. K. Misra). Since then, multifrequency EPR has developed extensively, and the parameters for the "EPR-silent" ions, for example, $Fe^{2+}$ and $Mn^{3+}$, have been determined at frequencies higher than X- and Q-band frequencies. The parameters listed here are taken from the published papers as found by an extensive search of the relevant databases, covering the period 1993–2012 (inclusive). It is possible that some relevant references may have been inadvertently missed. In addition, only the experimentally determined SHPs have been listed, leaving out the theoretically calculated ones.

Figure 4.1 shows the elements that have been detected by EPR, included in the data tabulation. The SHPs for the various ions are organized in the following order: **($3d^n$) iron group**: $3d^0$ ($Sc^{3+}$, $Ti^{4+}$, $V^{5+}$), $3d^1$ ($VO^{2+}$, $Ti^{3+}$, $V^{4+}$, $Cr^{5+}$, $Mn^{6+}$), $3d^2$ ($Ti^{2+}$, $V^{3+}$, $Cr^{4+}$, $Mn^{5+}$), $3d^3$ ($V^{2+}$, $Cr^{3+}$, $Mn^{4+}$), $3d^4$ ($Cr^{2+}$, $Mn^{3+}$, $Fe^{4+}$), $3d^5$ ($Cr^+$, $Mn^{2+}$, $Fe^{3+}$, $Co^{4+}$, $Cr^+$), $3d^6$ ($Mn^+$, $Fe^{2+}$, $Co^{3+}$), $3d^7$ ($Mn^0$, $Fe^+$, $Co^{2+}$, $Ni^{3+}$, $Cu^{4+}$), $3d^8$ ($Fe^0$, $Co^+$, $Ni^{2+}$, $Cu^{3+}$), $3d^9$ ($Ni^+$, $Cu^{2+}$), $3d^{10}$ ($Cu^+$), $3d^{10}4s^1$ ($Cu^0$, $Zn^+$); **($4d^n$) palladium group**: $4d^0$ ($Zr^{4+}$, $Mo^{6+}$), $4d^1$ ($Y^{2+}$, $Zr^{3+}$, $Nb^{4+}$, $Mo^{5+}$), $4d^2$ ($Nb^{3+}$, $Ru^{6+}$), $4d^3$ ($Nb^{2+}$, $Mo^{3+}$, $Tc^{4+}$), $4d^4$ ($Ru^{4+}$), $4d^5$ ($Ru^{3+}$, $Rh^{4+}$), $4d^6$ ($Rh^{3+}$), $4d^7$ ($Ru^+$, $Rh^{2+}$), $4d^8$ ($Rh^+$), $4d^9$ ($Rh^0$, $Pd^+$, $Ag^{2+}$), $4d^{10}$ ($Ag^+$, $Cd^{2+}$), $4d^{10}5s^1$ ($Ag^0$, $Cd^+$, $Sn^{3+}$); **($4f^n$) lanthanide group**: $4f^0$ ($Ce^{4+}$), $4f^1$ ($La^{2+}$, $Ce^{3+}$), $4f^2$ ($Pr^{3+}$), $4f^3$ ($Nd^{3+}$), $4f^5$ ($Sm^{3+}$), $4f^7$ ($Eu^{2+}$, $Gd^{3+}$), $4f^8$ ($Tb^{3+}$), $4f^9$ ($Dy^{3+}$), $4f^{10}$ ($Ho^{3+}$), $4f^{11}$ ($Ho^{2+}$, $Er^{3+}$), $4f^{12}$ ($Tm^{3+}$), $4f^{13}$ ($Tm^{2+}$, $Yb^{3+}$); **($5d^n$) platinum group**: $5d^1$ ($Ta^{4+}$, $W^{5+}$, $Re^{6+}$), $5d^2$ ($Re^{5+}$), $5d^3$ ($Re^{4+}$), $5d^4$ ($Pt^{6+}$), $5d^5$ ($Ir^{4+}$), $5d^7$ ($Os^+$, $Ir^{2+}$, $Pt^{3+}$), $5d^9$ ($Tl^{2+}$, $Pb^{3+}$), $5d^{10}$ ($Tl^1$, $Pb^{2+}$), $5d^{10}6s^1$ ($Pb^+$); and **($5f^n$) actinide group**: $5f^1$ ($U^{5+}$), $5f^2$ (($PuO_2$)$^{2+}$, $U^{4+}$), $5f^3$ ($U^{3+}$, $Np^{4+}$), $5f^4$($U^{2+}$), $5f^5$ ($Am^{4+}$, $Pu^{3+}$), $5f^7$ ($Am^{2+}$, $Cm^{3+}$), $5f^9$ ($Cf^{3+}$). SHPs for some ions in this list may not have been reported in the literature, in cases where they are missing.

---

*Multifrequency Electron Paramagnetic Resonance: Data and Techniques,* First Edition.
Edited by Sushil K. Misra.
© 2014 Wiley-VCH Verlag GmbH & Co. KGaA. Published 2014 by Wiley-VCH Verlag GmbH & Co. KGaA.

| | | 3 | 4 | 5 | 6 | 7 | 8 | 9 | 10 | 11 | 12 |
|---|---|---|---|---|---|---|---|---|---|---|---|
| | | $^{21}$Sc | $^{22}$Ti | $^{23}$V | $^{24}$Cr | $^{25}$Mn | $^{26}$Fe | $^{27}$Co | $^{28}$Ni | $^{29}$Cu | $^{30}$Zn |
| | | $^{39}$Y | $^{40}$Zr | $^{41}$Nb | $^{42}$Mo | $^{43}$Tc | $^{44}$Ru | $^{45}$Rh | $^{46}$Pd | $^{47}$Ag | $^{48}$Cd |
| | | | | $^{73}$Ta | $^{74}$W | $^{75}$Re | $^{76}$Os | $^{77}$Ir | $^{78}$Pt | | |

| | | 3 | 4 | 5 | 6 | 7 | 8 |
|---|---|---|---|---|---|---|---|
| | | | | | | | |
| | | | | | $^{90}$Sn | | |
| | | $^{81}$Tl | $^{82}$Pb | | | | |

| $^{57}$La | $^{58}$Ce | $^{59}$Pr | $^{60}$Nd | | $^{62}$Sm | $^{63}$Eu | $^{64}$Gd | $^{65}$Tb | $^{66}$Dy | $^{67}$Ho | $^{68}$Er | $^{69}$Tm | $^{70}$Yb |
|---|---|---|---|---|---|---|---|---|---|---|---|---|---|
| | | $^{92}$U | $^{93}$Np | $^{94}$Pu | $^{95}$Am | $^{96}$Cm | | $^{98}$Cf | | | | | |

**Figure 4.1** Metal ions detected by EPR that are included in the data tabulations. (Adapted from S. K. Misra, Chap. VIII, Handbook of Electron Spin Resonance, vol. 2, AIP Press, Springer Verlag, New York, 1999 (Eds. C. P. Poole, Jr. and H. A. Farach).)

The temperatures at which the parameters have been determined are given in most cases; it would be either in K, or indicated as RT (room temperature ~295 K), or LNT (liquid-nitrogen temperature ~77 K), or LHT (liquid-helium temperature ~4 K). In some cases no temperature was reported, for which the temperature column is either blank or contains the notation "NR." Either the microwave frequency, for example, 9.23 GHz, is specified, or just the band, which is X in this case. In the columns for $\widetilde{g}, \widetilde{A}$ matrices, if three values are specified they indicate the three principle values, while if only one value is given, it represents the isotropic value (sometimes expressed as $g_{iso}$ or $A_{iso}$), unless otherwise indicated. For the cases of axial symmetry, $\|$ and $\perp$ subscripts indicate the values along and perpendicular to the symmetry axis. The units of the various zero-field splitting parameters, $b_\ell^m$, are $10^{-4}$ cm$^{-1}$, unless otherwise specified in the particular references from which they are quoted, indicated within or without parentheses. For the other units chosen, the following abbreviations have been used: G = Gauss, kG = kilo Gauss, T = Tesla (= $10^4$ G), mT = milli Tesla (= 10 G), GHz = giga Hertz, MHz = mega Hertz. The following conversion factors should relate all units: (GHz) = 29.9792458 (cm$^{-1}$), (MHz) = 28.02494 (mT). In the listing, the hosts for the various ions are arranged in alphabetical order of the first letter of the important element or group in them. Sometimes, for a similar group of hosts, they are listed together, notwithstanding the alphabetical order. Sometimes, glass hosts for an ion are grouped together. One should be able to locate easily the hosts of interest following these guidelines.

Some common notations used for denoting zero-field SHP (ZFSHP)s are as follows:

$$D = b_2^0 = 3B_2^0, \; E = \frac{b_2^2}{3} = B_2^2, \; b_4^m = 60B_4^m, \; b_6^m = 1260B_6^m.$$

For cubic symmetry, $a = \frac{2}{5}b_4^4$, where "a" describes the fourth-order term in the spin Hamiltonian: $\frac{a}{6}\left[S_x^4 + S_y^4 + S_z^4 - \frac{1}{5}S(S+1)(3S^2 + 3S - 1)\right]$, while for axial-symmetry, $F = 3b_4^0$, where "F" describes the fourth-order term in the spin-Hamiltonian,

$$\frac{F}{180}[35S_z^4 - 30S(S+1)S_z^2 + 25S_z^2 - 6S(S+1) + 3S^2(S+1)^2].$$

Here $S$ denotes the electronic spin of the ion. As for the hyperfine parameters for axial symmetry, the notations are $A = A_z$; $B = A_x = A_y$.

It is noted that the number of parameters varies for different spins. Accordingly, the value of $\ell$ in the parameter $b_\ell^m$ is even and $\ell \leq 2S$, with $|m| \leq \ell$. The point-group symmetry about an ion determines which of the possible $b_\ell^m$ parameters is nonzero. These nonzero parameters are coefficients of real and imaginary operators; in practice, most frequently the former are used, which are listed here. [More details on coeficients of imaginary operators are given in S. K. Misra, Chapter 7, Multifrequency Electron Paramagnetic Resonance: Theory and Applications (Wiley-VCH, Weinheim, Germany, 2011; Ed. S. K. Misra).] These are given in the table below:

| Spin Hamiltonian | Nonzero coefficients ($B_\ell^m$) of real operators $O_\ell^m$ |
|---|---|
| Triclinic | $B_2^0, B_2^1, B_2^2, B_4^0, B_4^1, B_4^2, B_4^3, B_4^4,$ $B_6^0, B_6^1, B_6^2, B_6^3, B_6^4, B_6^5, B_6^6$ |
| Monoclinic: $C_2 \| Z^a$ | $B_2^0, B_2^2, B_4^0, B_4^2, B_4^4, B_6^0, B_6^2, B_6^4, B_6^6$ |
| Monoclinic: $C_2 \| Y^a$ | $B_2^0, B_2^2, B_4^0, B_4^2, B_4^4, B_6^0, B_6^2, B_6^4, B_6^6,$ $B_2^1, B_4^1, B_4^3, B_6^1, B_6^3, B_6^5$ |
| Monoclinic: $C_2 \| X^a$ | $B_2^0, B_2^2, B_4^0, B_4^2, B_4^4, B_6^0, B_6^2, B_6^4, B_6^6$ |
| Orthorhombic | $B_2^0, B_2^2, B_4^0, B_4^2, B_4^4, B_6^0, B_6^2, B_6^4, B_6^6$ |
| Tetragonal | $B_2^0, B_4^0, B_4^4, B_6^0, B_6^4$ |
| Trigonal | $B_2^0, B_1^0, B_1^1, B_4^3, B_6^0, B_6^1, B_6^3, B_6^6$ |
| Hexagonal | $B_2^0, B_4^0, B_6^0, B_6^6$ |
| Cubic (fourfold symmetry axis) | $B_4^0 \left(= B_4 = \frac{b_4}{60}\right), B_4^4 (= 5B_4^0),$ $B_6^0 \left(= B_6 = \frac{b_6}{1260}\right), B_6^2,$ $B_6^4 (= -21 B_6^0), B_6^6 (= -B_6^2)$ |
| Cubic (threefold symmetry axis) | $B_4^0, B_4^3 (= \pm 20\sqrt{2} B_4^0), B_6^0,$ $B_6^3 \left(= \mp \left(\frac{35\sqrt{2} B_6^0}{4}\right)\right)$ |

[a] $C_2$ axis is parallel to any one of the magnetic X, Y, Z axes for monoclinic symmetry.

## 4.2
## Listing of Spin-Hamiltonian Parameters

($3d^n$) iron group
$3d^1$ ($Cr^{5+}$, $Ti^{3+}$, $V^{4+}$, $VO^{2+}$), $S = 1/2$
$Cr^{5+}$. Data tabulation of SHPs
$Cr^{5+}(3d^1)$

| Host | Frequency (GHz)/band | T (K) | $\tilde{g}$ | $A_x$, $A_y$, $A_z$ ($10^{-4}$ cm$^{-1}$) | References |
|---|---|---|---|---|---|
| D-ribose 5′-monophosphate ($R_5P$)/Cr(VI) substrate | 9.78 | 295 | $g_{iso}$ = 1.9788 | $^{53}$Cr: $A_{iso}$ = 17.42 G | [1] |
| Adenosine 5′-monophosphate (AMP)/Cr(VI) substrate | 9.78 | 295 | $g_{iso}$ = 1.9789 | $^{53}$Cr: $A_{iso}$ = 17.37 G | [1] |
| Cytidine 5′-monophosphate (CMP)/Cr(VI) substrate | 9.78 | 295 | $g_{iso}$ = 1.9789 | $^{53}$Cr: $A_{iso}$ = 17.35 G | [1] |
| 2′-Deoxythymidine 5′-monophosphate (dTMP)/Cr(VI) substrate | 9.78 | 295 | $g_{iso}$ = 1.9790 | NR | [1] |
| Silica xerogels | X | RT | $g_\parallel$ = 1.953  $g_\perp$ = 1.976 | — | [2] |

$Ti^{3+}$. Data tabulation of SHPs
$Ti^{3+}(3d^1)$

| Host | Frequency (GHz)/band | T (K) | $\tilde{g}$ | $A_x$, $A_y$, $A_z$ ($10^{-4}$ cm$^{-1}$) | References |
|---|---|---|---|---|---|
| $Al_2O_3$ | X | — | 1.999, 1.958, 1.941 | — | [3] |
| $Al_2O_3$ | X | — | 1.985, 1987, 1.985 | — | [3] |
| $C(NH_3)_2Ga(SO_4)_2 \cdot 6H_2O$ (GuGaSH) | X | 1.9 | $g_\parallel$ = 0.965  $g_\perp$ = 0 | — | [4] |
| $C(NH_3)_2Al(SO_4)_2 \cdot 6H_2O$ (GuAlSH) | X | 1.9 | $g_\parallel$ = 0 : 984  $g_\perp$ = 0 | — | [4] |
| $TiO_2$ | 9.4 | 5 | 1.9732, 1.9765, 1.9405 | −0.401, 0.616, −0.338 (MHz) | [5] |
| $TiO_2$ | 9.57 | 4–15 | 1.9746, 1.9782, 1.9430 | −0.23, 0.47, 5.15 (MHz) | [6] |
| $TiO_2$ (self-trapped hole center) | 9.57 | 4–15 | 2.0040, 2.0129, 2.0277 | — | [7] |
| $TiO_2$ (extrinsic impurity-related hole center) | 9.57 | 4–15 | 2.0036, 2.0182, 2.0307 | — | [7] |

(continued)

## ($Ti^{3+}$($3d^1$) listing – contd.)

| Host | Frequency (GHz)/band | T (K) | $\tilde{g}$ | $A_x, A_y, A_z$ ($10^{-4}$ cm$^{-1}$) | References |
|---|---|---|---|---|---|
| YAl$_3$(BO$_3$)$_3$ | X | 4.2 | 1.860, 1.795, 1.455 | — | [8] |
| ZrSiO$_4$ | X | 10 | $g_\parallel$ = 1.9269, $g_\perp$ = 1.9408 | ($^{47}$Ti) $A_\parallel$ = 2.9714 mT, $A_\perp$ = 0.8986 mT | [9] |
| ZrSiO$_4$ | X | 10 | $g_\parallel$ = 1.9269, $g_\perp$ = 1.9408 | ($^{49}$Ti) $A_\parallel$ = 2.9638 mT, $A_\perp$ = 0.8554 mT | [9] |
| ZrSiO$_4$ ($^{46,48,50}$Ti) | X | 10 | $g_\parallel$ = 1.9269, $g_\perp$ = 1.9408 | — | [9] |

## $V^{4+}$. Data tabulation of SHPs
## $V^{4+}$($3d^1$)

| Host | Frequency (GHz)/band | T (K) | $\tilde{g}$ | $A_x, A_y, A_z$ ($10^{-4}$ cm$^{-1}$) | References |
|---|---|---|---|---|---|
| GeO$_2$ | X | 92 | 1.9467, 1.9211, 1.9638 | 18.39, 37.27, 134.66 | [10] |
| α-LiTiOPO$_4$ | X | 290–400 | 1.964, 1.964, 1.934 | 58.9, 58.9, 172.8 | [11] |
| β-LiTiOPO$_4$ | X | 290–400 | 1.967, 1.967, 1.928 | 56.4, 56.4, 172.2 | [11] |
| Mg$_2$InV$_3$O$_{11}$ | X | 4–300 | $g_\parallel$ = 1.924, $g_\perp$ = 1.965 | $A_\parallel$ = 162, $A_\perp$ = 50.8 | [12] |
| NaTiOPO$_4$ | X | 290–800 | 1.978, 1.978, 1.934 | 56.4, 56.4, 172.3 | [11] |
| Pb$_{10}$(PO$_4$)$_{5.5}$(VO$_4$)$_{0.5}$(OH)$_2$ | 9.5 | 573 | $g_\parallel$ = 1.93, $g_\perp$ = 1.97 | $A_\parallel$ = 175.625, $A_\perp$ = 58.853 | [13] |
| Pb$_{10}$(PO$_4$)$_5$(VO$_4$)$_1$(OH)$_2$ | 9.5 | 573 | $g_\parallel$ = 1.94, $g_\perp$ = 1.97 | $A_\parallel$ = 174.69, $A_\perp$ = 57.919 | [13] |
| Pb$_{10}$(PO$_4$)$_4$(VO$_4$)$_2$(OH)$_2$ | 9.5 | 573 | $g_\parallel$ = 1.94, $g_\perp$ = 1.97 | $A_\parallel$ = 174.69, $A_\perp$ = 57.919 | [13] |
| Pb$_{10}$(PO$_4$)$_3$(VO$_4$)$_3$(OH)$_2$ | 9.5 | 573 | $g_\parallel$ = 1.94, $g_\perp$ = 1.97 | $A_\parallel$ = 177.49, $A_\perp$ = 56.98 | [13] |

(continued)

## ($V^{4+}(3d^1)$) listing – contd.

| Host | Frequency (GHz)/band | T (K) | $\tilde{g}$ | $A_x, A_y, A_z$ ($10^{-4}$ cm$^{-1}$) | References |
|---|---|---|---|---|---|
| $Pb_{10}(PO_4)_6(OH)_2$ | 9.5 | 573 | $g_\parallel = 1.93$, $g_\perp = 1.97$ | $A_\parallel = 163.481$, $A_\perp = 53.248$ | [13] |
| $RbTiOPO_4$ | 9.378 | 77–300 | $g_\parallel = 1.9305$, $g_\perp = 1.9565$ | $A_\parallel = -168.2$, $A_\perp = -54.3$ | [14] |
| $RbTiOPO_4$ | 9.378 | 77–300 | $g_\parallel = 1.9340$, $g_\perp = 1.9523$ | $A_\parallel = -169$, $A_\perp = -55.2$ | [14] |
| $RbZnF_3$ (V-doped, $V^{2+}$-center) | X | RT | 1.9658, 1.9658, 1.9658 | −87.0, −87.0, −87.0 | [15] |
| $RbZnF_3$ (V-doped, center A) | X | RT | 1.9682, 1.9682, 1.9327 | −66.3, −66.3, −178.92 | [15] |
| $RbZnF_3$ (V, Li-doped, center B) | X | RT | 1.9696, 1.9708, 1.9339 | −74.1, −41, −176.9 | [15] |
| $RbZnF_3$ (V, Li-doped, center C) | X | RT | 1.971, 1.971, 1.936 | −52, −73, −174 | [15] |
| $SnO_2$ | X | 92 | 1.939, 1.903, 1.943 | 21.2, 41.8, 140.1 | [10] |
| $TiO_2$ | X | 92 | 1.915, 1.9135, 1.9565 | 31.5, 43, 142 | [10] |
| $\alpha$-$TeO_2$ | X | 92 | 1.9821, 1.9011, 1.9491 | 53.03, 160.13, 52.07 | [10] |
| ($TiO_2$-anatase) interstitial species A | X | 4–298 | 1.962, 1.962, 1.936 | 50, 50, 177 | [16] |
| ($TiO_2$-anatase) substitutional species B | X | 4–298 | 1.97, 1.97, 1.922 | 40, 40, 175 | [16] |
| ($TiO_2$-anatase) substitutional species C | X | 4–298 | 1.98, 1.98, 1.896 | 50, 50, 182 | [16] |
| ($TiO_2$-anatase) substitutional species D | X | 4–298 | 1.95, 1.95, 1.918 | 55, 55, 186 | [16] |
| 3,4,5-TMVOTPP (crystal) (TMVOTPP, trimenthyloxyte-traphenylporphyri-noxovanadium) | X | 295 | $g_\parallel = 1.965$, $g_\perp = 1.995$ | $A_\parallel = 160.68$, $A_\perp = 57.92$ | [17] |

(continued)

## ($V^{4+}$($3d^1$) listing – contd.)

| Host | Frequency (GHz)/band | T (K) | $\tilde{g}$ | $A_x, A_y, A_z$ ($10^{-4}$ cm$^{-1}$) | References |
|---|---|---|---|---|---|
| 3,4,5-TMVOTPP* (powder) | X | 295 | $g_\| = 1.964$, $g_\perp = 1.994$ | $A_\| = 161.61$, $A_\perp = 58.85$ | [17] |
| 3,4,5-TMVOTPP* (crystal) | X | 120 | $g_\| = 1.964$, $g_\perp = 1.994$ | $A_\| = 160.68$, $A_\perp = 58.85$ | [17] |
| 3,4,5-TMVOTPP* (powder) | X | 120 | $g_\| = 1.965$, $g_\perp = 1.996$ | $A_\| = 162.55$, $A_\perp = 57.92$ | [17] |
| 3,4,5-TMVOTPP* (crystal) | X | 77 | $g_\| = 1.964$, $g_\perp = 1.994$ | $A_\| = 160.8$, $A_\perp = 58.85$ | [17] |
| 3,4,5-TMVOTPP* (powder) | X | 77 | $g_\| = 1.965$, $g_\perp = 1.996$ | $A_\| = 162.55$, $A_\perp = 57.92$ | [17] |
| 3,4,5-TMVOTPP* (crystal) *TMVOTPP = trimenthyloxytetraphenylporphyrinoxovanadium | X | 295 | $g_\| = 1.965$, $g_\perp = 1.995$ | $A_\| = 160.68$, $A_\perp = 57.92$ | [17] |
| Vanadium oxide nanotube (site V1) | X | 4.2 | $g_\| = 1.944$, $g_\perp = 1.978$ | $A_\| = 478$, $A_\perp = 165$ (MHz) | [18] |
| Vanadium oxide nanotube (site V2) | X | 4.2 | $g_\| = 1.950$, $g_\perp = 1.991$ | $A_\| = 492$, $A_\perp = 163$ (MHz) | [18] |
| Vanadium oxide nanotube (site V1) | X | 298 | $g_\| = 1.944$, $g_\perp = 1.978$ | $A_\| = 470$, $A_\perp = 165$ (MHz) | [18] |
| Vanadium oxide nanotube (site V2) | X | 298 | $g_\| = 1.950$, $g_\perp = 1.989$ | $A_\| = 492$, $A_\perp = 163$ (MHz) | [18] |
| $VO_2$ | 9.479–9.510 | 5 | $g_\| = 1.936$, $g_\perp = 1.931$ | $A_\| = 427$, $A_\perp = 126$ (MHz) | [19] |
| $V_2O_5$ gel | X | 4–298 | 1.973, 1.973, 1.926 | 73, 73, 195 | [16] |
| $V_2O_5$ : 1.6 $H_2O$ | 9.5 | 65 | $g_\| = 1.9390$, $g_\perp = 1.9810$ | $A_\| = 203.3$ G, $A_\perp = 75.2$ G | [20] |
| 0.003$V_2O_5$ 0.997 [$P_2O_5$ $Li_2O$] | 9.4 | 295 | $g_\| = 1.93$, $g_\perp = 1.98$ | $A_\| = 166$, $A_\perp = 56.7$ $P = 127.5$ | [21] |
| 0.005$V_2O_5$ 0.995 [$P_2O_5$ $Li_2O$] | 9.4 | 295 | $g_\| = 1.92$, $g_\perp = 1.97$ | $A_\| = 165$, $A_\perp = 56.7$ $P = 126.4$ | [21] |

(continued)

## $V^{4+}(3d^1)$ listing – contd.

| Host | Frequency (GHz)/band | T (K) | $\tilde{g}$ | $A_x, A_y, A_z$ ($10^{-4}$ cm$^{-1}$) | References |
|---|---|---|---|---|---|
| 0.01V$_2$O$_5$ 0.99 [P$_2$O$_5$ Li$_2$O] | 9.4 | 295 | $g_\parallel = 1.92$, $g_\perp = 1.99$ | $A_\parallel = 165$, $A_\perp = 61.0$, $P = 121.4$ | [21] |
| 0.03V$_2$O$_5$ 0.97 [P$_2$O$_5$ Li$_2$O] | 9.4 | 295 | $g_\parallel = 1.92$, $g_\perp = 1.98$ | $A_\parallel = 165$, $A_\perp = 58.3$, $P = 124.5$ | [21] |
| 0.05V$_2$O$_5$ 0.95 [P$_2$O$_5$ Li$_2$O] | 9.4 | 295 | $g_\parallel = 1.92$, $g_\perp = 1.98$ | $A_\parallel = 165$, $A_\perp = 58.3$, $P = 124.5$ | [21] |
| 0.1V$_2$O$_5$ 0.9 [P$_2$O$_5$ Li$_2$O] | 9.4 | 295 | $g_\parallel = 1.92$, $g_\perp = 1.98$ | $A_\parallel = 165$, $A_\perp = 55.2$, $P = 128.2$ | [21] |
| YVO$_4$ (site A) | 9.45 | 10 | 1.9667, 1.8912, 1.8610 | 48, 229, 441 MHz | [22] |
| YVO$_4$ (site B) | 9.45 | 10 | 1.9876, 1.9235, 1.8604 | 45, 98, 332 MHz | [22] |
| YVO$_4$ (site C) | 9.45 | 10 | 1.9642, 1.8701, 1.7380 | 63, 128, 343 MHz | [22] |

## $V^{4+}(3d^1)$ Glasses and solutions

| Host | Frequency (GHz or band) | T (K) | $\tilde{g}$ | $A_x, A_y, A_z$ ($10^{-4}$ cm$^{-1}$) | References |
|---|---|---|---|---|---|
| Sol-solution | 9 | 300 | $g = 1.963$ | $A = 110$ | [23] |
| Sol-solution | 9 | 77 | $g_\parallel = 1.930$, $g_\perp = 1.981$ | $A_\parallel = 187$, $A_\perp = 71$ | [23] |
| G-i (i = 1–9) gels (fresh, dried and heat treated) | 9 | 300 | $g_\parallel = 1.93$, $g_\perp = 1.980$ | $A_\parallel = 185$, $A_\perp = 70$ | [23] |
| G-i (i = 1–9) gels (fresh, dried and heat treated) | 9 | 77 | $g_\parallel = 1.93$, $g_\perp = 1.981$ | $A_\parallel = 187$, $A_\perp = 71$ | [23] |
| G-10 G-11 heat treated gel | 9 | 300 | $g_\parallel = 1.930$, $g_\perp = 1.980$ | $A_\parallel = 185$, $A_\perp = 70$ | [23] |
| G-10 G-11 heat treated gel | 9 | 77 | $g_\parallel = 1.930$, $g_\perp = 1.981$ | $A_\parallel = 187$, $A_\perp = 71$ | [23] |

(continued)

## ($V^{4+}(3d^1)$) glasses listing – contd.

| Host | Frequency (GHz or band) | T (K) | $\tilde{g}$ | $A_x, A_y, A_z$ ($10^{-4}$ cm$^{-1}$) | References |
|---|---|---|---|---|---|
| Gl-14 heat treated gel | 9 | 300 | $g = 1.96$ | — | [23] |
| Gl-14 heat treated gel | 9 | 77 | $g = 1.96$ | — | [23] |
| Gl-1 glass | 9 | 300 | $g_\| = 1.930$, $g_\perp = 1.980$ | $A_\| = 185$, $A_\perp = 70$ | [23] |
| Gl-1 glass | 9 | 77 | $g_\| = 1.930$, $g_\perp = 1.981$ | $A_\| = 187$, $A_\perp = 71$ | [23] |
| Gl-2 glass | 9 | 300 | $g = 1.960$ | — | [23] |
| Gl-2 glass | 9 | 77 | $g = 1.96$ | — | [23] |
| Gl-3 glass | — | 77 | 1.995, 1.920, 1.934 | $A_z = 33$, $A_x = 112$, $A_y = 98$ | [23] |
| 5MgO-25Li$_2$O-68B$_2$O$_3$-2V$_2$O$_5$ | X | RT | $g_\| = 1.945$, $g_\perp = 1.986$ | $A_\| = 182$, $A_\perp = 75$ | [24] |
| 10MgO-20Li$_2$O-68B$_2$O$_3$-2V$_2$O$_5$ | X | RT | $g_\| = 1.942$, $g_\perp = 1.982$ | $A_\| = 180$, $A_\perp = 75$ | [24] |
| 12MgO-18Li$_2$O-68B$_2$O$_3$-2V$_2$O$_5$ | X | RT | $g_\| = 1.940$, $g_\perp = 1.982$ | $A_\| = 180$, $A_\perp = 75$ | [24] |
| 15MgO-15Li$_2$O-68B$_2$O$_3$-2V2O5 | X | RT | $g_\| = 1.935$, $g_\perp = 1.980$ | $A_\| = 180$, $A_\perp = 80$ | [24] |
| 17MgO-13Li$_2$O-68B$_2$O$_3$-2V$_2$O$_5$ | X | RT | $g_\| = 1.931$, $g_\perp = 1.980$ | $A_\| = 175$, $A_\perp = 80$ | [24] |
| 5MgO-25Na$_2$O-68B$_2$O$_3$-2V2O5 | X | RT | $g_\| = 1.952$, $g_\perp = 1.988$ | $A_\| = 165$, $A_\perp = 70$ | [24] |
| 10MgO-20Na$_2$O-68B$_2$O$_3$-2V$_2$O$_5$ | X | RT | $g_\| = 1.954$, $g_\perp = 1.985$ | $A_\| = 170$, $A_\perp = 70$ | [24] |
| 12MgO-18Na$_2$O-68B$_2$O$_3$-2V$_2$O$_5$ | X | RT | $g_\| = 1.950$, $g_\perp = 1.982$ | $A_\| = 168$, $A_\perp = 65$ | [24] |
| 15MgO-15Na$_2$O-68B$_2$O$_3$-2V$_2$O$_5$ | X | RT | $g_\| = 1.950$, $g_\perp = 1.982$ | $A_\| = 160$, $A_\perp = 65$ | [24] |
| 17MgO-13Na$_2$O-68B$_2$O$_3$-2V$_2$O$_5$ | X | RT | $g_\| = 1.950$, $g_\perp = 1.980$ | $A_\| = 165$, $A_\perp = 65$ | [24] |
| 3MgO-27K$_2$O-68B$_2$O$_3$-2V$_2$O$_5$ | X | RT | $g_\| = 1.967$, $g_\perp = 1.994$ | $A_\| = 164$, $A_\perp = 69$ | [24] |
| 6MgO-24K$_2$O-68B$_2$O$_3$-2V$_2$O$_5$ | X | RT | $g_\| = 1.968$, $g_\perp = 1.990$ | $A_\| = 175$, $A_\perp = 72$ | [24] |

(continued)

## ($V^{4+}$ ($3d^1$) glasses listing – contd.)

| Host | Frequency (GHz or band) | T (K) | $\tilde{g}$ | $A_x, A_y, A_z$ ($10^{-4}$ cm$^{-1}$) | References |
|---|---|---|---|---|---|
| 9MgO-21K$_2$O-68B$_2$O$_3$-2V$_2$O$_5$ | X | RT | $g_\parallel = 1.954$, $g_\perp = 1.989$ | $A_\parallel = 179$, $A_\perp = 69$ | [24] |
| 12MgO-18K$_2$O-68B$_2$O$_3$-2V$_2$O$_5$ | X | RT | $g_\parallel = 1.953$, $g_\perp = 1.985$ | $A_\parallel = 183$, $A_\perp = 67$ | [24] |
| 0.5V$_2$O$_5 \cdot$ (99.5)[2P$_2$O$_5 \cdot$ Na$_{2O}$] | 9.4 | RT | $g_\parallel = 1.933$, $g_\perp = 1.989$ | $A_\parallel = 178.5$, $A_\perp = 68.3$ | [25] |
| V$_2$O$_5 \cdot$ (99)[2P$_2$O$_5 \cdot$ Na$_{2O}$] | 9.4 | RT | $g_\parallel = 1.932$, $g_\perp = 1.988$ | $A_\parallel = 170.6$, $A_\perp = 70.9$ | [25] |
| 3V$_2$O$_5 \cdot$ (97)[2P$_2$O$_5 \cdot$ Na$_{2O}$] | 9.4 | RT | $g_\parallel = 1.933$, $g_\perp = 1.992$ | $A_\parallel = 175.1$, $A_\perp = 68.1$ | [25] |
| 5V$_2$O$_5 \cdot$ (95)[2P$_2$O$_5 \cdot$ Na$_{2O}$] | 9.4 | RT | $g_\parallel = 1.93$, $g_\perp = 1.99$ | $A_\parallel = 179.6$, $A_\perp = 72.0$ | [25] |
| 10V$_2$O$_5 \cdot$ (90)[2P$_2$O$_5 \cdot$ Na$_{2O}$] | 9.4 | RT | $g_\parallel = 1.931$, $g_\perp = 1.989$ | $A_\parallel = 192.9$, $A_\perp = 75.4$ | [25] |
| 20V$_2$O$_5 \cdot$ (80)[2P$_2$O$_5 \cdot$ Na$_{2O}$] | 9.4 | RT | $g_\parallel = 1.927$, $g_\perp = 1.988$ | $A_\parallel = 184.6$, $A_\perp = 68.9$ | [25] |

## $VO^{2+}$ Data tabulation of SHPs
## $VO^{2+}$ ($3d^1$)

| Host | Frequency (GHz)/band | T (K) | $\tilde{g}$ | $A_x, A_y, A_z$ ($10^{-4}$ cm$^{-1}$) | References |
|---|---|---|---|---|---|
| 10BaO $\cdot$ 10Li$_2$O $\cdot$ 10Na$_2$O $\cdot$ 10K$_2$O $\cdot$ 58B$_2$O$_3 \cdot$ 2V$_2$O$_5$ Glass | X | RT | $g_\parallel = 1.950$, $g_\perp = 1.978$ | $A_\parallel = 175.95$, $A_\perp = 78.20$ | [26] |
| Ba$_2$Zn(HCOO)$_6$(H$_2$O)$_4$ (powder) Ba$_2$Zn(HCOO)$_6$(H$_2$O)$_4$ (single crystal) | 9.8 | 295 | $g_\parallel = 1.948$, $g_\perp = 1.983$ (1.996, 1.994, 1.955) | $A_\parallel = 174.69$, $A_\perp = 63.524$ (6.166, 6.072, 16.628) | [27] |
| BeAlSiO$_4$(OH) (natural green euclase) | X | RT | 1.9740, 1.9669, 1.9447 | 150, 163, 502 (MHz) | [28] |

(continued)

## $(VO^{2+}(3d^1)$ listing – contd.)

| Host | Frequency (GHz)/ band | T (K) | $\tilde{g}$ | $A_x, A_y, A_z$ $(10^{-4} cm^{-1})$ | References |
|---|---|---|---|---|---|
| Bis(Glycinato) · Mg$^{2+}$ monohydrate | X | RT | 2.1447, 1.9974, 1.9131 | 49, 60, 82 | [29] |
| Ca(HCOO)$_2$-Sr(HCOO)$_2$ | X | RT | $g_\| = 1.935$, $g_\perp = 1.980$ | $A_\| = 19.6$ mT, $A_\perp = 7.4$ mT | [30] |
| Ca(HCOO)$_2$-Sr(HCOO)$_2$ | X | RT | $g_\| = 1.942$, $g_\perp = 1.986$ | $A_\| = 19.0$ mT, $A_\perp = 7.0$ mT | [30] |
| Ca(picrate)$_2$ (2,2′ bipyridyl)$_2$ (site I) | X | RT | 1.993, 1.998, 1.942 | 74.0, 150.6, 158.4 | [31] |
| Ca(picrate)$_2$ (2,2′ bipyridyl)$_2$ (site II) | X | RT | 1.995, 1.968, 1.954 | 47.6, 108.5, 188.1 | [31] |
| Cd(HCOO)$_2$ · 2H$_2$O | X | RT | $g_\| = 1.943$, $g_\perp = 1.989$ | $A_\| = 19.4$ mT, $A_\perp = 7.3$ mT | [32] |
| CdKPO$_4$ · 6H$_2$O | X | 295 | $g_\| = 1.943$, $g_\perp = 1.989$ | $A_\| = 172.36$, $A_\perp = 81.37$ | [33, 34] |
| CdNaPO$_4$ · 6H$_2$O (powder) (site I) | X | RT | $g_\| = 1.922$, $g_\perp = 1.995$ | $A_\| = 19.86$ mT, $A_\perp = 8.24$ mT | [35] |
| CdNaPO$_4$ · 6H$_2$O (powder) (site II) | X | RT | $g_\| = 1.933$, $g_\perp = 1.996$ | $A_\| = 20.72$ mT, $A_\perp = 7.77$ mT | [35] |
| CdNaPO$_4$ · 6H$_2$O (crystal) | X | RT | $g_\| = 1.922$, $g_\perp = 1.980$ | $A_\| = 19.72$ mT, $A_\perp = 8.63$ mT | [35] |
| Cd (Cadmium) Ammonium Phosphate Hexahydrate | X | RT, LNT | $g_\| = 1.931$, $g_\perp = 1.993$ | $A_\| = 183$, $A_\perp = 72$ | [36] |
| C$_4$H$_8$N$_2$O$_3$ · H$_2$O L-asparagine monohydrate (site I) | X | RT | 1.9633, 2.0274, 1.9797 | 88, 61, 161 | [37] |
| C$_4$H$_8$N$_2$O$_3$ · H$_2$O L-asparagine monohydrate (site II) | X | RT | 1.9627, 1.9880, 1.9425 | 90, 66, 167 | [37] |
| [C$_6$H$_4$AsNo (H$_2$O)$_2$] (complex I) (site I) | X | 295 | $g_\| = 1.919$, $g_\perp = 1.995$ | $A_\| = 191.2$, $A_\perp = 67.8$, $P = 144.0$ | [38] |
| [C$_6$H$_4$AsNo (H$_2$O)$_2$] (complex I) (site II) | X | 295 | $g_\| = 1.920$, $g_\perp = 1.995$ | $A_\| = 185.7$, $A_\perp = 70.5$, $P = 135.0$ | [38] |

(continued)

## (VO$^{2+}$(3d$^1$) listing – contd.)

| Host | Frequency (GHz)/band | T (K) | $\tilde{g}$ | $A_x, A_y, A_z$ ($10^{-4}$ cm$^{-1}$) | References |
|---|---|---|---|---|---|
| [C$_6$H$_4$AsNo (H$_2$O)$_2$] (complex II) (site I) | X | 295 | $g_\| = 1.921$, $g_\perp = 1.996$ | $A_\| = 167.2$, $A_\perp = 47.0$ $P = 140.2$ | [38] |
| [C$_6$H$_4$AsNo (H$_2$O)$_2$] (complex II) (site II) | X | 295 | $g_\| = 1.921$, $g_\perp = 1.995$ | $A_\| = 160.0$, $A_\perp = 55.1$ $P = 122.5$ | [38] |
| C$_6$H$_5$O$_7$ · Na$_3$ · 2H$_2$O (site I) | X | 300 | 1.9680, 2.0053, 1.9124 | 79, 76, 185 | [39] |
| C$_6$H$_5$O$_7$ · Na$_3$ · 2H$_2$O (site II) | X | 300 | 1.9650, 2.0067, 1.9418 | 78, 70, 186 | [39] |
| [C(NH$_2$)$_3$]$_2$ UO$_2$ (SO$_4$)$_2$ · 3H$_2$O | 9.4 | LNT | $g_\| = 1.938$, $g_\perp = 1.983$ | $A_\| = 193$, $A_\perp = 76$ | [40] |
| Co(C$_3$H$_4$N$_2$)$_6$SO$_4$ · 4H$_2$O (Hexaimidazole cobalt sulfate) (site I) | X | 300 | 1.941, 1.995, 1.985 | 18.9, 7.5, 7.7 (mT) | [41] |
| Co(C$_3$H$_4$N$_2$)$_6$SO$_4$ · 4H$_2$O (hexaimidazole cobalt sulfate) (site II) | X | 300 | 1.941, 1.986, 1.979 | 18.6, 7.3, 7.7 (mT) | [41] |
| Co(C$_3$H$_4$N$_2$)$_6$SO$_4$ · 4H$_2$O (hexaimidazole cobalt sulfate) (powder) | X | 300 | $g_\| = 1.939$, $g_\perp = 1.995$ | $A_\| = 19.2$, $A_\perp = 7.7$ (mT) | [41] |
| ([Co(H$_2$O)$_4$(py)$_2$] (sac)$_2$ · 4H$_2$O) (site I) | X | RT | 2.001, 1.989, 1.938 | 73, 52, 187 (G) | [42] |
| ([Co(H$_2$O)$_4$(py)$_2$] (sac)$_2$ · 4H$_2$O) (site II) | X | RT | 2.001, 1.987, 1.923 | 71.7, 73.5, 199.1 (G) | [42] |
| (COOK)$_2$ · H$_2$O | 9.5 | 295 | 2.0153, 1.9489, 1.9155 | 63, 92, 193 | [43] |
| CsLiSO$_4$ (as grown) (site I) | X | 295 | $g_\| = 1.9225$, $g_\perp = 1.9792$ | $A_\| = 180.296$, $A_\perp = 70.250$ | [44] |
| CsLiSO$_4$ (as grown) (site II) | X | 295 | $g_\| = 1.9332$, $g_\perp = 1.9827$ | $A_\| = 179.455$, $A_\perp = 71.091$ | [44] |
| CsLiSO$_4$ (annealed at 403 K) (site III) | X | 295 | 1.9250, 1.9746, 1.9771 | 180.856, 68.382, 68.568 | [44] |

(continued)

## ($VO^{2+}$ ($3d^1$) listing – contd.)

| Host | Frequency (GHz)/ band | T (K) | $\tilde{g}$ | $A_x, A_y, A_z$ ($10^{-4}$ cm$^{-1}$) | References |
|---|---|---|---|---|---|
| $Cs_2S_2O_7$–$Cs_2SO_4$(sat)– $V_2O_5/SO_2$ | X | RT | $g_\| = 1.981$, $g_\perp = 1.940$ | $A_\| = 187$ G, $A_\perp = 65.3$ G | [45] |
| $KC_6 \cdot H_{11}O_7 \cdot H_2O$ (complex I) (site I) | X | RT | $g_\| = 1.953$, $g_\perp = 1.991$, $g_{iso} = 1.978$ | $A_\| = 175.2$, $A_\perp = 52.3$, $A_{iso} = 93.3$ | [46] |
| $KC_6H_{11}O_7 \cdot H_2O$ (complex I) (site II) | X | RT | $g_\| = 1.954$, $g_\perp = 1.991$, $g_{iso} = 1.978$ | $A_\| = 169.6$, $A_\perp = 50.9$, $A_{iso} = 90.5$ | [46] |
| $KC_6H_{11}O_7 \cdot H_2O$ (complex II) (site I) | X | RT | $g_\| = 1.943$, $g_\perp = 1.996$, $g_{iso} = 1.978$ | $A_\| = 168.6$, $A_\perp = 55.9$, $A_{iso} = 93.5$ | [46] |
| $KC_6H_{11}O_7 \cdot H_2O$ (complex II) (site II) | X | RT | $g_\| = 1.944$, $g_\perp = 1.995$, $g_{iso} = 1.978$ | $A_\| = 169.3$, $A_\perp = 49.4$, $A_{iso} = 89.3$ | [46] |
| $KC_6H_{11}O_7 \cdot H_2O$ (powder) | X | RT | $g_\| = 1.947$, $g_\perp = 2.000$, $g_{iso} = 1.982$ | $A_\| = 177.6$, $A_\perp = 58.4$, $A_{iso} = 98.1$ | [46] |
| $KH_2PO_4$ complex I | X | RT | $g_\| = 1.923$, $g_\perp = 1.998$ | $A_\| = 19.39$, $A_\perp = 7.61$ | [47] |
| $KH_2PO_4$ complex II | X | RT | $g_\| = 1.924$, $g_\perp = 1.992$ | $A_\| = 18.80$, $A_\perp = 7.53$ | [47] |
| $KH_2PO_4$ complex III | X | RT | $g_\| = 1.933$, $g_\perp = 1.998$ | $A_\| = 18.00$, $A_\perp = 8.17$ | [47] |
| $KH_2PO_4$ complex IV | X | RT | $g_\| = 1.935$, $g_\perp = 1.993$ | $A_\| = 18.00$, $A_\perp = 8.51$ | [47] |
| $KH_2PO_4$ powder, site X | X | RT | $g_\| = 1.925$, $g_\perp = 1.997$ | $A_\| = 19.06$, $A_\perp = 7.24$ | [47] |
| $KH_2PO_4$ powder, site Y | X | RT | $g_\| = 1.935$, $g_\perp = 1.997$ | $A_\| = 18.69$, $A_\perp = 7.24$ | [47] |
| $KH_3C_4O_8 \cdot 2H_2O$ single crystal, site X | X | RT | $g_\| = 1.931$, $g_\perp = 1.997$ | $A_\| = 18.41$, $A_\perp = 7.24$ | [47] |
| $KH_3C_4O_8 \cdot 2H_2O$ single crystal, site Y | X | RT | $g_\| = 1.938$, $g_\perp = 1.994$ | $A_\| = 18.58$, $A_\perp = 6.85$ | [47] |
| $KH_3C_4O_8 \cdot 2H_2O$ powder | X | RT | $g_\| = 1.929$, $g_\perp = 1.995$ | $A_\| = 19.01$, $A_\perp = 7.45$ | [47] |
| $KRbB_4O_7$ | X | 295 | $g_\| = 1.948$, $g_\perp = 1.979$ | $A_\| = 157$, $A_\perp = 59$ | [48] |

(continued)

## ($VO^{2+}(3d^1)$ listing – contd.)

| Host | Frequency (GHz)/band | $T$ (K) | $\tilde{g}$ | $A_x, A_y, A_z$ ($10^{-4}$ cm$^{-1}$) | References |
|---|---|---|---|---|---|
| $KZnClSO_4 \cdot 3H_2O$ (site I) | X | 300 | 1.998, 1.972, 1.920 | 61, 108, 170 | [49] |
| $KZnClSO_4 \cdot 3H_2O$ (site II) | X | 300 | 1.995, 1.983, 1.920 | 52, 114, 174 | [49] |
| $KZnClSO_4 \cdot 3H_2O$ (site I) | X | RT | 1.998, 1.972, 1.920 | 61, 108, 173 | [50] |
| $KZnClSO_4 \cdot 3H_2O$ (site II) | X | RT | 1.995, 1.983, 1.920 | 52, 114, 174 | [50] |
| $10K_2O \cdot 25PbO \cdot 62B_2O_3 \cdot 3V_2O_5$ | X | RT | $g_\| = 1.923$, $g_\perp = 1.941$ | $A_\| = -185.4$, $A_\perp = -80.2$ | [51] |
| $K_2S_2O_7-K_2SO_4$(sat)$-V_2O_5/SO_2$ | X | RT | $g_\| = 1.980$, $g_\perp = 1.930$ | $A_\| = 202$ G, $A_\perp = 71.7$ G | [45] |
| $K_2[Zn(C_3H_2O_4)_2(H_2O)_2] \cdot 2H_2O$ | X | 300 | 1.978, 1.972, 1.936 | 66.513, 62.870, 170.393 | [52] |
| L-alanine $C_3H_7NO_2$ (complex I) | X | RT | $g_\| = 1.928$, $g_\perp = 1.998$ | $A_\| = 190.5$, $A_\perp = 67.8$ | [53] |
| L-alanine $C_3H_7NO_2$ (complex II) | X | RT | $g_\| = 1.937$, $g_\perp = 2.001$ | $A_\| = 187.2$, $A_\perp = 70.7$ | [53] |
| L-alanine $C_3H_7NO_2$ (complex III) | X | RT | $g_\| = 1.935$, $g_\perp = 2.000$ | $A_\| = 185.1$, $A_\perp = 72.5$ | [53] |
| $LiCsSO_4$ | X | RT | 1.987, 1.983, 1.932 | 7.8, 7.4, 19.8 (mT) | [54] |
| $LiHC_2O_4 \cdot H_2O$ (site I) | 9.5 | RT | $g_\| = 1.9304$, $g_\perp = 2.0002$ | $A_\| = 198$, $A_\perp = 81$ | [55] |
| $LiHC_2O_4 \cdot H_2O$ (site II) | 9.5 | RT | $g_\| = 1.9262$, $g_\perp = 2.0001$ | $A_\| = 195$, $A_\perp = 76$ | [55] |
| $LiHC_2O_4 \cdot H_2O$ (site III) | 9.5 | RT | $g_\| = 1.9287$, $g_\perp = 2.0000$ | $A_\| = 186$, $A_\perp = 71$ | [55] |
| $LiHC_2O_4 \cdot H_2O$ (site IV) | 9.5 | RT | $g_\| = 1.9312$, $g_\perp = 2.0001$ | $A_\| = 191$, $A_\perp = 69$ | [55] |
| $LiKSO_4$ (site I) | X | RT | 2.0015, 1.9835, 1.9211 | 48, 64, 169 | [56] |
| $LiKSO_4$ (site II) | X | RT | 2.0019, 1.9796, 1.9225 | 48, 83, 178 | [56] |

(continued)

## ($VO^{2+}$($3d^1$) listing – contd.)

| Host | Frequency (GHz)/ band | T (K) | $\tilde{g}$ | $A_x, A_y, A_z$ ($10^{-4}$ cm$^{-1}$) | References |
|---|---|---|---|---|---|
| $LiKSO_4$ (crystal) (site I) | X | 295 | $g_\parallel = 1.919$, $g_\perp = 1.984$ | $A_\parallel = 181$, $A_\perp = 70$ | [57] |
| $LiKSO_4$ (crystal) (site II) | X | 295 | $g_\parallel = 1.91$, $g_\perp = 1.982$ | $A_\parallel = 182$, $A_\perp = 77$ | [57] |
| $LiKSO_4$ (crystal) (site III) | X | 295 | $g_\parallel = 1.935$ | $A_\parallel = 174$ | [57] |
| $LiKSO_4$ (powder) (site I) | X | 295 | $g_\parallel = 1.919$, $g_\perp = 1.977$ | $A_\parallel = 182$, $A_\perp = 77$ | [57] |
| $LiKSO_4$ (powder) (site II) | X | 295 | $g_\parallel = 1.930$, $g_\perp = 1.975$ | $A_\parallel = 174$, $A_\perp = 68$ | [57] |
| $LiNaSO_4$ | 9.45 | 300 | 1.983, 1.983, 1.932 | 68, 68, 175 | [58] |
| $LiNH_3OHSO_4$ (site 1) | | | 2.0249, 1.9698, 1.9552 | 51, 93, 165 | [59] |
| $LiNH_3OHSO_4$ (site 2) | | | 2.0267, 1.9743, 1.9213 | 40, 80, 155 | [59] |
| $LiRbB_4O_7$ | X | 295 | $g_\parallel = 1.948$, $g_\perp = 1.976$ | $A_\parallel = 157$, $A_\perp = 58$ | [48] |
| $LiRbSO_4$-$LiCsSO_4$ (complex I) | X | RT | $g_\parallel = 1.929$, $g_\perp = 1.987$ | $A_\parallel = 181.0$, $A_\perp = 69.5$ | [60] |
| $LiRbSO_4$-$LiCsSO_4$ (edge angle $\theta_z = 66°$) (complex II) | X | RT | 1.985, 1.989, 1.938 | 71.3, 78.9, 180.9 | [60] |
| $LiRbSO_4$-$LiCsSO_4$ (edge angle $\theta_a = 85°$) (complex II) | X | RT | 1.994, 1.986, 1.945 | 68.8, 74.2, 177.9 | [60] |
| $10Li_2O \cdot 25PbO \cdot 62B_2O_3 \cdot 3V_2O_5$ | X | RT | $g_\parallel = 1.923$, $g_\perp = 1.941$ | $A_\parallel = -185.4$, $A_\perp = -80.2$ | [51] |
| $MgKPO_4 \cdot 6H_2O$ (site I) | 9.1 | 77 | 2.1130, 2.1556, 2.4225 | 24, 38, 75 | [61] |
| $MgKPO_4 \cdot 6H_2O$ (site II) | 9.1 | 77 | 2.0258, 2.0916, 2.3748 | 21, 31, 76 | [61] |

(continued)

## ($VO^{2+}(3d^1)$) listing – contd.

| Host | Frequency (GHz)/band | T (K) | $\tilde{g}$ | $A_x, A_y, A_z$ ($10^{-4}$ cm$^{-1}$) | References |
|---|---|---|---|---|---|
| $MgNH_4PO_4\ 6H_2O$ | 9.09169 | 296 | $g_\| = 1.930$, $g_\perp = 1.980$ | $A_\| = 176.56$, $A_\perp = 70.99$ | [62] |
| $MgNH_4PO_4 \cdot 6H_2O$ (site I) | X | RT | $g_\| = 1.941$, $g_\perp = 1.994$ | $A_\| = 19.23$ mT, $A_\perp = 7.149$ mT | [63] |
| $MgNH_4PO_4 \cdot 6H_2O$ (site II) | X | RT | $g_\| = 1.946$, $g_\perp = 1.997$ | $A_\| = 19.07$ mT, $A_\perp = 7.29$ mT | [63] |
| $MgTl_2(SO_4)_2\ 6H_2O$ | 9.09169 | 296 | $g_\| = 1.939$, $g_\perp = 1.987$ | $A_\| = 178.43$, $A_\perp = 67.26$ | [62] |
| Na (Sodium) Formate (complex I) | X | RT | $g_\| = 1.930$, $g_\perp = 1.984$ | $A_\| = 177$, $A_\perp = 67$ | [64] |
| Na (Sodium Formate) (complex II) | X | RT | $g_\| = 1.939$, $g_\perp = 1.989$ | $A_\| = 168$, $A_\perp = 67$ | [64] |
| $NaRbB_4O_7$ | X | 295 | $g_\| = 1.948$, $g_\perp = 1.977$ | $A_\| = 157$, $A_\perp = 58$ | [48] |
| $Na_2C_4H_4O_6$ single crystal (complex 1, site I) | X | RT | $g_\| = 1.894$, $g_\perp = 1.984$ | $A_\| = 180.2$, $A_\perp = 78.0$ | [65] |
| $Na_2C_4H_4O_6$ single crystal (complex 1, site II) | X | RT | $g_\| = 1.895$, $g_\perp = 1.984$ | $A_\| = 180.0$, $A_\perp = 78.2$ | [65] |
| $Na_2C_4H_4O_6$ single crystal (complex 2, site I) | X | RT | $g_\| = 1.894$, $g_\perp = 1.977$ | $A_\| = 184.6$, $A_\perp = 68.0$, | [65] |
| $Na_2C_4H_4O_6$ single crystal (complex 2, site II) | X | RT | $g_\| = 1.894$, $g_\perp = 1.977$ | $A_\| = 186.1$, $A_\perp = 70.5$ | [65] |
| $Na_2C_4H_4O_6$ single crystal (complex 3, site I) | X | RT | $g_\| = 1.902$, $g_\perp = 1.991$ | $A_\| = 160.0$, $A_\perp = 74.4$ | [65] |
| $Na_2C_4H_4O_6$ single crystal (complex 3, site II) | X | RT | $g_\| = 1.900$, $g_\perp = 1.990$ | $A_\| = 168.0$, $A_\perp = 74.2$, | [65] |
| $Na_2C_4H_4O_6$ powder | X | RT | $g_\| = 1.922$, $g_\perp = 1.984$ | $A_\| = 183.0$, $A_\perp = 70.7$ | [65] |
| $10Na_2O \cdot 25PbO \cdot 62B_2O_3 \cdot 3V_2O_5$ | X | RT | $g_\| = 1.926$, $g_\perp = 1.942$ | $A_\| = -183.4$, $A_\perp = -79.7$ | [51] |
| $Na_2S_2O_7 - Na_2SO_4$(sat) – $V_2O_5/SO_2$ | X | RT | $g_\| = 1.982$, $g_\perp = 1.930$ | $A_\| = 200$ G, $A_\perp = 71.7$ G | [45] |

(continued)

## $(VO^{2+}(3d^1)$ listing – contd.)

| Host | Frequency (GHz)/band | T (K) | $\tilde{g}$ | $A_x, A_y, A_z$ ($10^{-4}$ cm$^{-1}$) | References |
|---|---|---|---|---|---|
| $Na_2Zn(SO_4)_2 \cdot 4H_2O$ (site I) | X | RT | 2.0059, 1.9639, 1.9311 | 50, 69, 172 | [66] |
| $Na_2Zn(SO_4)_2 \cdot 4H_2O$ (site II) | X | RT | 2.0182, 1.9462, 1.9334 | 25, 61, 189 | [66] |
| $Na_2Zn(SO_4)_2 \cdot 4H_2O$ (site III) | X | RT | 2.0083, 1.9563, 1.9374 | 43, 66, 182 | [66] |
| $Na_3C_6H_5O_7 \cdot 2H_2O$ (site I) | X | 295 | $g_\parallel = 1.935$, $g_\perp = 1.997$ | $A_\parallel = 180.3$, $A_\perp = 60.2$ | [67] |
| $Na_3C_6H_5O_7 \cdot 2H_2O$ (site II) | X | 295 | $g_\parallel = 1.935$, $g_\perp = 1.9640$ | $A_\parallel = -160.4$, $A_\perp = -57.8$ | [67] |
| $[(NH_2CH_2COOH)_2 \cdot CaCl_2 \cdot 4H_2O]$ (site I) | X | RT | 1.9976, 1.9631, 1.9336 | 59, 81, 185 | [68] |
| $[(NH_2CH_2COOH)_2 \cdot CaCl_2 \cdot 4H_2O]$ (site II) | X | RT | 1.9969, 1.9689, 1.9306 | 49, 83, 175 | [68] |
| $[(NH_4)_2Cd_3(SO_4)_4 \cdot 5H_2O]$ (site I) | X | RT | 1.988, 1.990, 1.950 | 79, 75, 197 (G) | [69] |
| $[(NH_4)_2Cd_3(SO_4)_4 \cdot 5H_2O]$ (site II) | X | RT | 1.994, 1.992, 1.943 | 77, 75, 199 (G) | [69] |
| $(NH_4)_2C_4H_4O_6$ (site I) | X | RT | 2.00, 1.98, 1.94 | 74, 84, 197 | [70] |
| $(NH_4)_2C_4H_4O_6$ (site II) | X | RT | 2.00, 1.98, 1.94 | 71, 77, 195 | [70] |
| $(NH_4)_2C_4H_4O_6$ (site III) | X | RT | 1.99, 1.98, 1.94 | 70, 80, 190 | [70] |
| $(NH_4)_2C_4H_4O_6$ (site IV) | X | RT | 1.99, 1.99, 1.94 | 70, 75, 188 | [70] |
| $((NH_4)_2C_4H_4O_6)$ (complex I) (site I) | X | RT | $g_\parallel = 1.9124$, $g_\perp = 1.9894$ | $A_\parallel = 183.08$, $A_\perp = 61.05$ | [71] |
| $((NH_4)_2C_4H_4O_6)$ (complex I) (site II) | X | RT | $g_\parallel = 1.9105$, $g_\perp = 1.9883$ | $A_\parallel = 182.08$, $A_\perp = 55.39$ | [71] |

(continued)

## ($VO^{2+}(3d^1)$) listing – contd.

| Host | Frequency (GHz)/band | T (K) | $\tilde{g}$ | $A_x, A_y, A_z$ ($10^{-4}$ cm$^{-1}$) | References |
|---|---|---|---|---|---|
| $((NH_4)_2C_4H_4O_6)$ (complex II) (site I) | X | RT | $g_\| = 1.9160$, $g_\perp = 1.9828$ | $A_\| = 178.39$, $A_\perp = 63.81$ | [71] |
| $((NH_4)_2C_4H_4O_6)$ (complex II) (site II) | X | RT | $g_\| = 1.918$, $g_\perp = 1.9844$ | $A_\| = 178.03$, $A_\perp = 71.38$ | [71] |
| $((NH_4)_2C_4H_4O_6)$ (complex III) (site I) | X | RT | $g_\| = 1.9140$, $g_\perp = 1.9866$ | $A_\| = 181.10$, $A_\perp = 69.00$ | [71] |
| $((NH_4)_2C_4H_4O_6)$ (complex III) (site II) | X | RT | $g_\| = 1.9139$, $g_\perp = 1.9866$ | $A_\| = 181.32$, $A_\perp = 66.07$ | [71] |
| $((NH_4)_2C_4H_4O_6)$ (complex IV) (site I) | X | RT | $g_\| = 1.9110$, $g_\perp = 1.9117$ | $A_\| = 186.90$, $A_\perp = 47.29$ | [71] |
| $((NH_4)_2C_4H_4O_6)$ (complex IV) (site II) | X | RT | $g_\| = 1.91117$, $g_\perp = 1.9942$ | $A_\| = 186.23$, $A_\perp = 49.20$ | [71] |
| $[(NH_4)HC_2O_4 \cdot 1/2H_2O]$ (complex I) (site I) | X | 295 | 1.999, 1.995, 1.947 | 59.79, 67.26, 194.31 | [72] |
| $[(NH_4)HC_2O_4 \cdot 1/2H_2O]$ (complex I) (site II) | X | 295 | 2.001, 1.991, 1.945 | 57.92, 78.47, 190.57 | [72] |
| $[(NH_4)HC_2O_4 \cdot 1/2H_2O]$ (complex II) (site I) | X | 295 | 2.001, 1.989, 1.944 | 46.71, 70.997, 187.77 | [72] |
| $[(NH_4)HC_2O_4 \cdot 1/2H_2O]$ (complex II) (site II) | X | 295 | 2.002, 1.971, 1.946 | 59.79, 74.73, 179.36 | [72] |
| $(NH_4)_2C_6H_6O_7$ (single crystal) (site I) | 9.52 | 295 | $g_\| = 1.933$, $g_\perp = 1.992$ (average) | $A_\| = 184.43$, $A_\perp = 80.833$, $A_{iso} = 115.3$ | [73] |
| $(NH_4)_2C_6H_6O_7$ (single crystal) (site II) | 9.52 | 295 | $g_\| = 1.933$, $g_\perp = 1.998$ (average) | $A_\| = 180.63$, $A_\perp = 83.73$, $A_{iso} = 116.0$ | [73] |
| $(NH_4)_2C_6H_6O_7$ (powder) | 9.52 | 295 | $g_\| = 1.937$, $g_\perp = 2.000$ (average) | $A_\| = 180.3$, $A_\perp = 60.7$, $A_{iso} = 99.6$ | [73] |
| $Pb(HCOO)_2$ | — | RT | $g_\| = 1.931$, $g_\perp = 1.996$ | $A_\| = 20.0$ mT, $A_\perp = 7.3$ mT | [74] |
| $Pb(HCOO)_2 \cdot 2H_2O$ | X | RT | $g_\| = 1.931$, $g_\perp = 1.996$ | $A_\| = 20.0$ mT, $A_\perp = 7.3$ mT | [32] |
| $35PbO \cdot B_2O_3 \cdot 3V_2O_5$ | X | RT | $g_\| = 1.924$, $g_\perp = 1.943$ | $A_\| = 81.6$, $A_\perp = -79.0$ | [51] |

(continued)

## $(VO^{2+}(3d^1))$ listing – contd.

| Host | Frequency (GHz)/band | T (K) | $\tilde{g}$ | $A_x, A_y, A_z$ $(10^{-4} \text{ cm}^{-1})$ | References |
|---|---|---|---|---|---|
| Poly(vinyl alcohol) (PVA): $VO^{2+}$ (2 mol% $V_2O_5$ content) | X | 295 | $g_\| = 1.927$, $g_\perp = 1.983$ | $A_\| = 63.0$, $A_\perp = 24.7$ $P = 124$ | [75] |
| PVA: $VO^{2+}$ (4 mol% $V_2O_5$ content) | X | 295 | $g_\| = 1.928$, $g_\perp = 19.84$ | $A_\| = 63.3$, $A_\perp = 25.0$ $P = 125$ | [75] |
| PVA: $VO^{2+}$ (5 mol% $V_2O_5$ content) | X | 295 | $g_\| - 1.928$, $g_\perp = 1.985$ | $A_\| - 62.7$, $A_\perp = 24$ $P = 126$ | [75] |
| PVA: $VO^{2+}$ (6 mol% $V_2O_5$ content) | X | 295 | $g_\| = 1.928$, $g_\perp = 1.984$ | $A_\| = 63.3$, $A_\perp = 25$ $P = 124$ | [75] |
| PVA: $VO^{2+}$ (7 mol% $V_2O_5$ content) | X | 295 | $g_\| = 1.928$, $g_\perp = 1.983$ | $A_\| = 63.0$, $A_\perp = 25.3$ $P = 123$ | [75] |
| PVA: $VO^{2+}$ (9 mol% $V_2O_5$ content) | X | 295 | $g_\| = 1.928$, $g_\perp = 1.982$ | $A_\| = 188$, $A_\perp = 76$ $P = 121$ | [75] |
| $RbHC_2O_4$ (site I) | X | 290 | 1.976, 1.978, 1.933 | 72, 66, 176 | [76] |
| $RbHC_2O_4$ (site II) | X | 290 | 1.976, 1.978, 1.936 | 72, 66, 175 | [76] |
| $RbHC_2O_4$ (site III) | X | 290 | 1.990, —, 1.935 | 74, —, 172 | [76] |
| $Rb_2Mg(SO_4)_2 \cdot 6H_2O$ | 9.09169 | 296 | $g_\| = 1.938$, $g_\perp = 1.986$ | $A_\| = 177.50$, $A_\perp = 67.26$ | [62] |
| $Rb_2S_2O_7-Rb_2SO_4$(sat)– $V_2O_5/SO_2$ | X | RT | $g_\| = 1.980$, $g_\perp = 1.930$ | $A_\| = 194$ G, $A_\perp = 67.1$ G | [45] |
| SHOMH ($NaHC_2O_4 \cdot H_2O$) (complex I) single crystal (SC) | X | 295 | $g_\| = 1.931$, $g_\perp = 1.999$ $g_{iso} = 1.976$ | $A_\| = 183.2$, $A_\perp = 65.4$ $A_{iso} = 104.6$ | [77] |
| SHOMH (complex II) SC | X | 295 | $g_\| = 1.935$, $g_\perp = 2.000$ $g_{iso} = 1.978$ | $A_\| = 188.8$, $A_\perp = 55.6$ $A_{iso} = 99.7$ | [77] |

(continued)

## (VO$^{2+}$(3d$^1$) listing – contd.)

| Host | Frequency (GHz)/band | T (K) | $\tilde{g}$ | $A_x, A_y, A_z$ ($10^{-4}$ cm$^{-1}$) | References |
|---|---|---|---|---|---|
| SHOMH (complex III) SC | X | 295 | $g_\| = 1.947$, $g_\perp = 1.987$, $g_{iso} = 1.973$ | $A_\| = 166.4$, $A_\perp = 44.8$, $A_{iso} = 85.2$ | [77] |
| SHOMH (complex IV) SC | X | 295 | $g_\| = 1.950$, $g_\perp = 1.987$, $g_{iso} = 1.974$ | $A_\| = 161.7$, $A_\perp = 42.5$, $A_{iso} = n74.0$ | [77] |
| SHOMH (powder complex I) | X | 295 | $g_\| = 1.932$, $g_\perp = 2.000$, $g_{iso} = 1.976$ | $A_\| = 182.2$, $A_\perp = 64.8$, $A_{iso} = 103.9$ | [77] |
| SHOMH (powder complex II) | X | 295 | $g_\| = 1.938$, $g_\perp = 2.000$, $g_{iso} = 1.977$ | $A_\| = 186.9$, $A_\perp = 67.0$, $A_{iso} = 196.9$ | [77] |
| SHOMH (powder complex III) | X | 295 | $g_\| = 1.942$, $g_\perp = 1.998$, $g_{iso} = 1.971$ | $A_\| = 166.4$, $A_\perp = 44.9$, $A_{iso} = 85.6$ | [77] |
| SHOMH (powder complex IV) | X | 295 | $g_\| = 1.945$, $g_\perp = 1.998$, $g_{iso} = 1.975$ | $A_\| = 158.9$, $A_\perp = 32.6$, $A_{iso} = 74.7$ | [77] |
| TiO$_2$ (anatase) | X | 290–800 | 1.980, 1.980, 1.935 | 74, 74, 202 (G) | [16] |
| Triaqua(1,10-phenanthroline-k$^2$N,N′)(sulfato-kO) magnesium(II) complex | — | — | 1.973, 1.972, 1.930 | 7.15, 6.77, 18.92 (mT) | [78] |
| [(VO)L$_I^{II}$] (powder) (L = thiosemicarbazone ligand) | X | 295 | $g_\| = 1.985$, $g_\perp = 1.993$, $g_0 = 1.990$ | — | [79] |
| [(VO)L$_{II}^{I}$] (powder) (L = thiosemicarbazone ligand) | X | 295 | $g_\| = 1.985$, $g_\perp = 1.996$, $g_0 = 1.992$ | — | [79] |
| [(VO)L$_I^{III}$] (powder) (L = thiosemicarbazone ligand) | X | 295 | $g_\| = 1.995$, $g_\perp = 2.000$, $g_0 = 1.998$ | — | [79] |

(*continued*)

## ($VO^{2+}(3d^1)$) listing – contd.)

| Host | Frequency (GHz)/band | T (K) | $\tilde{g}$ | $A_x, A_y, A_z$ ($10^{-4}$ cm$^{-1}$) | References |
|---|---|---|---|---|---|
| [(VO)L$_{III}$$^I$] (powder) (L = thiosemicarbazone ligand) | X | 295 | $g_{\|} = 1.995$, $g_\perp = 1.996$ $g_0 = 1.996$ | — | [79] |
| [(VO)L$_I$$^{II}$] (solution) (L = thiosemicarbazone ligand) | X | 295 | $g_0 = 1.995$ | — | [79] |
| [(VO)L$_{II}$$^I$] (solution) (L = thiosemicarbazone ligand) | X | 295 | $g_0 = 2.001$ | — | [79] |
| [(VO)L$_I$$^{III}$] (solution) (L = thiosemicarbazone ligand) | X | 295 | $g_0 = 1.997$ | — | [79] |
| [(VO)L$_{III}$$^I$] (solution) (L = thiosemicarbazone ligand) | X | 295 | $g_0 = 1.998$ | — | [79] |
| VO(acac)$_2$ coordinated supramolecular phthalocyanine in powder form | X | 295 | $g_{\|} = 1.955$, $g_\perp = 1.985$ | $A_{\|} = 23.354$, $A_\perp = 163.48$ | [80] |
| Zn[CS(NH$_2$)$_2$]$_3$SO$_4$ | X | RT | $g_{\|} = 1.969$, $g_\perp = 1.979$ | $A_{\|} = 166.08$, $A_\perp = 69.94$ | [81] |
| [Zn(CH$_3$CHOHCOO)$_2$ · 3H$_2$O)] | X | 295 | 1.9771, 2.0229, 1.9236 | 76, 104, 197 | [82] |
| [Zn(C$_3$H$_2$O$_4$) (C$_{10}$H$_8$N$_2$) (H$_2$O)$_2$] single crystal | X | RT | 1.968, 1.964, 1.928 | 7.54, 6.36, 18.81 (mT) | [83] |
| [Zn(C$_3$H$_2$O$_4$) (C$_{10}$H$_8$N$_2$) (H$_2$O)$_2$] powder | X | RT | $g_{\|} = 1.930$, $g_\perp = 1.973$ | $A_{\|} = 18.89$, $A_\perp = 7.56$ (mT) | [83] |
| [Zn(C$_3$H$_2$O$_4$)(H$_2$O)$_2$] (single crystal) | 9.084 | 295 | 1.998, 1.960, 1.924 | 69.129, 61.936, 170.674 | [84] |
| [Zn(C$_3$H$_2$O$_4$)(H$_2$O)$_2$] (powder) | 9.416 | 195 | 1.982, 1.982, 1.939 | 68.008, 68.008, 167.498 | [84] |

(continued)

## (VO$^{2+}$(3d$^1$) listing – contd.)

| Host | Frequency (GHz)/band | T (K) | $\tilde{g}$ | $A_x, A_y, A_z$ ($10^{-4}$ cm$^{-1}$) | References |
|---|---|---|---|---|---|
| Zn(C$_3$H$_3$O$_4$)$_2$(H$_2$O)$_2$ | X | 300 | 1.980, 1.972, 1.937 | 7.847, 5.698, 16.909 | [85] |
| [Zn(H$_2$O)$_6$] · [Zn(C$_3$H$_2$O$_4$)$_2$(H2O)$_2$] | X | RT | 1.981, 1.976, 1.941 | 7.96, 6.09, 17.8 (mT) | [86] |
| ZnKPO$_4$ · 6H$_2$O (site I) | X | RT | $g_\parallel = 1.9664$, $g_\perp = 1.9973$ | $A_\parallel = 150$, $A_\perp = 60$ | [87] |
| ZnKPO$_4$ · 6H$_2$O (site II) | X | RT | $g_\parallel = 1.9276$, $g_\perp = 1.9921$ | $A_\parallel = 155$, $A_\perp = 62$ | [87] |
| ZnKPO$_4$ 6H$_2$O | 9.09169 | 296 | $g_\parallel = 1.936$, $g_\perp = 1.976$ | $A_\parallel = 187.77$, $A_\perp = 70.99$ | [62] |
| ZnNH$_4$PO$_4$ 6H$_2$O | 9.09169 | 296 | $g_\parallel = 1.929$, $g_\perp = 1.979$ | $A_\parallel = 186.83$, $A_\perp = 74.73$ | [62] |
| ZnNH$_4$PO$_4$ · 6H$_2$O (site I) | 9.5 | RT | — | 62, 70, 172 | [88] |
| ZnNH$_4$PO$_4$ · 6H$_2$O (site II) | 9.5 | RT | — | 54, 60, 149 | [88] |
| 10ZnO · 10Li$_2$O · 10Na$_2$O · 10K$_2$O · 58B$_2$O$_3$ · 2V$_2$O$_5$ glass | X | RT | $g_\parallel = 1.941$, $g_\perp = 1.996$ | $A_\parallel = 181.50$, $A_\perp = 67.03$ | [26] |
| [Zn(sac)$_2$(py)$_2$] complex I (site I) | X | RT | 2.001, 1.993, 1.947 | 71.46, 52.89, 185.61 (G) | [89] |
| [Zn(sac)$_2$(py)$_2$] complex I (site II) | X | RT | 1.997, 2.002, 1.940 | 74.89, 37.64, 191.34 (G) | [89] |
| [Zn(sac)$_2$(py)$_2$] complex II (site I) | X | RT | 1.997, 2.002, 1.935 | 70.82, 55.09, 184.89 (G) | [89] |
| [Zn(sac)$_2$(py)$_2$] complex II (site II) | X | RT | 1.989, 1.999, 1.934 | 78.96, 55.94, 187.04 (G) | [89] |
| [Zn(sac)$_2$(py)$_2$] complex III (site I) | X | RT | 1.996, 2.002, 1.933 | 57.02, 80.35, 184.08 (G) | [89] |
| [Zn(sac)$_2$(py)$_2$] complex III (site II) | X | RT | 1.995, 2.002, 1.934 | 57.01, 80.36, 184.07 (G) | [89] |
| ZnTl$_2$(SO$_4$)$_2$ 6H$_2$O | 9.09169 | 296 | $g_\parallel = 1.939$, $g_\perp = 1.989$ | $A_\parallel = 178.42$, $A_\perp = 67.26$ | [62] |

## $VO^{2+}(3d^1)$ Glasses

| Host | Frequency (GHz)/band | T (K) | $\tilde{g}$ | $A_x, A_y, A_z$ ($10^{-4}$ cm$^{-1}$) | References |
|---|---|---|---|---|---|
| 40AgI · 21Ag$_2$O · 18V$_2$O$_5$ · 21P$_2$O$_5$ glass | X | 40 | 1.968, 1.968, 1.9255 | 6.3352, 6.3352, 1.7517 | [90] |
| 40Ag$_2$O · 40V$_2$O$_5$ · 20P$_2$O$_5$ glass | X | 40 | 1.968, 1.968, 1.9255 | 6.1516, 6.1516, 1.7517 | [90] |
| 10AgI · 40Ag$_2$O · 40V$_2$O$_5$ · 10P$_2$O$_5$ glass | X | 40 | 1.970, 1.970, 1.9300 | 5.744, 5.744, 1.6748 | [90] |
| 16AgI · 40Ag$_2$O · 40V$_2$O$_5$ · 4P$_2$O$_5$ glass | X | 40 | 1.966, 1.966, 1.9301 | 5.595, 5.595, 1.6718 | [90] |
| 0.3BaO · 0.7B$_2$O$_3$ (+2.0 mol% V$_2$O$_5$) glass | X | RT | $g_\parallel = 1.9398$, $g_\perp = 1.9712$ | $A_\parallel = 166.6$, $A_\perp = 58.4$ | [91] |
| 0.02Bi$_2$O$_3$ · 0.28BaO · 0.7B$_2$O$_3$ (+2.0 mol% V$_2$O$_5$) glass | X | RT | $g_\parallel = 1.9398$, $g_\perp = 1.9712$ | $A_\parallel = 166.6$, $A_\perp = 58.4$ | [91] |
| 0.04Bi$_2$O$_3$ · 0.26BaO · 0.7B$_2$O$_3$ (+2.0 mol% V$_2$O$_5$) glass | X | RT | $g_\parallel = 1.94$, $g_\perp = 1.972$ | $A_\parallel = 167.1$, $A_\perp = 58.5$ | [91] |
| 0.06Bi$_2$O$_3$ · 0.24BaO · 0.7B$_2$O$_3$ (+2.0 mol% V$_2$O$_5$) glass | X | RT | $g_\parallel = 1.941$, $g_\perp = 1.9732$ | $A_\parallel = 167.2$, $A_\perp = 59.4$ | [91] |
| 0.08Bi$_2$O$_3$ · 0.22BaO · 0.7B$_2$O$_3$ (+2.0 mol% V$_2$O$_5$) glass | X | RT | $g_\parallel = 1.9414$, $g_\perp = 1.9740$ | $A_\parallel = 167.2$, $A_\perp = 59.4$ | [91] |
| 0.1Bi$_2$O$_3$ · 0.2BaO · 0.7B$_2$O$_3$ (+2.0 mol% V$_2$O$_5$) glass | X | RT | $g_\parallel = 1.9414$, $g_\perp = 1.9740$ | $A_\parallel = 167.2$, $A_\perp = 59.9$ | [91] |
| 0.12Bi$_2$O$_3$ · 0.18BaO · 0.7B$_2$O$_3$ (+2.0 mol% V$_2$O$_5$) glass | X | RT | $g_\parallel = 1.9421$, $g_\perp = 1.9748$ | $A_\parallel = 167.3$, $A_\perp = 59.9$ | [91] |
| 0.15Bi$_2$O$_3$ · 0.15BaO · 0.7B$_2$O$_3$ (+2.0 mol% V$_2$O$_5$) glass | X | RT | $g_\parallel = 1.9421$, $g_\perp = 1.9750$ | $A_\parallel = 167.3$, $A_\perp = 59.9$ | [91] |
| $x$Bi2O$_3$ · $(70-x)$B2O$_3$ · 30Li$_2$O glasses | | | | | |
| 1. $x=0$ mol%; V$_2$O$_5$ = 2 mol% | 9.13 | 295 | 1. $g_\parallel = 1.9391$, $g_\perp = 1.9735$ | 1. $|A_\parallel| = 176.30$, $|A_\perp| = 61.27$ | [92] |
| 2. $x=2$ mol%; V$_2$O$_5$ = 2 mol% | | | 2. $g_\parallel = 1.9390$, $g_\perp = 1.9725$ | 2. $A_\parallel = 167.20$, $|A_\perp| = 61.24$ | |

(continued)

## ($VO^{2+}$($3d^1$) glasses listing – contd.)

| Host | Frequency (GHz)/band | T (K) | $\tilde{g}$ | $A_x, A_y, A_z$ ($10^{-4}$ cm$^{-1}$) | References |
|---|---|---|---|---|---|
| 3. $x = 4$ mol%; $V_2O_5 = 2$ mol% |  |  | 3. $g_\| = 1.9376$, $g_\perp = 1.9711$ | 3. $\|A_\|\| = 166.98$, $\|A_\perp\| = 61.19$ |  |
| 4. $x = 6$ mol%; $V_2O_5 = 2$ mol% |  |  | 4. $g_\| = 1.9361$, $g_\perp = 1.9696$ | 4. $\|A_\|\| = 166.77$, $\|A_\perp\| = 61.15$ |  |
| 5. $x = 8$ mol%; $V_2O_5 = 2$ mol% |  |  | 5. $g_\| = 1.9347$, $g_\perp = 1.9681$ | 5. $\|A_\|\| = 166.65$, $\|A_\perp\| = 61.10$ |  |
| 6. $x = 10$ mol%; $V_2O_5 = 2$ mol% |  |  | 6. $g_\| = 1.9347$, $g_\perp = 1.9681$ | 6. $\|A_\|\| = 166.65$, $\|A_\perp\| = 61.10$ |  |
| 7. $x = 12$ mol%; $V_2O_5 = 2$ mol% |  |  | 7. $g_\| = 1.9347$, $g_\perp = 1.9681$ | 7. $\|A_\|\| = 166.65$, $\|A_\perp\| = 61.10$ |  |
| 8. $x = 15$ mol%; $V_2O_5 = 2$ mol% |  |  | 8. $g_\| = 1.9347$, $g_\perp = 1.9681$ | 8. $\|A_\|\| = 166.65$, $\|A_\perp\| = 61.10$ |  |
| 9. $x = 20$ mol%; $V_2O_5 = 2$ mol% |  |  | 9. $g_\| = 1.9333$, $g_\perp = 1.9667$ | 9. $\|A_\|\| = 166.53$, $\|A_\perp\| = 61.06$ $A_\|$ and $A_\perp$ are negative (above) |  |
| $xBi_2O_3 \cdot (30-x)K_2O \cdot 70B_2O_3$ glasses |  |  |  |  |  |
| 1. $x = 0$ mol%; $V_2O_5 = 2$ mol% | 9.3 | 295 | 1. $g_\| = 1.9437$, $g_\perp = 1.9738$ | 1. $\|A_\|\| = 167.97$, $\|A_\perp\| = 58.98$ | [93] |
| 2. $x = 2$ mol%; $V_2O_5 = 2$ mol% |  |  | 2. $g_\| = 1.9437$, $g_\perp = 1.9738$ | 2. $A_\| = 167.97$, $\|A_\perp\| = 58.98$ |  |
| 3. $x = 4$ mol%; $V_2O_5 = 2$ mol% |  |  | 3. $g_\| = 1.9436$, $g_\perp = 1.9738$ | 3. $\|A_\|\| = 166.50$, $\|A_\perp\| = 58.98$ |  |
| 4. $x = 6$ mol%; $V_2O_5 = 2$ mol% |  |  | 4. $g_\| = 1.9429$, $g_\perp = 1.9738$ | 4. $\|A_\|\| = 166.45$, $\|A_\perp\| = 58.98$ |  |
| 5. $x = 8$ mol%; $V_2O_5 = 2$ mol% |  |  | 5. $g_\| = 1.9430$, $g_\perp = 1.9738$ | 5. $\|A_\|\| = 166.27$, $\|A_\perp\| = 58.98$ |  |
| 6. $x = 10$ mol%; $V_2O_5 = 2$ mol% |  |  | 6. $g_\| = 1.9431$, $g_\perp = 1.9738$ | 6. $\|A_\|\| = 166.01$, $\|A_\perp\| = 58.98$ |  |
| 7. $x = 12$ mol%; $V_2O_5 = 2$ mol% |  |  | 7. $g_\| = 1.9424$, $g_\perp = 1.9738$ | 7. $\|A_\|\| = 165.95$, $\|A_\perp\| = 58.98$ |  |
| 8. $x = 15$ mol%; $V_2O_5 = 2$ mol% |  |  | 8. $g_\| = 1.9424$, $g_\perp = 1.9738$ | 8. $\|A_\|\| = 165.95$, $\|A_\perp\| = 59.16$ $A_\|$ and $A_\perp$ are negative (above) |  |
| $0.02Bi_2O_3 \cdot 0.28PbO \cdot 0.70B_2O_3$ glass | 9.14 | 295 | $g_\| = 1.9365$, $g_\perp = 1.9780$ | $A_\| = 168.6$, $A_\perp = 64.6$ | [94] |
| $0.04Bi_2O_3 \cdot 0.26PbO \cdot 0.70B_2O_3$ glass | 9.14 | 295 | $g_\| = 1.9371$, $g_\perp = 1.9783$ | $A_\| = 168.7$, $A_\perp = 64.6$ | [94] |

(continued)

## ($VO^{2+}(3d^1)$) glasses listing – contd.

| Host | Frequency (GHz)/band | T (K) | $\tilde{g}$ | $A_x, A_y, A_z$ ($10^{-4}$ cm$^{-1}$) | References |
|---|---|---|---|---|---|
| $0.06Bi_2O_3 \cdot 0.24PbO \cdot 0.70B_2O_3$ glass | 9.14 | 295 | $g_\| = 1.9377$, $g_\perp = 1.9789$ | $A_\| = 168.7$, $A_\perp = 64.7$ | [94] |
| $0.08Bi_2O_3 \cdot 0.22PbO \cdot 0.70B_2O_3$ glass | 9.14 | 295 | $g_\| = 1.9382$, $g_\perp = 1.9794$ | $A_\| = 168.8$, $A_\perp = 64.7$ | [94] |
| $0.10Bi_2O_3 \cdot 0.20PbO \cdot 0.70B_2O_3$ glass | 9.14 | 295 | $g_\| = 1.9291$, $g_\perp = 1.9797$ | $A_\| = 168.8$, $A_\perp = 64.7$ | [94] |
| $0.12Bi_2O_3 \cdot 0.18PbO \cdot 0.70B_2O_3$ glass | 9.14 | 295 | $g_\| = 1.9397$, $g_\perp = 1.9806$ | $A_\| = 168.9$, $A_\perp = 64.7$ | [94] |
| $0.15Bi_2O_3 \cdot 0.15PbO \cdot 0.70B_2O_3$ glass | 9.14 | 295 | $g_\| = 1.9399$, $g_\perp = 1.9815$ | $A_\| = 168.9$, $A_\perp = 64.8$ | [94] |
| Calcium aluminum borate glass | X | 295 | $g_\| = 1.928$, $g_\perp = 1.948$ | $A_\| = 185$, $A_\perp = 76$ | [95] |
| | X | 273 | $g_\| = 1.928$, $g_\perp = 1.947$ | $A_\| = 185$, $A_\perp = 77$ | [95] |
| | X | 243 | $g_\| = 1.929$, $g_\perp = 1.947$ | $A_\| = 184$, $A_\perp = 77$ | [95] |
| | X | 213 | $g_\| = 1.928$, $g_\perp = 1.947$ | $A_\| = 185$, $A_\perp = 78$ | [95] |
| | X | 183 | $g_\| = 1.929$, $g_\perp = 1.947$ | $A_\| = 185$, $A_\perp = 77$ | [95] |
| | X | 153 | $g_\| = 1.929$, $g_\perp = 1.947$ | $A_\| = 184$, $A_\perp = 77$ | [95] |
| | X | 123 | $g_\| = 1.929$, $g_\perp = 1.947$ | $A_\| = 183$, $A_\perp = 76$ | [95] |
| $5CaO \cdot 15SrO \cdot 19.9Na_2O \cdot 60B_2O_3 \cdot 0.1V_2O_5$ glass | X | RT | $g_\| = 1.9468$, $g_\perp = 1.9875$ | $A_\| = 165$, $A_\perp = 52$ | [96] |
| $10CaO \cdot 10SrO \cdot 19.9Na_2O \cdot 60B_2O_3 \cdot 0.1V_2O_5$ glass | X | RT | $g_\| = 1.9431$, $g_\perp = 1.9879$ | $A_\| = 163$, $A_\perp = 57$ | [96] |
| $15CaO \cdot 5SrO \cdot 19.9Na_2O \cdot 60B_2O_3 \ 0.1V_2O_5$ glass | X | RT | $g_\| = 1.9443$, $g_\perp = 1.9871$ | $A_\| = 155$, $A_\perp = 47$ | [96] |
| $CoO.(0.4-x) \cdot PbO.0.6B_2O_3$ (x = 0-0.2) | X | RT | $g_\| = 1.922\text{-}1.927$, $g_\perp = 1.968\text{-}1.973$ | $A_\| = 166.9\text{-}165.1$, $A_\perp = 58.8\text{-}62.6$ | [97] |

*(continued)*

## ($VO^{2+}$ ($3d^1$) glasses listing – contd.)

| Host | Frequency (GHz)/ band | T (K) | $\tilde{g}$ | $A_x, A_y, A_z$ ($10^{-4}$ cm$^{-1}$) | References |
|---|---|---|---|---|---|
| $x$CoO-$(0.03-x)$Li$_2$O-0.7B$_2$O$_3$ glass with 1.0 mol% V$_2$O$_5$ ($x=0.0$–0.6) glasses | X | RT | $g_\| = 1.9404$–1.9422, $g_\perp = 1.9741$–1.9760 | $A_\| = 169.9$–170.1, $A_\perp = 58.6$–60.0 | [98] |
| $x$CoO-$(0.03-x)$Li$_2$O-0.7B$_2$O$_3$ glass with 2.0 mol% V$_2$O$_5$ ($x=0.0$–0.1) glasses | X | RT | $g_\| = 1.9401$–1.9413, $g_\perp = 1.9721$–1.9764 | $A_\| = 169.5$–169.8, $A_\perp = 58.8$–59.7 | [98] |
| $x$CoO-$(0.03-x)$K$_2$O-0.7B$_2$O$_3$ glass with 1.0 mol% V$_2$O$_5$ ($x=0.0$) glass | X | RT | $g_\| = 1.9422$, $g_\perp = 1.9751$ | $A_\| = 168.9$, $A_\perp = 59.0$ | [98] |
| $x$CoO-$(0.03-x)$K$_2$O-0.7B$_2$O$_3$ glass with 2.0 mol% V$_2$O$_5$ ($x=0.0$) glass | X | RT | $g_\| = 1.9422$, $g_\perp = 1.9734$ | $A_\| = 168.5$, $A_\perp = 57.9$ | [98] |
| 0.02GeO$_2\cdot$0.28K$_2$O$\cdot$0.70B$_2$O$_3$ glass | 9.14 | 295 | $g_\| = 1.9370$, $g_\perp = 1.9657$ | $A_\| = -161.0$, $A_\perp = -57.8$ | [99] |
| 0.04GeO$_2\cdot$0.26K$_2$O$\cdot$0.70B$_2$O$_3$ glass | 9.14 | 295 | $g_\| = 1.9355$, $g_\perp = 1.9660$ | $A_\| = -160.8$, $A_\perp = -57.8$ | [99] |
| 0.06GeO$_2\cdot$0.24K$_2$O$\cdot$0.70B$_2$O$_3$ glass | 9.14 | 295 | $g_\| = 1.9338$, $g_\perp = 1.9651$ | $A_\| = -160.7$, $A_\perp = -57.8$ | [99] |
| 0.08GeO$_2\cdot$0.22K$_2$O$\cdot$0.70B$_2$O$_3$ glass | 9.14 | 295 | $g_\| = 1.9327$, $g_\perp = 1.9645$ | $A_\| = -160.6$, $A_\perp = -57.8$ | [99] |
| 0.10GeO$_2\cdot$0.20K$_2$O$\cdot$0.70B$_2$O$_3$ glass | 9.14 | 295 | $g_\| = 1.9313$, $g_\perp = 1.9643$ | $A_\| = -160.5$, $A_\perp = -57.8$ | [99] |
| 0.12GeO$_2\cdot$0.18K$_2$O$\cdot$0.70B$_2$O$_3$ glass | 9.14 | 295 | $g_\| = 1.9298$, $g_\perp = 1.9643$ | $A_\| = -160.4$, $A_\perp = -57.8$ | [99] |
| 0.15GeO$_2\cdot$0.15K$_2$O$\cdot$0.70B$_2$O$_3$ glass | 9.14 | 295 | $g_\| = 1.9298$, $g_\perp = 1.998$ | $A_\| = 181.3$, $A_\perp = 59.1$ | [99] |
| 0.02GeO$_2\cdot$0.28Li$_2$O$\cdot$0.70B$_2$O$_3$ glass | 9.14 | 295 | $g_\| = 1.9327$, $g_\perp = 1.9634$ | $A_\| = -160.6$, $A_\perp = -59.6$ | [99] |
| 0.04GeO$_2\cdot$0.26Li$_2$O$\cdot$0.70B$_2$O$_3$ glass | 9.14 | 295 | $g_\| = 1.9327$, $g_\perp = 1.9634$ | $A_\| = -160.6$, $A_\perp = -59.6$ | [99] |
| 0.06GeO$_2\cdot$0.24Li$_2$O$\cdot$0.70B$_2$O$_3$ glass | 9.14 | 295 | $g_\| = 1.9327$, $g_\perp = 1.9634$ | $A_\| = -160.6$, $A_\perp = -59.6$ | [99] |

(continued)

## ($VO^{2+}(3d^1)$) glasses listing – contd.)

| Host | Frequency (GHz)/band | T (K) | $\tilde{g}$ | $A_x, A_y, A_z$ ($10^{-4}$ cm$^{-1}$) | References |
|---|---|---|---|---|---|
| $0.08GeO_2 \cdot 0.22Li_2O \cdot 0.70B_2O_3$ glass | 9.14 | 295 | $g_\parallel = 1.9321$, $g_\perp = 1.9628$ | $A_\parallel = -160.6$, $A_\perp = -59.6$ | [99] |
| $0.10GeO_2 \cdot 0.20Li_2O \cdot 0.70B_2O_3$ glass | 9.14 | 295 | $g_\parallel = 1.9321$, $g_\perp = 1.9628$ | $A_\parallel = -160.6$, $A_\perp = -59.6$ | [99] |
| $0.12GeO_2 \cdot 0.28Li_2O \cdot 0.70B_2O_3$ glass | 9.14 | 295 | $g_\parallel = 1.9321$, $g_\perp = 1.9628$ | $A_\parallel = -160.6$, $A_\perp = -59.6$ | [99] |
| $0.15GeO_2 \cdot 0.15Li_2O \cdot 0.70B_2O_3$ glass | 9.14 | 295 | $g_\parallel = 1.9315$, $g_\perp = 1.9628$ | $A_\parallel = -160.5$, $A_\perp = -59.3$ | [99] |
| KBaP glass (10 $K_2O$ + 27$BaCO_3$ + 60 $P_2O_5$ + 3$V_2O_5$) | X | RT | $g_\parallel = 1.924$, $g_\perp = 1.977$ | $A_\parallel = 178$, $A_\perp = 68$ | [100] |
| KCaB glass | 9.205 | RT | $g_\parallel = 1.939$, $g_\perp = 1.969$ | $|A_\parallel| = 161$, $|A_\perp| = 50$ | [101] |
| KPbBTe: (potassium lead borotellurite) glass | X | 295 | $g_\parallel = 1.979$, $g_\perp = 1.988$ | $A_\parallel = 153$, $A_\perp = 63$ | [102] |
| 30$K_2O \cdot$ 69$B_2O_3$ glass | 9.14 | RT | $g_\parallel = 1.943$, $g_\perp = 1.981$ | $A_\parallel = 166$, $A_\perp = 61.5$ | [103] |
| 0.30$K_2O \cdot$ 0.70$B_2O_3$ glass | 9.14 | 295 | $g_\parallel = 1.9370$, $g_\perp = 1.9663$ | $A_\parallel = -161.0$, $A_\perp = -57.8$ | [99] |
| 10$K_2SO_4$ – 40 $Na_2SO_4$ – 50$ZnSO_4$ | X | RT | $g_\parallel = 1.927$, $g_\perp = 1.977$ | $A_\parallel = 181$, $A_\perp = 71$ | [104] |
| 25$K_2SO_4$ – 25 $Na_2SO_4$ – 50$ZnSO_4$ | X | RT | $g_\parallel = 1.928$, $g_\perp = 1.977$ | $A_\parallel = 181$, $A_\perp = 71$ | [104] |
| 30$K_2SO_4$ – 20 $Na_2SO_4$ – 50$ZnSO_4$ | X | RT | $g_\parallel = 1.929$, $g_\perp = 1.976$ | $A_\parallel = 180$, $A_\perp = 70$ | [104] |
| 40$K_2SO_4$ – 10 $Na_2SO_4$ – 50$ZnSO_4$ | X | RT | $g_\parallel = 1.930$, $g_\perp = 1.975$ | $A_\parallel = 180$, $A_\perp = 70$ | [104] |
| LiBaP glass (10$Li_2O$ + 27$BaCO_3$ + 60 $P_2O_5$+3$V_2O_5$) | X | RT | $g_\parallel = 1.920$, $g_\perp = 1.974$ | $A_\parallel = 180$, $A_\perp = 68$ | [100] |
| 20LiF + 79 $B_2O_3$ + 1 $V_2O_5$ | X | RT | $g_\parallel = 1.9565$, $g_\perp = 1.9890$ | $A_\parallel = 176$, $A_\perp = 59$ | [105] |
| 30LiF + 69 $B_2O_3$ + 1 $V_2O_5$ | X | RT | $g_\parallel = 1.9510$, $g_\perp = 1.9873$ | $A_\parallel = 175$, $A_\perp = 58$ | [105] |
| 40LiF + 59 $B_2O_3$ + 1$V_2O_5$ | X | RT | $g_\parallel = 1.9470$, $g_\perp = 1.9849$ | $A_\parallel = 174$, $A_\perp = 58$ | [105] |

(continued)

## ($VO^{2+}$ ($3d^1$) glasses listing – contd.)

| Host | Frequency (GHz)/ band | T (K) | $\tilde{g}$ | $A_x, A_y, A_z$ ($10^{-4}$ cm$^{-1}$) | References |
|---|---|---|---|---|---|
| 50LiF + 49 $B_2O_3$ + 1$V_2O_5$ | X | RT | $g_\parallel = 1.9444$<br>$g_\perp = 1.9824$ | $A_\parallel = 172$,<br>$A_\perp = 57$ | [105] |
| LiCa B glass | 9.205 | RT | $g_\parallel = 1.941$,<br>$g_\perp = 1.966$ | $\|A_\parallel\| = 158$,<br>$\|A_\perp\| = 54$ | [101] |
| LiPbBTe (lithium lead borotellurite) glass | X | 295 | $g_\parallel = 1.940$,<br>$g_\perp = 1.953$ | $A_\parallel = 164$,<br>$A_\perp = 59$ | [102] |
| 15$Li_2O$ · 5BaO · 80$B_2O_3$ glass | 9.12 | 295 | $g_\parallel = 1.937$,<br>$g_\perp = 1.978$ | $A_\parallel = 171.8$,<br>$A_\perp = 64.6$ | [107] |
| 15$Li_2O$ · 10BaO · 75$B_2O_3$ glass | 9.12 | 295 | $g_\parallel = 1.938$,<br>$g_\perp = 1.977$ | $A_\parallel = 170.7$,<br>$A_\perp = 63.8$ | [107] |
| 15$Li_2O$ · 15BaO · 70$B_2O_3$ glass | 9.12 | 295 | $g_\parallel = 1.939$,<br>$g_\perp = 1.977$ | $A_\parallel = 168.3$,<br>$A_\perp = 61.4$ | [107] |
| 15$Li_2O$ · 20BaO · 65$B_2O_3$ glass | 9.12 | 295 | $g_\parallel = 1.941$,<br>$g_\perp = 1.977$ | $A_\parallel = 166.7$,<br>$A_\perp = 60.9$ | [107] |
| 30$Li_2O$ · 5BaO · 65$B_2O_3$ glass | 9.12 | 295 | $g_\parallel = 1.940$,<br>$g_\perp = 1.976$ | $A_\parallel = 166.6$,<br>$A_\perp = 60.9$ | [107] |
| 25$Li_2O$ · 10BaO · 65$B_2O_3$ glass | 9.12 | 295 | $g_\parallel = 1.940$,<br>$g_\perp = 1.976$ | $A_\parallel = 166.6$,<br>$A_\perp = 60.9$ | [107] |
| 20$Li_2O$ · 15BaO · 65$B_2O_3$ glass | 9.12 | 295 | $g_\parallel = 1.940$,<br>$g_\perp = 1.976$ | $A_\parallel = 166.6$,<br>$A_\perp = 60.9$ | [107] |
| 10$Li_2O$ · 25BaO · 65$B_2O_3$ glass | 9.12 | 295 | $g_\parallel = 1.940$,<br>$g_\perp = 1.976$ | $A_\parallel = 166.6$,<br>$A_\perp = 60.9$ | [107] |
| 5$Li_2O$ · 30BaO · 65$B_2O_3$ glass | 9.12 | 295 | $g_\parallel = 1.940$,<br>$g_\perp = 1.976$ | $A_\parallel = 166.6$,<br>$A_\perp = 60.9$ | [107] |
| 15$Li_2O$ · 15$K_2O$ · 10$Bi_2O_3$ · 55$B_2O_3$: 5$V_2O_5$ glass | 9.3 | RT | $g_\parallel = 1.952$,<br>$g_\perp = 1.992$ | $A_\parallel = 157.9$,<br>$A_\perp = 64.30$ | [108] |
| 15$Li_2O$ · 15$K_2O$ · 10$Bi_2O_3$ · 53$B_2O_3$: 7$V_2O_5$ glass | 9.3 | RT | $g_\parallel = 1.959$,<br>$g_\perp = 1.992$ | $A_\parallel = 152.1$,<br>$A_\perp = 38.33$ | [108] |
| 15$Li_2O$ · 15$K_2O$ · 10$Bi_2O_3$ · 51$B_2O_3$: 9$V_2O_5$ glass | 9.3 | RT | $g_\parallel = 1.958$,<br>$g_\perp = 1.991$ | $A_\parallel = 182.9$,<br>$A_\perp = 43.05$ | [108] |
| 15$Li_2O$ · 15$K_2O$ · 3$Bi_2O_3$ · 62$B_2O_3$/5$V_2O_5$ glass | 9.3 | RT | $g_\parallel = 1.928$,<br>$g_\perp = 1.985$ | $A_\parallel = 176.27$,<br>$A_\perp = 65.85$ | [109] |
| 15$Li_2O$ · 15$K_2O$ · 5$Bi_2O_3$ · 60$B_2O_3$/5$V_2O_5$ glass | 9.3 | RT | $g_\parallel = 1.956$,<br>$g_\perp = 1.989$ | $A_\parallel = 159.18$,<br>$A_\perp = 65.04$ | [109] |

(continued)

## ($VO^{2+}(3d^1)$) glasses listing – contd.

| Host | Frequency (GHz)/band | T (K) | $\tilde{g}$ | $A_x, A_y, A_z$ ($10^{-4}$ cm$^{-1}$) | References |
|---|---|---|---|---|---|
| $15Li_2O \cdot 15K_2O \; 7Bi_2O_3 \cdot 58B_2O_3/5V_2O_5$ glass | 9.3 | RT | $g_\parallel = 1.962$, $g_\perp = 1.994$ | $A_\parallel = 156.60$, $A_\perp = 63.97$ | [109] |
| $15Li_2O \cdot 15K_2O \cdot 10Bi_2O_3 \cdot 55B_2O_3/5V_2O_5$ glass | 9.3 | RT | $g_\parallel = 1.952$, $g_\perp = 1.992$ | $A_\parallel = 157.90$, $A_\perp = 61.30$ | [109] |
| $15Li_2O \cdot 15K_2O \cdot 12Bi_2O_3 \cdot 53B_2O_3/5V_2O_5$ glass | 9.3 | RT | $g_\parallel = 1.950$, $g_\perp = 1.990$ | $A_\parallel = 159.19$, $A_\perp = 64.99$ | [109] |
| $15Li_2O \cdot 15K_2O \cdot 15Bi_2O_3 \cdot 50B_2O_3/5V_2O_5$ glass | 9.3 | RT | $g_\parallel = 1.950$, $g_\perp = 1.990$ | $A_\parallel = 159.19$, $A_\perp = 65.46$ | [109] |
| $15Li_2O \cdot 85B_2O_3$ glass | 9.12 | 295 | $g_\parallel = 1.936$, $g_\perp = 1.980$ | $A_\parallel = 172.2$, $A_\perp = 64.7$ | [107] |
| $30Li_2O \cdot 69B_2O_3$ glass | 9.14 | 300 | $g_\parallel = 1.939$, $g_\perp = 1.974$ | $|A_\parallel| = 164.8$, $|A_\perp| = 59.9$ | [110] |
| $30Li_2O \cdot 70\,B_2O_3$ glass | X | RT | $g_\parallel = 1.930$, $g_\perp = 1.963$ | $A_\parallel = 163.0$, $A_\perp = 59.6$ | [111] |
| $35Li_2O \cdot 65B_2O_3$ glass | 9.12 | 295 | $g_\parallel = 1.940$, $g_\perp = 1.977$ | $A_\parallel = 166.6$, $A_\perp = 60.9$ | [107] |
| $0.3Li_2O \cdot 0.70\,B_2O_3$ (glass) | X | RT | $g_\parallel = 1.9409$, $g_\perp = 1.9725$ | $A_\parallel = 169$, $A_\perp = 58.5$ | [112] |
| $5Li_2O-25MoO_3-70\,B_2O_3$ glass | X | RT | $g_\parallel = 1.915$, $g_\perp = 1.972$ | $A_\parallel = 171.68$, $A_\perp = 47.99$ | [113] |
| $10Li_2O-20MoO_3-70\,B_2O_3$ glass | X | RT | $g_\parallel = 1.918$, $g_\perp = 1.975$ | $A_\parallel = 173.52$, $A_\perp = 48.92$ | [113] |
| $15Li_2O-15MoO_3-70\,B_2O_3$ glass | X | RT | $g_\parallel = 1.929$, $g_\perp = 1.977$ | $A_\parallel = 173.52$, $A_\perp = 55.38$ | [113] |
| $20Li_2O-10MoO_3-70\,B_2O_3$ glass | X | RT | $g_\parallel = 1.932$, $g_\perp = 1.979$ | $A_\parallel = 175.37$, $A_\perp = 55.38$ | [113] |
| $25Li_2O-5MoO_3-70\,B_2O_3$ glass | X | RT | $g_\parallel = 1.936$, $g_\perp = 1.983$ | $A_\parallel = 176.29$, $A_\perp = 56.3$ | [113] |
| $xLi_2O \cdot (20-x)Na_2O \cdot 20CdO \cdot 59.5P_2O_5$ glasses ($5 \le x \le 15$; 3 varying $x$) | X | RT | $g_\parallel = 1.95254$ to $1.9237$, $g_\perp = 1.9772-1.9785$ | $A_\parallel = 183-185$, $A_\perp = 61-62$ | [114] |
| $[Mg(H_2O)_6] \cdot [MgC_6H_5O_7(H_2O)]_2 \cdot 2H_2O$ (site I) | X | RT | 2.0976, 1.9093, 1.9505 | 73, 115, 237 (absolute values) | [115] |

(continued)

## ($VO^{2+}$ ($3d^1$)) glasses listing – contd.)

| Host | Frequency (GHz)/band | T (K) | $\tilde{g}$ | $A_x, A_y, A_z$ ($10^{-4}$ cm$^{-1}$) | References |
|---|---|---|---|---|---|
| [Mg(H$_2$O)$_6$] · [MgC$_6$H$_5$O$_7$(H$_2$O)]$_2$ · 2H$_2$O (site II) | X | RT | 2.0735, 1.9235, 1.9699 | 72, 111, 233 (absolute values) | [115] |
| 5MgO–25Li$_2$O–68 B$_2$O$_3$–2V$_2$O$_5$ glass | X | RT | $g_\| = 1.945$, $g_\perp = 1.986$ | $A_\| = 182$, $A_\perp = 75$ | [24] |
| 10MgO–20Li$_2$O–68 B$_2$O$_3$–2V$_2$O$_5$ glass | X | RT | $g_\| = 1.942$, $g_\perp = 1.982$ | $A_\| = 180$, $A_\perp = 75$ | [24] |
| 12MgO–18Li$_2$O–68 B$_2$O$_3$–2V$_2$O$_5$ glass | X | RT | $g_\| = 1.940$, $g_\perp = 1.982$ | $A_\| = 180$, $A_\perp = 75$ | [24] |
| 15MgO–15Li$_2$O–68 B$_2$O$_3$–2V$_2$O$_5$ glass | X | RT | $g_\| = 1.935$, $g_\perp = 1.980$ | $A_\| = 180$, $A_\perp = 80$ | [24] |
| 17MgO–13Li$_2$O–68 B$_2$O$_3$–2V$_2$O$_5$ glass | X | RT | $g_\| = 1.931$, $g_\perp = 1.980$ | $A_\| = 175$, $A_\perp = 80$ | [24] |
| 5MgO–25Na$_2$O–68 B$_2$O$_3$–2V$_2$O$_5$ glass | X | RT | $g_\| = 1.952$, $g_\perp = 1.988$ | $A_\| = 165$, $A_\perp = 70$ | [24] |
| 10MgO–20Na$_2$O–68 B$_2$O$_3$–2V$_2$O$_5$ glass | X | RT | $g_\| = 1.954$, $g_\perp = 1.985$ | $A_\| = 170$, $A_\perp = 70$ | [24] |
| 12MgO–18Na$_2$O–68 B$_2$O$_3$–2V$_2$O$_5$ glass | X | RT | $g_\| = 1.950$, $g_\perp = 1.982$ | $A_\| = 168$, $A_\perp = 65$ | [24] |
| 15MgO–15Na$_2$O–68 B$_2$O$_3$–2V$_2$O$_5$ glass | X | RT | $g_\| = 1.950$, $g_\perp = 1.982$ | $A_\| = 160$, $A_\perp = 65$ | [24] |
| 17MgO–13Na$_2$O–68 B$_2$O$_3$–2V$_2$O$_5$ glass | X | RT | $g_\| = 1.950$, $g_\perp = 1.980$ | $A_\| = 165$, $A_\perp = 65$ | [24] |
| 3MgO–27K$_2$O–68 B$_2$O$_3$–2V$_2$O$_5$ glass | X | RT | $g_\| = 1.967$, $g_\perp = 1.994$ | $A_\| = 164$, $A_\perp = 69$ | [24] |
| 6MgO–24K$_2$O–68 B$_2$O$_3$–2V$_2$O$_5$ glass | X | RT | $g_\| = 1.968$, $g_\perp = 1.990$ | $A_\| = 175$, $A_\perp = 72$ | [24] |
| 9MgO–21K$_2$O–68 B$_2$O$_3$–2V$_2$O$_5$ glass | X | RT | $g_\| = 1.954$, $g_\perp = 1.989$ | $A_\| = 179$, $A_\perp = 69$ | [24] |
| 12MgO–18K$_2$O–68 B$_2$O$_3$–2V$_2$O$_5$ glass | X | RT | $g_\| = 1.953$, $g_\perp = 1.985$ | $A_\| = 183$, $A_\perp = 67$ | [24] |
| 10MgO · 10Li$_2$O · 10Na$_2$O · 10K$_2$O · 58B$_2$O$_3$ · 2V$_2$O$_5$ glass | X | RT | $g_\| = 1.957$, $g_\perp = 1.983$ | $A_\| = 173.16$, $A_\perp = 75.41$ | [26] |
| 30MoO$_3$–70B$_2$O$_3$ glass | X | RT | $g_\| = 1.913$, $g_\perp = 1.969$ | $A_\| = 167.98$, $A_\perp = 47.07$ | [113] |

(continued)

## ($VO^{2+}$($3d^1$)) glasses listing – contd.)

| Host | Frequency (GHz)/band | $T$ (K) | $\tilde{g}$ | $A_x, A_y, A_z$ ($10^{-4}$ cm$^{-1}$) | References |
|---|---|---|---|---|---|
| 98$K_2B_4O_7$+2$VOSO_4$ (alkali tetraborate glass) | X | RT | $g_\parallel = 1.947$<br>$g_\perp = 1.971$ | $A_\parallel = -165$,<br>$A_\perp = -50$ | [116] |
| 98$Li_2B_4O_7$ + 2 $VOSO_4$ (alkali tetraborate glass) | X | RT | $g_\parallel = 1.948$,<br>$g_\perp = 1.966$ | $A_\parallel = -158$,<br>$A_\perp = -55$ | [116] |
| NaBaP glass (10 $Na_2O$ + 27$BaCO_3$ ∣ 60 $P_2O_5$ ∣ 3$V_2O_5$) | X | RT | $g_\parallel = 1.925$,<br>$g_\perp = 1.977$ | $A_\parallel = 178$,<br>$A_\perp = 67$ | [100] |
| NaCaB glass | 9.205 | RT | $g_\parallel = 1.935$,<br>$g_\perp = 1.963$ | $\|A_\parallel\| = 161$,<br>$\|A_\perp\| = 53$ | [101] |
| 20NaF+79 $B_2O_3$+1$V_2O_5$ | X | RT | $g_\parallel = 1.9496$<br>$g_\perp = 1.9849$ | $A_\parallel = 174$,<br>$A_\perp = 57$ | [105] |
| 30NaF+69 $B_2O_3$+1$V_2O_5$ | X | RT | $g_\parallel = 1.9469$,<br>$g_\perp = 1.9838$ | $A_\parallel = 174$,<br>$A_\perp = 56$ | [105] |
| 40NaF+59 $B_2O_3$+1$V_2O_5$ | X | RT | $g_\parallel = 1.9439$,<br>$g_\perp = 1.9819$ | $A_\parallel = 174$,<br>$A_\perp = 56$ | [105] |
| 50NaF+49 $B_2O_3$+1$V_2O_5$ | X | RT | $g_\parallel = 1.9411$<br>$g_\perp = 1.9797$ | $A_\parallel = 174$,<br>$A_\perp = 57$ | [105] |
| 98$Na_2B_4O_7$+2$VOSO_4$ (alkali tetraborate glass) | X | RT | $g_\parallel = 1.944$<br>$g_\perp = 1.972$ | $A_\parallel = -162$,<br>$A_\perp = -50$ | [116] |
| 15NaI–12$Na_2$O–3$K_2$O–70$B_2O_3$ glass | X | 310 | $g_\parallel = 1.932$,<br>$g_\perp = 1.986$ | $A_\parallel = 176$,<br>$A_\perp = 55$ | [117] |
| 15NaI–9$Na_2$O–6$K_2$O–70$B_2O_3$ glass | X | 310 | $g_\parallel = 1.930$,<br>$g_\perp = 1.983$ | $A_\parallel = 175$,<br>$A_\perp = 57$ | [117] |
| 15NaI–6$Na_2$O–9$K_2$O–70$B_2O_3$ glass | X | 310 | $g_\parallel = 1.930$,<br>$g_\perp = 1.981$ | $A_\parallel = 175$,<br>$A_\perp = 57.5$ | [117] |
| 15NaI–3$Na_2$O–12$K_2$O–70$B_2O_3$ glass | X | 310 | $g_\parallel = 1.931$,<br>$g_\perp = 1.980$ | $A_\parallel = 172$,<br>$A_\perp = 55.5$ | [117] |
| 15NaI–0$Na_2$O–15$K_2$O–70$B_2O_3$ glass | X | 310 | $g_\parallel = 1.930$,<br>$g_\perp = 1.985$ | $A_\parallel = 171$,<br>$A_\perp = 54.5$ | [117] |
| NaPbBTe (sodium lead borotellurite) glass | X | 295 | $g_\parallel = 1.950$,<br>$g_\perp = 1.954$ | $A_\parallel = 153$,<br>$A_\perp = 64$ | [102] |
| 30$Na_2O$ · 69$B_2O_3$ glass | 9.14 | 300 | $g_\parallel = 1.945$,<br>$g_\perp = 1.980$ | $\|A_\parallel\| = 165.3$,<br>$\|A_\perp\| = 59.6$ | [110] |
| 30$Na_2O$ · 70$B_2O_3$ glass | 9.14 | 300 | $g_\parallel = 1.924$,<br>$g_\perp = 1.957$ | $A_\parallel = 160$,<br>$A_\perp = 50.4$ | [118] |

(*continued*)

## ($VO^{2+}$ ($3d^1$) glasses listing – contd.)

| Host | Frequency (GHz)/band | T (K) | $\tilde{g}$ | $A_x, A_y, A_z$ ($10^{-4}$ cm$^{-1}$) | References |
|---|---|---|---|---|---|
| $Na_2O \cdot P_2O_5$ glass | X | RT | $g_\parallel = 1.937$, $g_\perp = 1.979$ | $A_\parallel = 175$, $A_\perp = 59$ | [119] |
| $(30-x)(NaPO_3)_6 + 30 PbO + 40B_2O_3 + xV_2O_5$ (x = 1, 2, 3, 4, 5, 6, and 7 mol%) | X | 93–333 | $g_\parallel = 1.943$, $g_\perp = 1.995$ | $A_\parallel = 183$, $A_\perp = 68$ | [120] |
| $4Nb_2O_5 \cdot 26K_2O\ 69B_2O_3$ glass | 9.14 | 300 | $g_\parallel = 1.946$, $g_\perp = 1.978$ | $|A_\parallel| = 163.5$, $|A_\perp| = 59.1$ | [110] |
| $8Nb_2O_5 \cdot 22K_2O\ 69B_2O_3$ glass | 9.14 | 300 | $g_\parallel = 1.952$, $g_\perp = 1.976$ | $|A_\parallel| = 161.3$, $|A_\perp| = 57.2$ | [110] |
| $4Nb_2O_5 \cdot 26Li_2O\ 69B_2O_3$ glass | 9.14 | 300 | $g_\parallel = 1.942$, $g_\perp = 1.966$ | $|A_\parallel| = 163.2$, $|A_\perp| = 57.8$ | [110] |
| $8Nb_2O_5 \cdot 22Li_2O\ 69B_2O_3$ glass | 9.14 | 300 | $g_\parallel = 1.945$, $g_\perp = 1.962$ | $|A_\parallel| = 161.6$, $|A_\perp| = 55.0$ | [110] |
| $4Nb_2O_5 \cdot 26Na_2O\ 69B_2O_3$ glass | 9.14 | 300 | $g_\parallel = 1.948$, $g_\perp = 1.977$ | $|A_\parallel| = 163.7$, $|A_\perp| = 57.2$ | [110] |
| $8Nb_2O_5 \cdot 22Na_2O\ 69B_2O_3$ glass | 9.14 | 300 | $g_\parallel = 1.951$, $g_\perp = 1.974$ | $|A_\parallel| = 162.1$, $|A_\perp| = 55.3$ | [110] |
| $0.02NiO \cdot 0.28\ Li_2O \cdot 0.70B_2O_3$ (glass) | X | RT | $g_\parallel = 1.9410$, $g_\perp = 1.9731$ | $A_\parallel = 169$, $A_\perp = 58.5$ | [112] |
| $0.04NiO \cdot 0.26\ Li_2O \cdot 0.70B_2O_3$ (glass) | X | RT | $g_\parallel = 1.9409$, $g_\perp = 1.9731$ | $A_\parallel = 69$, $A_\perp = 58.5$ | [112] |
| $0.05NiO \cdot 0.25Li_2O \cdot 0.70B_2O_3$ (glass) | X | RT | $g_\parallel = 1.9410$, $g_\perp = 1.9731$ | $A_\parallel = 69$, $A_\perp = 58.5$ | [112] |
| $0.06NiO \cdot 0.24Li_2O \cdot 0.70B_2O_3$ (glass) | X | RT | $g_\parallel = 1.9390$, $g_\perp = 1.9740$ | $A_\parallel = 170.6$, $A_\perp = 59.0$ | [112] |
| $0.3PbO \cdot 0.70B_2O_3$ glass | 9.14 | 295 | $g_\parallel = 1.9361$, $g_\perp = 1.9772$ | $A_\parallel = 168.6$, $A_\perp = 64.6$ | [94] |
| $Pb_3O_4 \cdot ZnO \cdot P_2O_5$ glass | X | RT | $g_\parallel = 1.9275$, $g_\perp = 1.9760$ | $A_\parallel = 184$, $A_\perp = 63$ | [121] |
| $SrB_4O_7$ glass | — | — | $g_\parallel = 1.9359$, $g_\perp = 1.9967$ | $A_\parallel = 172$, $A_\perp = 56$ | [122] |
| $10SrO \cdot 10Li_2O \cdot 10Na_2O \cdot 10K_2O \cdot 58B_2O_3 \cdot 2V_2O_5$ glass | X | RT | $g_\parallel = 1.944$, $g_\perp = 1.976$ | $A_\parallel = 178.75$, $A_\perp = 83.79$ | [26] |
| $10SrO \cdot 29.9ZnO\ 60B_2O_3 \cdot 0.1V_2O_5$ glass | 9.305 | RT | $g_\parallel = 1.9852$, $g_\perp = 2.002$ | $A_\parallel = 167$, $A_\perp = 67$ | [123] |

(continued)

## ($VO^{2+}(3d^1)$) glasses listing – contd.)

| Host | Frequency (GHz)/band | T (K) | $\tilde{g}$ | $A_x, A_y, A_z$ ($10^{-4}$ cm$^{-1}$) | References |
|---|---|---|---|---|---|
| $10SrO \cdot 29.7ZnO \cdot 60B_2O_3 \cdot 0.3V_2O_5$ glass | 9.305 | RT | $g_\parallel = 1.9855$, $g_\perp = 2.002$ | $A_\parallel = 164$, $A_\perp = 67$ | [123] |
| $10SrO \cdot 29.5ZnO \cdot 60B_2O_3 \cdot 0.5V_2O_5$ glass | 9.305 | RT | $g_\parallel = 1.9861$, $g_\perp = 2.002$ | $A_\parallel = 162$, $A_\perp = 68$ | [123] |
| $10SrO \cdot 29.3ZnO \cdot 60B_2O_3 \cdot 0.7V_2O_5$ glass | 9.305 | RT | $g_\parallel = 1.9864$, $g_\perp = 2.002$ | $A_\parallel = 164$, $A_\perp = 68$ | [123] |
| $10SrO \cdot 29.1ZnO \cdot 60B_2O_3 \cdot 0.9V_2O_5$ glass | 9.305 | RT | $g_\parallel = 1.9869$, $g_\perp = 2.002$ | $A_\parallel = 161$, $A_\perp = 67$ | [123] |
| $10SrO \cdot 29.1ZnO \cdot 60B_2O_3 \cdot 0.9V_2O_5$ glass | 9.305 | 93 | $g_\parallel = 1.9884$, $g_\perp = 2.002$ | $A_\parallel = 163$, $A_\perp = 68$ | [123] |
| $10SrO \cdot 29.1ZnO \cdot 60B_2O_3 \cdot 0.9V_2O_5$ glass | 9.305 | 123 | $g_\parallel = 1.9866$, $g_\perp = 2.002$ | $A_\parallel = 164$, $A_\perp = 68$ | [123] |
| $10SrO \cdot 29.1ZnO \cdot 60B_2O_3 \cdot 0.9V_2O_5$ glass | 9.305 | 153 | $g_\parallel = 1.9861$, $g_\perp = 2.002$ | $A_\parallel = 164$, $A_\perp = 67$ | [123] |
| $10SrO \cdot 29.1ZnO \cdot 60B_2O_3 \cdot 0.9V_2O_5$ glass | 9.305 | 183 | $g_\parallel = 1.9871$, $g_\perp = 2.002$ | $A_\parallel = 163$, $A_\perp = 68$ | [123] |
| $10SrO \cdot 29.1ZnO \cdot 60B_2O_3 \cdot 0.9V_2O_5$ glass | 9.305 | 213 | $g_\parallel = 1.9861$, $g_\perp = 2.002$ | $A_\parallel = 164$, $A_\perp = 68$ | [123] |
| $10SrO \cdot 29.1ZnO \cdot 60B_2O_3 \cdot 0.9V_2O_5$ glass | 9.305 | 243 | $g_\parallel = 1.9868$, $g_\perp = 2.002$ | $A_\parallel = 163$, $A_\perp = 68$ | [123] |
| $10SrO \cdot 29.1ZnO \cdot 60B_2O_3 \cdot 0.9V_2O_5$ glass | 9.305 | 273 | $g_\parallel = 1.9861$, $g_\perp = 2.002$ | $A_\parallel = 163$, $A_\perp = 67$ | [123] |
| $xTiO_2 \cdot (70-x)B_2O_3 \cdot 30Na_2O$ glasses <br> 1. $x = 0$ mol%; $V_2O_5 = 2$ mol% <br> 2. $x = 2$ mol%; $V_2O_5 = 2$ mol% <br> 3. $x = 5$ mol%; $V_2O_5 = 2$ mol% <br> 4. $x = 7$ mol%; $V_2O_5 = 2$ mol% | 9.13 | 295 | 1. $g_\parallel = 1.9376$, $g_\perp = 1.9746$ <br> 2. $g_\parallel = 1.9370$, $g_\perp = 1.9740$ <br> 3. $g_\parallel = 1.9370$, $g_\perp = 1.9734$ <br> 4. $g_\parallel = 1.9370$, $g_\perp = 1.9734$ | 1. $\lvert A_\parallel \rvert = 166.89$, $\lvert A_\perp \rvert = 60.38$ <br> 2. $A_\parallel = 166.75$, $\lvert A_\perp \rvert = 60.36$ <br> 3. $\lvert A_\parallel \rvert = 165.94$, $\lvert A_\perp \rvert = 59.70$ <br> 4. $\lvert A_\parallel \rvert = 165.85$, $\lvert A_\perp \rvert = 59.70$ <br> (above) $A_\parallel$ and $A_\perp$ are negative | [124] |
| $xTiO_2 \cdot (30-x)Na_2O \cdot 70B_2O_3$ glasses | | | | | |

*(continued)*

## ($VO^{2+}(3d^1)$) glasses listing – contd.

| Host | Frequency (GHz)/band | T (K) | $\tilde{g}$ | $A_x, A_y, A_z$ ($10^{-4}$ cm$^{-1}$) | References |
|---|---|---|---|---|---|
| 1. $x = 0$ mol%; $V_2O_5 = 2$ mol% <br> 2. $x = 2$ mol%; $V_2O_5 = 2$ mol% <br> 3. $x = 5$ mol%; $V_2O_5 = 2$ mol% <br> 4. $x = 7$ mol%; $V_2O_5 = 2$ mol% | 9.13 | 295 | 1. $g_\| = 1.9376$, $g_\perp = 1.9746$ <br> 2. $g_\| = 1.9347$, $g_\perp = 1.9734$ <br> 3. $g_\| = 1.9342$, $g_\perp = 1.9734$ <br> 4. $g_\| = 1.9336$, $g_\perp = 1.9734$ | 1. $\|A_\|\| = 166.89$, $\|A_\perp\| = 60.38$ <br> 2. $A_\| = 166.65$, $\|A_\perp\| = 60.35$ <br> 3. $\|A_\|\| = 167.14$, $\|A_\perp\| = 60.81$ <br> 4. $\|A_\|\| = 167.00$, $\|A_\perp\| = 61.27$ <br> $A_\|$ and $A_\perp$ are negative (above) | [124] |
| $V_2O_5 \cdot (99)[P_2O_5 \cdot 1.5Na_2O]$ glass | 9.4 | RT | $g_\| = 1.954$, $g_\perp = 2.005$ | $A_\| = 171.1$, $A_\perp = 59.9$ | [25] |
| $V_2O_5 \cdot (99)[P_2O_5 \cdot 2Na_2O]$ glass | 9.4 | RT | $g_\| = 1.948$, $g_\perp = 2.004$ | $A_\| = 163.4$, $A_\perp = 65.7$ | [25] |
| $3V_2O_5 \cdot (97)[P_2O_5 \cdot 2Na_2O]$ glass | 9.4 | RT | $g_\| = 1.945$, $g_\perp = 2.002$ | $A_\| = 160.6$, $A_\perp = 67.5$ | [25] |
| $0.7V_2O_5$–$0.3P_2O_5$ glass | X | 300, 4 | $g_\| = 1.959$, $g_\perp = 1.987$ (at 4 K) | $A_\| = 156.6$, $A_\perp = 53.8$ (at 300 K) | [125] |
| $0.7V_2O_5$–$0.3P_2O_5$ glass | X | 77 | $g_{iso} = 1.963$ | — | [125] |
| $2.5WO_3 \cdot 27.5Li_2O \cdot 70 B_2O_3$ glass | X | RT | $g_\| = 1.917$, $g_\perp = 1.957$ | $A_\| = 162.0$, $A_\perp = 59.4$ | [111] |
| $5.0WO_3 \cdot 25.0Li_2O \cdot 70 B_2O_3$ glass | X | RT | $g_\| = 1.912$, $g_\perp = 1.953$ | $A_\| = 162.2$, $A_\perp = 59.7$ | [111] |
| $7.5WO_3 \cdot 22.5Li_2O \cdot 70 B_2O_3$ glass | X | RT | $g_\| = 1.911$, $g_\perp = 1.953$ | $A_\| = 162.6$, $A_\perp = 61.6$ | [111] |
| $10.0WO_3 \cdot 20.0Li_2O \cdot 70 B_2O_3$ glass | X | RT | $g_\| = 1.903$, $g_\perp = 1.956$ | $A_\| = 161.8$, $A_\perp = 62.0$ | [111] |
| $12.5WO_3 \cdot 17.5Li_2O \cdot 70 B_2O_3$ glass | X | RT | $g_\| = 1.902$, $g_\perp = 1.957$ | $A_\| = 160.0$, $A_\perp = 62.6$ | [111] |
| $15.0WO_3 \cdot 1.0Li_2O \cdot 70 B_2O_3$ glass | X | RT | $g_\| = 1.930$, $g_\perp = 1.963$ | $A_\| = 158.9$, $A_\perp = 63.6$ | [111] |
| $2.5WO_3 \cdot 27.5Na_2O \cdot 70B_2O_3$ glass | 9.14 | 300 | $g_\| = 1.920$, $g_\perp = 1.963$ | $A_\| = 161.1$, $A_\perp = 60.6$ | [118] |
| $5WO_3 \cdot 25Na_2O \cdot 70B_2O_3$ glass | 9.14 | 300 | $g_\| = 1.919$, $g_\perp = 1.963$ | $A_\| = 162.9$, $A_\perp = 60.3$ | [118] |

(continued)

## ($VO^{2+}(3d^1)$) glasses listing – contd.)

| Host | Frequency (GHz)/band | T (K) | $\tilde{g}$ | $A_x, A_y, A_z$ ($10^{-4}$ cm$^{-1}$) | References |
|---|---|---|---|---|---|
| 7.5WO$_3$ · 22.5Na$_2$O · 70B$_2$O$_3$ glass | 9.14 | 300 | $g_\parallel = 1.913$, $g_\perp = 1.962$ | $A_\parallel = 163.9$, $A_\perp = 60.6$ | [118] |
| 10WO$_3$ · 20Na$_2$O · 70B$_2$O$_3$ glass | 9.14 | 300 | $g_\parallel = 1.907$, $g_\perp = 1.962$ | $A_\parallel = 163.9$, $A_\perp = 62.1$ | [118] |
| 12.5WO$_3$ · 17.5Na$_2$O · 70B$_2$O$_3$ glass | 9.14 | 300 | $g_\parallel = 1.907$, $g_\perp = 1.961$ | $A_\parallel = 163.9$, $A_\perp = 62.3$ | [118] |
| 15WO$_3$ · 15Na$_2$O · 70B$_2$O$_3$ glass | 9.14 | 300 | $g_\parallel = 1.892$, $g_\perp = 1.961$ | $A_\parallel = 163.9$, $A_\perp = 64.1$ | [118] |
| Zinc lead borate glasses (varying Pb content) | | | | | |
| xZnCO$_3$ + (40 – x) PbO + 56.5H$_3$BO$_3$ + 3.5 V$_2$O$_5$ | X | 295 | | | [126] |
| 1. x = 0 mol% | | | 1. $g_\parallel = 1.914$, $g_\perp = 1.960$ | 1. $A_\parallel = -162.8$, $A_\perp = -59.6$ | |
| 2. x = 10 mol% | | | 2. $g_\parallel = 1.918$, $g_\perp = 1.964$ | 2. $A_\parallel = -162.1$, $A_\perp = -60.4$ | |
| 3. x = 20 mol% | | | 3. $g_\parallel = 1.934$, $g_\perp = 1.966$ | 3. $A_\parallel = -160.6$, $A_\perp = -60.7$ | |
| 4. x = 30 mol% | | | 4. $g_\parallel = 1.938$, $g_\perp = 1.967$ | 4. $A_\parallel = -159.9$, $A_\perp = -61.1$ | |
| 5. x = 40 mol% | | | 5. $g_\parallel = 1.940$, $g_\perp = 1.968$ | 5. $A_\parallel = -159.2$, $A_\perp = -61.1$ | |
| 19.9ZnO · 5Li$_2$O · 25Na$_2$O · 50B$_2$O$_3$ · 0.1V$_2$O$_5$ glass | X | RT | $g_\parallel = 1.9355$, $g_\perp = 1.9806$ | $A_\parallel = 164$, $A_\perp = 56$ | [127] |
| 19.9ZnO · 10Li$_2$O · 20Na$_2$O · 50B$_2$O$_3$ · 0.1V$_2$O$_5$ glass | X | RT | $g_\parallel = 1.9444$, $g_\perp = 1.9807$ | $A_\parallel = 164$, $A_\perp = 56$ | [127] |
| 19.9ZnO · 15Li$_2$O · 15Na$_2$O · 50B$_2$O$_3$ · 0.1V$_2$O$_5$ glass | X | RT | $g_\parallel = 1.9417$, $g_\perp = 1.9810$ | $A_\parallel = 167$, $A_\perp = 56$ | [127] |
| 19.9ZnO · 20Li$_2$O · 10Na$_2$O · 50B$_2$O$_3$ · 0.1V$_2$O$_5$ glass | X | RT | $g_\parallel = 1.9387$, $g_\perp = 1.9809$ | $A_\parallel = 169$, $A_\perp = 56$ | [127] |
| 19.9ZnO · 25Li$_2$O · 5Na$_2$O · 50B$_2$O$_3$ · 0.1V$_2$O$_5$ glass | X | RT | $g_\parallel = 1.9357$, $g_\perp = 1.9808$ | $A_\parallel = 169$, $A_\perp = 56$ | [127] |
| 1.667ZnO · (0.443)P$_2$O$_5$ glass | X | 77 | $g_\parallel = 1.9329$, $g_\perp = 1.9824$ | $A_\parallel = 157$, $A_\perp = 52$ | [128] |

## 3d² (Cr⁴⁺, Mn⁵⁺, Ti²⁺, V³⁺), S = 1

### Cr⁴⁺. Data tabulation of SHPs
$Cr^{4+}(3d^2)$

| Host | Frequency (GHz)/ band | T (K) | $\tilde{g}$ | $b_2^0 (= D), b_2^2 (= 3E)$ $(10^{-4}\,cm^{-1})$ | References |
|---|---|---|---|---|---|
| $Bi_4Ge_3O_{12}$ | 9.5, 35.5 | 77 | 1.915, 1.915, 1.932 | 550, 30 (G) | [129] |
| $La_2Ga_5SiO_{14}$ | 9.4 | 77, 293 | 1.971 | 10 000.0, 0.0 | [130] |

### Mn⁵⁺. Data tabulation of SHPs
$Mn^{5+}(3d^2)$

| Host | Frequency (GHz)/ band | T (K) | $\tilde{g}$ | $(b_2^0 = D),$ $(b_2^2 = 3E)$ $(10^{-4}\,cm^{-1})$ | $A_x, A_y, A_z$ $(10^{-4}\,cm^{-1})$ | References |
|---|---|---|---|---|---|---|
| $Ba_3(VO_4)_2$ | 9.09 | RT | $g_\parallel = 1.9608$, $g_\perp = 1.9722$ | $D = 5.81$ (GHz) | $A_\parallel = 70$, $A_\perp = 60$ | [131] |
| $Ca_2MnO_4Cl$ | 9.6 | 295 | $g_\parallel = 1.9796$, $g_\perp = 1.9722$ | 13.194, −5.448 (GHz) | $A_\parallel = 0.207$, $A_\perp = 0.201$ (GHz) | [132] |
| $Ca_2MnO_4Cl$ | 9.6 | 190 | $g_\parallel = 1.9736$, $g_\perp = 1.9716$ | 14.072, −5.532 (GHz) | $A_\parallel = 0.213$, $A_\perp = 0.200$ (GHz) | [132] |
| $Ca_2MnO_4Cl$ | 9.6 | 120 | $g_\parallel = 1.9797$, $g_\perp = 1.9728$ | 14.520, −5.634 (GHz) | $A_\parallel = 0.188$, $A_\perp = 0.194$ (GHz) | [132] |
| $Sr_5(MnO_4)_3Cl$ | X | 296 | $g_\parallel = 1.9658$, $g_\perp = 1.9774$ | 11.79, −3.915 (GHz) | $A_\parallel = 0.203$, $A_\perp = 0.201$ (GHz) | [133] |
| $Sr_5(MnO_4)_3Cl$ | X | 190 | $g_\parallel = 1.9625$, $g_\perp = 1.9785$ | 12.41, −4.494 (GHz) | $A_\parallel = 0.196$, $A_\perp = 0.201$ (GHz) | [133] |
| $Sr_5(MnO_4)_3Cl$ | X | 120 | $g_\parallel = 1.9608$, $g_\perp = 1.9768$ | 12.77, −4.797 (GHz) | $A_\parallel = 0.197$, $A_\perp = 0.200$ (GHz) | [133] |

## Ti$^{2+}$. Data tabulation of SHPs
## Ti$^{2+}$ (3d$^2$)

| Host | Frequency (GHz)/band | T (K) | $\tilde{g}$ | $A_x, A_y, A_z$ ($10^{-4}$ cm$^{-1}$) | References |
|---|---|---|---|---|---|
| CsNO$_3$ | X | 77 | 1.9930, 1.9928, 1.9965 | 93.778, 93.778, 94.179 (GHz) | [134] |
| CsTi(SO$_4$)$_2$ · 12H$_2$O | X | <8 | $g_\parallel = 1.168, g_\perp \sim 0.0$ | — | [135] |
| NaNO$_3$ | X | 77 | 1.9923, 1.9923, 1.9970 | 93.781, 93.781, 94.185 (GHz) | [134] |
| KNO$_3$ | X | 77 | 1.9949, 1.9947, 1.9975 | 122.655, 122.744, 123.024 (GHz) | [134] |
| RbNO$_3$ (site 1) | X | 77 | 1.9967, 1.9992, 1.9970 | 109.295, 109.376, 109.480 (GHz) | [134] |
| RbNO$_3$ (site 2) | X | 77 | 1.9843, 1.9987, 1.9985 | 108.196, 108.451, 108.553 (GHz) | [134] |

## V$^{3+}$. Data tabulation of SHPs
## V$^{3+}$ (3d$^2$)

| Host | Frequency (GHz)/band | T (K) | $\tilde{g}$ | $b_2^0(=D)$, $b_2^2(=3E)$ ($10^{-4}$ cm$^{-1}$) | $A_x, A_y, A_z$ ($10^{-4}$ cm$^{-1}$) | References |
|---|---|---|---|---|---|---|
| Al$_2$O$_3$ | 95–700 HFEPR | Variable | $g_\parallel = 1.921$, $g_\perp = 1.74$ | — | $A_\parallel = 98$, $A_\perp = 78$ | [136] |
| CdTe | X | 4–100 | 1.962 | — | 60 | [137] |
| CsGa(SO$_4$)$_2$ · 12D$_2$O | 95–700 HFEPR | Variable | $g_\parallel = 1.9549$, $g_\perp = 1.8690$ | 47 735, 0 | $A_\parallel = 99$, $A_\perp = 78$ | [136] |
| CsGa(SO$_4$)$_2$ · 12H$_2$O | 95–700 HFEPR | Variable | $g_\parallel = 1.9500$, $g_\perp = 1.8656$ | 48 581, 0 | $A_\parallel = 98$, $A_\perp = 78$ | [136] |
| Na[V(trdta)] · 3H$_2$O trdta= trimethylenediamine-N,N,N',N'-tetraacetate | Multi-frequency HFEPR | variable | $g = 1.95$ | 5.60, 2.55 | | [138] |
| Na[V(edta)] · 3H$_2$O edta=ethylene diamine tetraacetic acid | Multi-frequency HFEPR | variable | $g = 1.95$ | 1.4, 0.42 | | [138] |

*(continued)*

## ($V^{3+}$ ($3d^2$) listing – contd.)

| Host | Frequency (GHz)/ band | T (K) | $\tilde{g}$ | $b_2^0 (= D)$, $b_2^2 (= 3E)$ ($10^{-4}$ cm$^{-1}$) | $A_x, A_y, A_z$ ($10^{-4}$ cm$^{-1}$) | References |
|---|---|---|---|---|---|---|
| Rb(V,Ga)(SO$_4$)$_2$ · 12H$_2$O | 94.9685, 189.937, 284.9055 | 5–20 | $g_\parallel = 1.944$, $g_\perp = 1.863$ | 49 060, — | $A_\parallel = 111$ G | [139] |
| VBr$_3$(thf)$_3$ (thf = tetrahydrofuran) | 337 | 10 | | −161 620, −36 940 | — | [140] |
| V(acac)$_3$ (acac = anion of 2,4-penta-nedione) | 95–700 | 5 | | 74 700, 57 480 | — | [141] |
| VCl$_3$(thf)$_3$ (thf = tetrahydrofuran) | 95–700 | 4.5 | — | 118 500, 118 500 (absolute values) | — | [141] |
| VBr$_3$(thf)$_3$ (thf = tetrahydrofuran) | 95–700 | 7/10 | 1.86, 1.90, 1.710 | −161 620, −50 820 | — | [141] |

## $3d^3$ ($Cr^{3+}$, $Fe^{5+}$, $Mn^{4+}$, $V^{2+}$), S = 3/2
$Cr^{3+}$. Data tabulation of SHPs
$Cr^{3+}$ ($3d^3$)

| Host | Frequency (GHz)/ band | T (K) | $\tilde{g}$ | $(b_2^0 = D)$, $(b_2^2 = 3E)$ ($10^{-4}$ cm$^{-1}$) | $A_x, A_y, A_z$ ($10^{-4}$ cm$^{-1}$) | References |
|---|---|---|---|---|---|---|
| Al$_2$O$_3$ | 90–250 | RT | $g_\parallel = 1.9803$, $g_\perp = 1.9813$ | $b_2^0 = -5.742$ | — | [142] |
| α-Al$_2$O$_3$ | 94.9 | 295 | $g_\parallel = 1.9812$, $g_\perp = 1.9814$ | $|D| = 5.738$ GHz, — | — | [143] |
| Al$_2$O$_3$ (β signal) | X, Q | 150, 300 | 2.05, 2.05, 2.05 | 0, 0 | — | [144] |
| Al$_2$O$_3$ (δ signal) | X, Q | 150, 300 | 1.970, 1.565, 1.565 | 4900, 4890 | — | [144] |
| Al$_2$O$_3$ (natural pink sapphire) | X | RT | 1.976, 1.976, 1.967 | $B_2^0 = 92.099$ (mT) | — | [145] |
| | X | 1473 | 1.9791.979, 1.975 | $B_2^0 = 92.648$ (mT) | — | [145] |

*(continued)*

## ($Cr^{3+}$(3d³) listing – contd.)

| Host | Frequency (GHz)/ band | T (K) | $\tilde{g}$ | ($b_2^0 = D$), ($b_2^2 = 3E$) ($10^{-4}$ cm$^{-1}$) | $A_x, A_y, A_z$ ($10^{-4}$ cm$^{-1}$) | References |
|---|---|---|---|---|---|---|
| | X | 1573 | 1.985, 1.985, 1.984 | $B_2^0 = 91.445$ (mT) | | [145] |
| | X | 1673 | 1.983, 1.983, 1.971 | $B_2^0 = 88.215$ (mT) | | [145] |
| | X | 1773 | 1.987 1.987, 1.983 | $B_2^0 = 91.845$ (mT) | | [145] |
| | X | 1873 | 1.967 1.967, 1.963 | $B_2^0 = 94.527$ (mT) | | [145] |
| 6H-$BaTiO_3$ (site I) | 9.34, 94 | 300 | 1.9857, 1.9857, 1.9797 | 1050, — | — | [146] |
| 6H-$BaTiO_3$ (site II) | 9.34, 94 | 300 | 1.9756, 1.0756, 1.9736 | 3220, — | — | [146] |
| 6H-$BaTiO_3$ (site I) | 9.34, 94 | 170 | 1.9860, 1.9860, 1.9795 | 980, 50 | — | [146] |
| 6H-$BaTiO_3$ (site II) | 9.34, 94 | 170 | 1.9756, 1.9756, 1.9736 | 3170, — | — | [146] |
| 6H-$BaTiO_3$ (site I) | 9.34, 94 | 15 | 1.9867, 1.9867, 1.9795 | 930, 80 | — | [146] |
| 6H-$BaTiO_3$ (site II) | 9.34, 94 | 15 | 1.9756, 1.9756, 1.9736 | 3170, — | — | [146] |
| $BeAlSiO_4(OH)$ (natural green Euclase) | | | 2.018, 2.001, 1.956 | $D = -8.27$, $F = 1.11$ (GHz) | — | [28] |
| $Bi_{12}SiO_{20}$ | X, Q, W | 12 | 1.983 | $B_2^0 = 1950$ | — | [147] |
| $Ca_3Ga_2Ge_4O_{14}$ (CGGO) line A | X | RT | $g_\| = 1.980$, $g_\perp = 1.98$ | $b_2^0 = 8000$ | — | [148] |
| $Ca_3Ga_2Ge_4O_{14}$ (CGGO) line B | X | RT | $g_\| = 1.97$, $g_\perp = 1.97$ | $b_2^0 = 8000$, $b_2^2 = 2400$ | — | [148] |
| $(CH_3)_2NH_2Ga(SO_4)_2 \cdot 6H_2O$ | X | RT | 1.982 | 890, 386 | — | [149] |

(continued)

## ($Cr^{3+}$ ($3d^3$) listing – contd.)

| Host | Frequency (GHz)/ band | T (K) | $\tilde{g}$ | ($b_2^0 = D$), ($b_2^2 = 3E$) ($10^{-4}$ cm$^{-1}$) | $A_x, A_y, A_z$ ($10^{-4}$ cm$^{-1}$) | References |
|---|---|---|---|---|---|---|
| $(CH_3)_2NH_2Ga$ $(SO_4)_2 \cdot 6H_2O$ | X | 125 | 1.982 | 1088, 39 | — | [149] |
| $(CH_3)_2NH_2Ga$ $(SO_4)_2 \cdot 6H_2O$ | X | 40 | 1.976 | 1425, 1120 | — | [149] |
| $(CH_3)_4NCdCl_3$ | X | 295 | g = 1.9741 | 553, — | — | [150] |
| $C_6H_{12}O_7, H_2O$ | X | 77 | $g_{iso}$ = 1.9919 | $|D|$ = 349, $|E|$=113 | — | [151] |
| CrAPO-5 (signal C) (APO = $Cr_{0.24}Al_{11.76}$ $P_{12}O_{48}$) | X, Q | 150, 300 | 1.987, 1.987, 1.987 | 0,0 | — | [144] |
| CrAPO-5 (signal D) (APO = $Cr_{0.24}Al_{11.76}$ $P_{12}O_{48}$) | X, Q | 150, 300 | 1.970, 1.865, 1.865 | 4900, 4890 | — | [144] |
| $CsLiSO_4$ | X | RT | 1.9674, 1.9673, 1.9684 | D = 1529, E = 414 | — | [152] |
| $CsLiSO_4$ | X | RT | 1.9674, 1.9673, 1.9684 | D = 1529, E = 414 | — | [152] |
| $CsMgCl_3$ (center I) | X, Q | 1.6 | 1.9841, 1.9841, 1.9838 | D = 1049.4 | — | [153] |
| $CsMgCl_3$ (center I) | X, Q | 290 | 1.9868, 1.9868, 1.9868 | D = 1253.3 | — | [153] |
| $CsMgCl_3$ (center II) | X, Q | 1.6 | 1.9893, 1.9896, 1.9832 | D = 1334.3, E = −193.3 | — | [153] |
| $CsMgCl_3$ (center II) | X, Q | 300 | 1.984, 1.985, 1.985 | D = 1546, E = −184.0 | — | [153] |
| $CsMgCl_3$ (center III) | X, Q | 1.6 | 1.995, 1.995, 1.990 | D = −2807 | — | [153] |
| $CsMgCl_3$ (center III) | X, Q | 300 | 1.9831, 1.9831, 1.9861 | D = −2811.8 | — | [153] |

(continued)

## ($Cr^{3+}$($3d^3$) listing – contd.)

| Host | Frequency (GHz)/band | T (K) | $\tilde{g}$ | ($b_2^0 = D$), ($b_2^2 = 3E$) ($10^{-4}$ $cm^{-1}$) | $A_x, A_y, A_z$ ($10^{-4}$ $cm^{-1}$) | References |
|---|---|---|---|---|---|---|
| CsMgCl$_3$ (center IV) | X, Q | 1.6 | 1.9835, 1.9835, 1.9828 | $D = -170.1$ | — | [153] |
| CsMgCl$_3$ (center IV) | X, Q | 300 | 1.9864, 1.9864, 1.9858 | $D = 86.6$ | — | [153] |
| CuAlO$_2$ | X | 294 | 1.979 | $B_2^0 = -2601$ | — | [154] |
| CuAlO$_2$ | X | 110 | 1.979 | $B_2^0 = -2581$ | — | [154] |
| CuAlO$_2$ | Q | 123 | 1.979 | $B_2^0 = -2606$ | — | [154] |
| [Cu(bpy)$_3$]$_2$ [Cr(C$_2$O$_4$)$_3$]NO$_3$ · 9H$_2$O (bpy = 2,2'-bipyridine) | 9.6 | — | $g = 1.963$ | $D = 6300$, $|E| = 210$ | — | [155] |
| GaAs | X | 4 | 1.522, 2.002, 2.357 | $B_2^0 = 376$, $B_2^2 = 108$ | — | [156] |
| H$_2$Ni(C$_4$H$_2$O$_4$)$_2$ 4H$_2$O | X | RT | 2.0980, 1.9750, 1.9680 | $-954, -864$ | — | [157] |
| KAl(MoO$_4$)$_2$ | X | RT | $g_\parallel = 1.9781$, $g_\perp = 1.9727$ | $|D| = 4798.0$ | — | [158] |
| KGaF$_4$ (phase II) | 9.75 | 493 | 1.978, 1.978, 1.965 | $b_2^0 = 1711$, $b_2^1 = 2949$, $b_2^2 = 469$ | — | [159] |
| KGaF$_4$ (phase III) | 9.75 | RT | 1.983, 1.965, 1.965 | $b_2^{-2} = 217.3$, $b_2^{-1} = -1453$, $b_2^0 = 1585.2$, $b_2^1 = -2920$, $b_2^2 = 464$ | — | [159] |
| KNaC$_4$H$_4$O$_6$ · 4H$_2$O | X | 77 | 1.9257, 1.9720, 2.0102 | $|D| = 313$, $|E| = 101$ | — | [160] |
| KTiOPO$_4$ (center A) | Q | RT | 1.9758, 1.9713, 1.9637 | 1675.7, 2797.2 | — | [161] |

(continued)

## $(Cr^{3+}(3d^3))$ listing – contd.

| Host | Frequency (GHz)/band | T (K) | $\tilde{g}$ | $(b_2^0 = D)$, $(b_2^2 = 3E)$ $(10^{-4}\,cm^{-1})$ | $A_x, A_y, A_z$ $(10^{-4}\,cm^{-1})$ | References |
|---|---|---|---|---|---|---|
| KTiOPO$_4$ (center B) | Q | RT | 1.9741, 1.9708, 1.9674 | 1364.5, 3584.7 | — | [161] |
| KTiOPO$_4$ (center C) | Q | RT | 2.1207, 1.9769, 1.8228 | $\tilde{D}=$ −1779.8, 529.8, 1249.6 | — | [162] |
| KTiOPO$_4$ (center D) | Q | RT | 1.9826, 1.9777, 1.9733 | $\tilde{D}=$ −3625.1, −602.7, 4228.1 | — | [162] |
| K$_2$MgCl$_4$ (center I) | X | RT | 1.9846, 1.9846, 1.9831 | −583.5, — | — | [163] |
| K$_2$MgCl$_4$ (center II) | X | RT | 1.9752, 1.9643, 1.9832 | −851.3, 578.3 | — | [163] |
| K$_2$MgCl$_4$ (center III) | X | RT | 1.9845, 1.9828, 1.9867 | 455.1, 387.8 | — | [163] |
| K$_2$MgCl$_4$ (center IV) | X | RT | 1.9862, 1.9862, 1.9821 | −855.5, — | — | [163] |
| K$_3$H(SO$_4$)$_2$ | X | RT | 1.959 | D = 4450, E = 1460 (G) | — | [164] |
| L-histidine · HCl · H$_2$O (site I) | X | RT | 1.9108, 1.9791, 2.0389 | D = 300, E = 96 | 252, 254, 304 | [165] |
| L-histidine · HCl · H$_2$O (site II) | X | RT | 1.8543, 1.9897, 2.0793 | D = 300, E = 96 | 251, 257, 309 | [165] |
| La$_3$Ga$_5$GeO$_{14}$ (LGGO) line A | X | RT | $g_\parallel = 1.970$, $g_\perp = 1.97$ | $b_2^0 = 8000$ | — | [148] |
| La$_3$Ga$_5$SiO$_{14}$ (LGS) line A | X | RT | $g_\parallel = 1.97$, $g_\perp = 1.97$ | $b_2^0 = 9000$ | — | [148] |

(continued)

## ($Cr^{3+}(3d^3)$) listing – contd.)

| Host | Frequency (GHz)/ band | T (K) | $\tilde{g}$ | ($b_2^0 = D$), ($b_2^2 = 3E$) ($10^{-4}$ cm$^{-1}$) | $A_x, A_y, A_z$ ($10^{-4}$ cm$^{-1}$) | References |
|---|---|---|---|---|---|---|
| LiCaAlF$_6$ | 9.5 | 300 | $g_\| = 1.974$, $g_\perp = 1.974$ | −1010, — | — | [166] |
| LiCaGaF$_6$ | 9.5 | 300 | $g_\| = 1.974$, $g_\perp = 1.974$ | — | — | [166] |
| LiKSO$_4$ | X | 300 | 2.0763, 1.9878, 1.8685 | $D = 549$, $E = 183$ | — | [167] |
| LiNbO$_3$ | 9.5 | 300 | 1.972 | −3869.34, — | — | [168] |
| LiNbO$_3$ | 9.4 | 4.2 | 1.97 | Range = (−3500 to −4300), — | — | [169] |
| LiNbO$_3$ (site Li) | X | RT | 1.957 | $D = \|0.393\|$ | — | [170] |
| LiNbO$_3$ (site Nb) | X | RT | 1.96 | $D = \|0.1\|$ | — | [170] |
| LiScGeO$_4$ | X, Q, 36–656 | 4.2 | 1.97, 1.97, 1.97 | −6298.8, −1032.4 | — | [171] |
| LiSrAlF$_6$ | 9.5 | 300 | $g_\| = 1.974$, $g_\perp = 1.974$ | 135, — | — | [166] |
| LiSrGaF$_6$ | 9.5 | 300 | $g_\| = 1.974$, $g_\perp = 1.974$ | — | — | [166] |
| LiTaO$_3$ | 9.5 | 300 | 1.995 | −4439.7, — | — | [168] |
| Mg$_2$SiO$_4$ (forsterite) (octahedral site) | 65–535 | 4.2 | 1.978, 1.969, 1.965 | −30.97, −8.59 (GHz) | — | [172] |
| NH$_4$LiSO$_4$ (site I) | X | RT | 2.0003 | $D = 269$, $E = 82$ | — | [173] |
| NH$_4$LiSO$_4$ (site II) | X | RT | 1.9904 | $D = 251$, $E = 79$ | — | [173] |
| NaNH$_4$SO$_4 \cdot 2H_2O$ (site α) | Q | RT | $g_{iso} = 1.981$ | $\tilde{D} =$ −1779.8, 529.8, 1249.6(MHz) | — | [174] |
| NaNH$_4$SO$_4 \cdot 2H_2O$ (site β) | Q | RT | $g_{iso} = 1.987$ | $\tilde{D} =$ −3625.1, −602.7, 4228.1(MHz) | — | [174] |

(continued)

## ($Cr^{3+}$ ($3d^3$) listing – contd.)

| Host | Frequency (GHz)/band | T (K) | $\tilde{g}$ | ($b_2^0 = D$), ($b_2^2 = 3E$) ($10^{-4}$ cm$^{-1}$) | $A_x, A_y, A_z$ ($10^{-4}$ cm$^{-1}$) | References |
|---|---|---|---|---|---|---|
| $(NH_4)_2Co(SO_4)_2 \cdot 6H_2O$ (site I) | Q | 295 | 2.13528, 1.44974, 1.96175 | 1564.7, 276.2 | — | [175] |
| $(NH_4)_2Co(SO_4)_2 \cdot 6H_2O$ (site II) | Q | 295 | 2.13662, 1.62653, 1.91655 | −1447.9, −272.0 | — | [175] |
| $(NH_4)_2Mg(SO_4)_2 \cdot 6H_2O$ | X | 295 | g = 1.9763 | D = 611 | — | [176] |
| $PbWO_4$ | X | 10 | 1.8050, 1.8050, 1.9231 | $B_2^0 = -287.6$, $B_2^2 = 0$ | — | [177] |
| Silica xerogels | X | RT | 1.975 | — | — | [2] |
| $SiO_2$ (δ signal) | X, Q | 150, 300 | 1.970, 1.725, 1.725 | 4900, 4890 | — | [144] |
| $SiO_2 \cdot Al_2O_3$ (δ signal) | X, Q | 150, 300 | 1.970, 1.570, 1.570 | 4900, 4890 | — | [144] |
| $SnO_2$ (Center 1) | 9.423 | 5 | $g_{iso} = 1.997$ | −2315, 1680 | | [178] |
| $SnO_2$ (Center 2) | 9.423 | 5 | $g_{iso} = 1.997$ | −2310, 1500 | | [178] |
| $SnO_2$ (Center 3) | 9.423 | 5 | $g_{iso} = 1.997$ | −24205, 1680 | | [178] |
| $SnO_2$ (Center 4) | 9.423 | 5 | $g_{iso} = 1.997$ | −300, 0 | | [178] |
| $SrB_4O_7$ | X | RT | 4.448 | — | — | [122] |
| $Sr_3Ga_2Ge_4O_{14}$ | 9.4 | 300 | 1.973 | 9000.0, 0.0 | — | [179] |
| $Sr_3Ga_2Ge_4O_{14}$ (SGGO) line A | X | RT | $g_\parallel = 1.980$, $g_\perp = 1.98$ | $b_2^0 = 9000$ | — | [148] |
| $Sr_3Ga_2Ge_4O_{14}$ (SGGO) line B | X | RT | $g_\parallel = 1.97$, $g_\perp = 1.97$ | $b_2^0 = 8000$, $b_2^2 = 2400$ | — | [148] |
| Stishovite | 9.390 | 294 | 1.9799, 1.9799, 1.9799 | — | 1.79, 1.68, 1.71 (mT) | [180] |
| $TiO_2$ | X | RT | 1.970, 1.970, 1.970 | D = −6858, E = −1352 | — | [181] |
| $TlZnF_3$ (site E) | 9.5 | 297 | 1.9711, 1.9728, 1.9721 | −1954.1, 541.6 | — | [182] |

*(continued)*

## ($Cr^{3+}(3d^3)$) listing – contd.

| Host | Frequency (GHz)/ band | T (K) | $\tilde{g}$ | ($b_2^0 = D$), ($b_2^2 = 3E$) ($10^{-4}$ cm$^{-1}$) | $A_x, A_y, A_z$ ($10^{-4}$ cm$^{-1}$) | References |
|---|---|---|---|---|---|---|
| TlZnF$_3$ (site F) | 9.5 | 297 | 1.9714, 1.9716, 1.9738 | 3462.4, −1037.0 | — | [182] |
| YAlO$_3$ | X | RT | 1.981, 1.981, 1.981 | 445, −278 | — | [183] |
| Y$_3$Al$_5$O$_{12}$ | X, Q | — | — | $B_2^0 = 26.97$, $B_4^0 = -20.13$, $B_4^3 = 535.53$ | — | [184] |
| Zn(C$_4$H$_3$O$_4$)$_2$ · 4H$_2$O | 9.6 | 295 | $g_\parallel = 2.265$, $g_\perp = 2.202$ | 15.25, 0.010, $b_2^{-2} = -0.005$ (GHz) | — | [185] |
| Zn(C$_4$H$_3$O$_4$)$_2$ · 4H$_2$O | 9.6 | 79 | $g_\parallel = 2.255$, $g_\perp = 2.198$ | 15.45, 0.015, $b_2^{-2} = -0.006$ (GHz) | — | [185] |
| Zn(C$_4$H$_3$O$_4$)$_2$ · 4H$_2$O | 9.6 | 4.2 | $g_\parallel = 2.252$, $g_\perp = 2.192$ | 15.62, 0.016, $b_2^{-2} = -0.006$ (GHz) | — | [185] |
| ZnKPO$_4$ · 6H$_2$O | X | 295 | 2.1065, 1.9952, 1.9687 | $D_{xx}, D_{yy}, D_{zz} = 66.2, 91.7, -157.9$ mT | — | [186] |

## $Cr^{3+}(3d^3)$ Glasses

| Host | Frequency (GHz or band) | T (K) | $\tilde{g}$ | ($b_2^0 = D$), ($b_2^2 = 3E$) ($10^{-4}$ cm$^{-1}$) | $A_x, A_y, A_z$ ($10^{-4}$ cm$^{-1}$) | References |
|---|---|---|---|---|---|---|
| CdO · P$_2$O$_5$ glass (site I) | X | RT | $g_{iso} = 4.8223$ | — | — | [187] |
| CdO · P$_2$O$_5$ glass (site II) | X | RT | $g_{iso} = 1.9630$ | — | — | [187] |

## $Fe^{5+}$. Data tabulation of SHPs
### $Fe^{5+}(3d^3)$

| Host | Frequency (GHz)/ band | T (K) | $\tilde{g}$ | $(b_2^0 = D)$, $(b_2^2 = 3E)$ $(10^{-4}\ cm^{-1})$ | $A_x, A_y, A_z$ $(10^{-4}\ cm^{-1})$ | References |
|---|---|---|---|---|---|---|
| [(Cyclam-acetato)Fe(V)(N)]$^+$ | X | 10 | $g_{iso}=2.0$ | −2900, −1131 | $\tilde{A}\,(A_x, A_y, A_z)/g_N\mu_N(T) = -12.8,$ $-11.4, 1.9$ | [188] |
| [(Cyclam)(N)$_3$Fe(V)(N)]$^+$ | X | 10 | $g_{iso}=2.0$ | −3700, −1053 | $\tilde{A}\,(A_x, A_y, A_z)/g_N\mu_N(T) =$ $-13.3, -10.6, 2.5$ | [188] |

## $Mn^{4+}$. Data tabulation of SHPs
### $Mn^{4+}(3d^3)$

| Host | Frequency (GHz)/ band | T (K) | $\tilde{g}$ | $(b_2^0 = D)$, $(b_2^2 = 3E)$ $(10^{-4}\ cm^{-1})$ | $A_x, A_y, A_z$ $(10^{-4}\ cm^{-1})$ | References |
|---|---|---|---|---|---|---|
| $Ba_3(VO_4)_2$ | 9.09 | RT | $g_\| = 1.977,$ $g_\perp = 1.985$ | $D = 28$ (GHz) | $A_\| = 80, A_\perp = 19$ | [131] |
| $LaGa_{1-x}Mn_xO_3$ | X | RT | 1.990, 1.988, 1.991 | 3556, −1650 (MHz) | −212, −112, −215 MHz | [189] |
| $PbTiO_3$ | X, Q | RT | $g_\| = 1.990,$ $g_\perp = 1.987$ | $|b_2^0| = 3166$ | $A_\| = 79.46,$ $A_\perp = 71.05$ | [190] |
| $PbTiO_3$ | 9.4 | 10 | $g_\| = 1.98205,$ $g_\perp = 1.9868$ | $B_2^0 = -3.0833$ (GHz) | $A^{55}{}_\| = -213.5$ (MHz), $A^{55}{}_\perp = -212$ (MHz) | [191] |
| $TiO_2$ | X | RT | 1.988, 1.985, 1.985 | $D = 4003,$ $E = -1324$ | 72 | [181] |

## $V^{2+}$. Data tabulation of SHPs
### $V^{2+}(3d^3)$

| Host | Frequency (GHz)/ band | T (K) | $\tilde{g}$ | $b_2^0(= D)$, $b_2^2(= 3E)$ $(10^{-4}\ cm^{-1})$ | $A_x, A_y, A_z$ $(10^{-4}\ cm^{-1})$ | References |
|---|---|---|---|---|---|---|
| $CsMgCl_3$ | X, Q | 302 | 1.9747, 1.9747, 1.9729 | $D = 965$ | — | [153] |
| $Zn_xCd_{1-x}Te$ ($X = 0.04, 0.10$) | X | 4 | 1.976, 1.959, 1.974 | −9330, −2190 | −65, −57, −65 | [192] |

## 3d⁴ ($Cr^{2+}$, $Mn^{3+}$), $S = 2$
## $Cr^{2+}$. Data tabulation of SHPs
## $Cr^{2+}$ ($3d^4$)

| Host | Frequency (GHz)/band | T (K) | $\tilde{g}$ | $b_2^0 (= D)$, $b_2^2 (= 3E)$ ($10^{-4}$ cm$^{-1}$) | $b_4^0, b_4^2, b_4^4$ ($10^{-4}$ cm$^{-1}$) | $A_x, A_y, A_z$ ($10^{-4}$ cm$^{-1}$) | References |
|---|---|---|---|---|---|---|---|
| AgGaSe₂ (orthorhombic center) | 65–530 | 4.2 | 1.91, 1.98, 1.98 | $\|B_2^0\|$ = 21.87 GHz, $B_2^2$ = 4.77 GHz | $B_4^0$ = 0.0025 GHz, $B_4^2$ = 0.0789 GHz, $B_4^4$ = 0.0226 GHz | — | [193] |
| AgGaSe₂ (tetragonal center) | 65–530 | 4.2 | $g_\|$ = 1.92, $g_\perp$ = 1.98 | $\|B_2^0\|$ = 22.16 GHz | $B_4^0$ = 0.0039 GHz, $B_4^4$ = 0.292 GHz | — | [193] |
| AgGaS₂ (orthorhombic center) | 65–530 | 4.2 | $g_{zz}$ = 1.90 | $\|B_2^0\|$ = 23.05 GHz, $B_2^2$ = 3.24 GHz | $B_4^0$ = −0.0167 GHz, $B_4^2$ = 1.2248 GHz, $B_4^4$ = 0.1719 GHz | — | [193] |
| AgGaS₂ (tetragonal center) | 65–530 | 4.2 | $g_\|$ = 1.96 | — | $B_4^4$ = 0.0333 GHz | — | [193] |
| BaF₂ | X, Q | 4–300 | $g_\|$ = 1.974, $g_\perp$ = 1.997 | $D - a$ = −21 140, $\|a\|$ = 183 | — | $A_\perp(^{53}Cr)$ = 35 MHz | [194] |
| CaF₂ | X, Q | 77–300 | $g_\|$ = 1.965, $g_\perp$ = 1.995 | $b_2^0$ = −22370, $b_2^2$ = 1428 cm$^{-1}$ | — | $A_\perp(^{53}Cr)$ = 33.3 MHz, $A_\|(^{53}Cr)$ = 36 MHz | [195] |
| CdGa₂S₄ | 65–240 | | $g_\|$ = 1.93, $g_\perp$ = 1.99 | 70 977, — (MHz) | 114, —, 3252 (MHz) | — | [196] |
| [Cr·(H₂O)₆](SO₄) | 109.5, 330 | 10 | $g_\| = g_\perp$ = 1.98 | −22 000, 0 | — | — | [197] |
| CrSO₄ · 5H₂O | 90–440 | 10 | $g_\|$ = 1.98, $g_\perp$ = 1.98 | −22 000, 100 | — | — | [197] |
| Cu(SO₄) · 5H₂O | ~95–440 | 10 | 1.98, 1.98, 1.98 | −22 400, 0 | — | — | [197] |
| SrCl₂ | X, Q | 4–300 | $g_\|$ = 1.974, $g_\perp$ = 1.996 | $D - a$ = −19790, $\|a\|$ = 213 | — | $A_\perp(^{53}Cr)$ = 43.5 MHz, $A_\|(^{53}Cr)$ = 31.2 MHz | [194] |

## $Mn^{3+}$. Data tabulation of SHPs
## $Mn^{3+}$ (3d$^4$)

| Host | Frequency (GHz)/ band | T (°K) | $\tilde{g}$ | $b_2^0 (= D)$, $b_2^2 (= 3E)$ ($10^{-4}$ cm$^{-1}$) | $b_4^0, b_4^2, b_4^4$ ($10^{-4}$ cm$^{-1}$) | $A_x, A_y, A_z$ ($10^{-4}$ cm$^{-1}$) | References |
|---|---|---|---|---|---|---|---|
| 8,12-Diethyl-2,3,7,13,17,18-hexamethylcorrolato (immobilized solid1) | 276.62 | 4.2 | $g_\| = 2.00$, $g_\perp = 2.02$ | −26 400, 450 | — | — | [198] |
| $C_{101}H_{196}Mn_{12}O_{49}$ | 115 | 5 | $g_\| = 2.00$, $g_\perp = 1.93$ | $D = -459$ | 14.04, —, 12.0 | — | [199] |
| $4 \cdot 3CH_2Cl_2$ ($C_{33}H_{36}N_6O_{14}C_{17}Mn_3$) | 129 | 5 | $g = 2.00$ | −3 000, ≤150 | −0.3, —, — | — | [200] |
| $CsMn(SO_4)_2 \cdot 12D_2O$ (deuterated salt) | 344.7 | <30 | 1.981, 1.993, 1.988 | −44 910, 2 480 | — | — | [201] |
| $CsMn(SO_4)_2 \cdot 12H_2O$ (protonated salt) | 344.7 | <30 | 2.001, 1.997, 1.966 | −44 310, 2 580 | — | — | [201] |
| $[Ga(H_2O)_6](SO_4)_2 \cdot 6H_2O$ | X and 190 GHz | 5 | 2.000, 2.000, 1.9844 | −45 140, −1 610 | −41, −1.4, −57.7 | −87, −53.1 | [202] |
| $Mn(acac)_3$ (Hacac = 2,4 pentanedione) | 190–575 | | 1.99, 1.99, 1.99 | −45 200, 7 500 | — | 156 MHz | [203] |
| $[Mn(bpea)(F)_3]$ (bpea, N,N-bis(2-pyridylmethyl)-etylamine) | 190–575 | 5–15 | 1.96, 1.98, 1.98 | −36 700, 21 000 | — | — | [204] |
| $[Mn(bpea)(N3)_3]$ (bpea = N,N-bis(2-pyridylmethyl)-etylamine) | 190–575 | 5–15 | 2.02, 1.98, 1.95 | −35 000, 24 600 | — | — | [204] |
| MnCor | 95–575 | 4.2 | $g_\| = 2.00$, $g_\perp = 2.02$ | −26 400, 150 | — | — | [205] |
| MnCor(py) | 95–575 | 4.2 | $g_\| = 2.00$, $g_\perp = 2.02$ | −27 800, 300 | — | — | [205] |
| Mn(dbm)$_3$ dbm, 1,3-diphenyl-1,3-propanedionate | ~95, 110 | 10 | 1.99, 1.99, 1.97 | $b_2^0 = $ −43 500, $b_2^2 = -7800$ | — | — | [197] |
| Mn(dbm)$_3$ dbm = 1,3-diphenyl-1,3-propanedionate | 245–349.3 | 4.2–30 | 1.99, 1.99, 1.97 | $b_2^0 = $ −45 500, $b_2^2 = -8400$ | — | — | [206] |
| [Mn(dbm)$_3$] | 245–349.3 | 15 | 1.99, 1.99, 1.96 | −45 500, 2 800 | — | — | [207] |

(continued)

## ($Mn^{3+}$ ($3d^4$) listing – contd.)

| Host | Frequency (GHz)/ band | $T$ (°K) | $\tilde{g}$ | $b_2^0 (= D)$, $b_2^2 (= 3E)$ ($10^{-4}$ cm$^{-1}$) | $b_4^0, b_4^2, b_4^4$ ($10^{-4}$ cm$^{-1}$) | $A_x, A_y, A_z$ ($10^{-4}$ cm$^{-1}$) | References |
|---|---|---|---|---|---|---|---|
| [Mn(dbm)$_2$(py)$_2$](ClO$_4$) | 95–440 | 4.2–77 | 1.993, 1.994, 1.983 | −45 040, −4 250 | −108, 420, 2 880 | — | [208] |
| MnDPDME Br | 95–575 | 5 | $g_\parallel = 2$, $g_\perp = 2$ | −11 000, ∼0 | — | — | [205] |
| MnDPDME Cl | 95–575 | 5 | $g_\parallel = 2$, $g_\perp = 2$ | −25 300, ∼100 | — | — | [205] |
| Mn(DP-IX-DME)Cl | X, Q | 4 | $g_\parallel = 2$ | −25 300 | — | — | [209] |
| MnMo$_6$Se$_8$ | 208 | 10 | $g_\parallel = 2.145$, $g_\perp = 2.186$ | $D = -480.1$ (G) | $B_4^0 = 42.3$, $B_4^4 = 275.2$ | — | [210] |
| [Mn$^{3+}$(NCTPP)(py)$_2$] | 192 | 10 | 2.000, 2.000, 2.006 | −30 840, −6 080 | — | — | [211] |
| Mn(ODMAPz)Cl (ODMAPz = 2,3,7,8, 12,13,17,18-octakis (dimethyl-amino) porphyrazinato) | X, Q | 4 | $g_\parallel = 1.984$ | $b_2^0 = -2.33$ cm$^{-1}$ | — | — | [212] |
| Mn(ODMAPz)DTC (ODMAPz = 2,3,7,8, 12,13,17,18-octakis (dimethyl-amino) porphyrazinato; DTC = diethyldithio-carbamato) | X, Q | 4 | $g_\parallel = 1.983$ | $b_2^0 = -2.62$ cm$^{-1}$ | — | — | [212] |
| MnPcCl | 95–575 | 5 | $g_\parallel = 2.00$, $g_\perp = 2.005$ | −23 100, 0 | — | — | [205] |
| Mn(salen) (solid) (salen = [(R-R)-(−)-N, N′-bis(3,5-di-tert-butylsalicylidene)-1,2-cyclohexanediamino-manganese) | 192.8 | 30 | 2.00 | −22 400, — | — | — | [213] |
| Mn(salen) (solution) (salen = [(R-R)-(−)-N, N′-bis(3,5-di-tert-butylsalicylidene)-1,2-cyclohexanediamino-manganese) | 192.8 | 30 | 2.00 | −24 700, 5 100 | — | — | [213] |

(continued)

## ($Mn^{3+}$($3d^4$) listing – contd.)

| Host | Frequency (GHz)/ band | $T$ (°K) | $\tilde{g}$ | $b_2^0 (= D)$, $b_2^2 (= 3E)$ ($10^{-4}$ cm$^{-1}$) | $b_4^0, b_4^2, b_4^4$ ($10^{-4}$ cm$^{-1}$) | $A_x, A_y, A_z$ ($10^{-4}$ cm$^{-1}$) | References |
|---|---|---|---|---|---|---|---|
| [Mn(terpy)(F)$_3$] (terpy = 2,2′:6′,2″-terpyridine) | 190–575 | 5–15 | 1.97, 2.04, 1.96 | −38 200, 22 500 | — | — | [204] |
| [Mn(terpy)(N$_3$)$_3$] (terpy = 2,2′:6′,2″-terpyridine) | 190–575 | 5–15 | 2.00, 1.98, 2.01 | −32 900, 14 400 | — | — | [204] |
| Mn(TPP)Cl TPP = 5,10,15,20-tetraphenyl-porphyrinato | X, Q | 4 | $g_\parallel = 1.822$ | $b_2^0 = -2.27$ cm$^{-1}$ | — | — | [212] |
| MnTPPCl | 95–575 | 5 | $g_\parallel = 1.98$, $g_\perp = 2.005$ | −25 000, 0 | — | — | [205] |
| MnTPPCl | 95–575 | 12 | $g_\parallel = 1.98$, $g_\perp = 2.005$ | −22 900, 0 | — | — | [205] |
| MnTPP(ClO$_4$) | 95–575 | 5 | $g_\parallel = 2$, $g_\perp = 2$ | −20 000, — | — | — | [205] |
| MnTPP(py)Cl | 95–575 | 5 | $g_\parallel = 2$, $g_\perp = 2$ | −30 000, — | — | — | [205] |
| Mn(TSP)Cl | 493.1 | 8 | 1.996, 1.996, 2.01 | −31 160, −7 | — | — | [214] |
| Mn(TSP) (solid) (TSP, mesotetrasulfanato-porphyrinatomanganese) | 108.41 | 20 | 2 | −31 200, — | — | — | [213] |
| Mn(TSP) (solution) (TSP, mesotetrasulfanato-porphyrinatomanganese) | 220.1 | 30 | 2 | −31 600, — | — | — | [213] |
| SrTiO$_3$ (Mn$^{3+}$-Vo(I) center) | X, Q | 295 | $g_\parallel = 1.986$, $g_\perp = 1.997$ | $b_2^0 = -28 440.0$ | $b_4^0 = -170.0$, $b_4^4 = -3800.0$ | $A_\parallel = -38.0$, $A_\perp = -64.0$ | [215] |
| SrTiO$_3$ (Mn$^{3+}$-X center) | X, Q | 295 | $g_\parallel = 1.986$, $g_\perp = 1.997$ | $b_2^0 = -27 550.0$ | $b_4^0 = -170.0$, $b_4^4 = -1350.0$ | $A_\parallel = -10.4$, $A_\perp = -29.8$ | [215] |
| SrTiO$_3$ (Mn$^{3+}$-Vo(II) center) | X, Q | 295 | $g_\parallel = 1.986$, $g_\perp = 1.997$ | $b_2^0 = -26 340.0$ | $b_4^0 = -170.0$, $b_4^4 = -1350.0$ | $A_\parallel = -28.0$, $A_\perp = -45.0$ | [215] |
| [(tpfc)Mn(OPPh$_3$)] | 95–575 | 4.2 | $g_\parallel = 1.980$, $g_\perp = 1.994$ | −26 900, 300 | — | — | [205] |

## 3d$^5$ (Cr$^+$, Fe$^{3+}$, Mn$^{2+}$), S = 5/2

### Cr$^+$. Data tabulation of SHPs

**Cr$^+$ (3d$^5$)**

| Host | Frequency (GHz)/band | T (K) | $\tilde{g}$ | $b_4^0, b_4^2, b_4^4$ ($10^{-4}$ cm$^{-1}$) | References |
|---|---|---|---|---|---|
| KTaO$_3$ | X | 77 | $g_{iso} = 1.997$ | $b_4^0 = 171$, $b_4^4 = 830$ | [216] |

### Fe$^{3+}$. Data tabulation of SHPs

**Fe$^{3+}$ (3d$^5$)**

| Host | Frequency (GHz)/band | T (°K) | $\tilde{g}$ | $b_2^0 (= D)$, $b_2^2 (= 3E)$ ($10^{-4}$ cm$^{-1}$) | $b_4^0, b_4^2, b_4^4$ ($10^{-4}$ cm$^{-1}$) | $A_x, A_y, A_z$ ($10^{-4}$ cm$^{-1}$) | References |
|---|---|---|---|---|---|---|---|
| AlF$_3 \cdot$ 3H$_2$O (center 1) | X | 77 | $g_{iso} = 2$ | $b_2^0 = 9250$, $b_2^2 = 9249$ MHz | — | — | [217] |
| AlF$_3 \cdot$ 3H$_2$O (center 2) | X | 77 | $g_{iso} = 2$ | $b_2^0 = 14390$, $b_2^2 = 10101$ MHz | — | — | [217] |
| $\alpha$-Al$_2$O$_3$ | 94.9 | 295 | $g_{\parallel} = 2.0034$ | $\lvert D \rvert = 5.033$ GHz | $\lvert a - F \rvert = 0.986$ GHz | — | [143] |
| Al$_2$O$_3$ | X | 4.2 | 2.0 | 5.032 GHz, — | — | — | [218] |
| Al$_2$O$_3$ | X | RT | 2.00 | 1705, — | $-108$, —, —, $b_4^3 = 2181$ | — | [219] |
| Al$_2$O$_3$ (unannealed) (site I) | 9.45 | 297 | $g_{\parallel} = 2.00$, $g_{\perp} = 2.00$ | $D = 64.650$ mT | $B_4^0 = 0.190$ mT, $B_4^3 = 0.005$ mT | — | [220] |
| Al$_2$O$_3$ (unannealed) (site II) | 9.45 | 297 | $g_{\parallel} = 2.00$, $g_{\perp} = 2.00$ | $D = 73.120$ mT | $B_4^0 = 0.261$ mT, $B_4^3 = 0.009$ mT | — | [220] |
| Al$_2$O$_3$ (annealed at 1473 K) (site I) | 9.45 | 297 | $g_{\parallel} = 1.99$, $g_{\perp} = 1.98$ | $D = 63.545$ mT | $B_4^0 = 0.191$ mT, $B_4^3 = 0.006$ mT | — | [220] |
| Al$_2$O$_3$ (annealed at 1473 K) (site II) | 9.45 | 1473 | $g_{\parallel} = 1.99$, $g_{\perp} = 1.99$ | $D = 73.226$ mT | $B_4^0 = 0.212$ mT, $B_4^3 = 0.008$ mT | — | [220] |
| Al$_2$O$_3$ (annealed at 1573 K) (site I) | 9.45 | 297 | $g_{\parallel} = 1.97$, $g_{\perp} = 1.96$ | $D = 67.861$ mT | $B_4^0 = 0.197$ mT, $B_4^3 = 0.008$ mT | — | [220] |
| Al$_2$O$_3$ (annealed at 1573 K) (site II) | 9.45 | 1573 | $g_{\parallel} = 2.00$, $g_{\perp} = 1.99$ | $D = 75.631$ mT | $B_4^0 = 0.259$ mT, $B_4^3 = 0.009$ mT | — | [220] |

(continued)

## ($Fe^{3+}(3d^5)$) listing – contd.

| Host | Frequency (GHz)/ band | $T$ (°K) | $\tilde{g}$ | $b_2^0 (= D)$, $b_2^2 (= 3E)$ ($10^{-4}$ cm$^{-1}$) | $b_4^0, b_4^2, b_4^4$ ($10^{-4}$ cm$^{-1}$) | $A_x, A_y, A_z$ ($10^{-4}$ cm$^{-1}$) | References |
|---|---|---|---|---|---|---|---|
| $Al_2O_3$ (annealed at 1673 K) (site I) | 9.45 | 297 | $g_\| = 1.98$, $g_\perp = 1.97$ | $D = 69.377$ mT | $B_4^0 = 0.198$ mT, $B_4^3 = 0.008$ mT | — | [220] |
| $Al_2O_3$ (annealed at 1673 K) (site II) | 9.45 | 1673 | $g_\| = 2.00$, $g_\perp = 1.99$ | $D = 77.621$ mT | $B_4^0 = 0.294$ mT, $B_4^3 = 0.008$ mT | — | [220] |
| $Al_2O_3$ (annealed at 1773 K) (site I) | 9.45 | 297 | $g_\| = 1.99$, $g_\perp = 1.93$ | $D = 69.394$ mT | $B_4^0 = 0.199$ mT, $B_4^3 = 0.009$ mT | — | [220] |
| $Al_2O_3$ (annealed at 1773 K) (site II) | 9.45 | 1773 | $g_\| = 2.00$, $g_\perp = 2.00$ | $D = 78.357$ mT | $B_4^0 = 0.296$ mT, $B_4^3 = 0.009$ mT | — | [220] |
| $Al_2O_3$ (annealed at 1873 K) (site I) | 9.45 | 297 | $g_\| = 2.08$, $g_\perp = 2.07$ | $D = 69.406$ mT | $B_4^0 = 0.199$ mT, $B_4^3 = 0.010$ mT | — | [220] |
| $Al_2O_3$ (annealed at 1873 K) (site II) | 9.45 | 1873 | $g_\| = 2.00$, $g_\perp = 2.00$ | $D = 78.401$ mT | $B_4^0 = 0.323$ mT, $B_4^3 = 0.019$ mT | — | [220] |
| $\beta''$-(BEDT-TTF)$_4$[(H$_3$O)Fe(C$_2$O$_4$)3]C$_6$H$_5$CN | 55, 63, 72, 95, 100 | 1–7 | $g_\| = 1.98$, $g_\perp = 1.96$ | $\|D\| = 1300$, $\|E\| = 150$ | — | 89 | [221] |
| $BiVO_4$ | Q | RT | $g_{xx} = 1.976$, $g_{xz} = 0.0033$, $g_{yy} = 1.995$, $g_{zz} = 1.994$ | $B_2^0 = 467.3$, $B_2^1 = 1693$, $B_2^2 = 2339$ | $B_4^0 = 0.12$, $B_4^1 = -2.3$, $B_4^2 = -0.8$, $B_4^3 = 4.0$, $B_4^4 = -0.3$ | — | [222] |
| $BiVO_4$ | X | 3.8–300 | 1.976, 1.995, 1.995 | 4209, 2904 | — | — | [223] |
| $Bi_4Ge_3O_{12}$ single crystal | X | 90 | $g_\| = 2.015$, $g_\perp = 2.003$ | $b_2^0 = 10300$ | $b_4^0 = 10$, $b_4^4 = -360$ | — | [224] |
| $CaWO_4$ (site 1) | 8.9 | 10 | 2.09509, 2.02054, 1.90750 | $D_{xx} = 835.36$, $D_{yy} = 4.95$, $D_{zz} = -840.31$ (mT) | — | — | [225] |
| $CaYAlO_4$ | 249.9 | 253 | $g_\| = 1.992$, $g_\perp = 1.987$ | 9787.10, — | −1325.92, —, 1371.62 | — | [226] |
| $CaYAlO_4$ | 9.79, 35.67 | 295 | $g_\| = 1.992$, $g_\perp = 1.987$ | 9787.10, — | −1325.92, —, 1371.62 | — | [226] |

(continued)

## ($Fe^{3+}$ ($3d^5$) listing – contd.)

| Host | Frequency (GHz)/band | T (°K) | $\tilde{g}$ | $b_2^0$ (= D), $b_2^2$ (= 3E) ($10^{-4}$ cm$^{-1}$) | $b_4^0, b_4^2, b_4^4$ ($10^{-4}$ cm$^{-1}$) | $A_x, A_y, A_z$ ($10^{-4}$ cm$^{-1}$) | References |
|---|---|---|---|---|---|---|---|
| $Ca_{9.4}Fe_{0.4}Mn_{0.2}$ $(PO_4)_6(OH)_2$ hydroxyapatite | X | — | 4.3 | — | — | — | [227] |
| $CdCl_2$ | 9.50 | 26 | g = 2.008 | 815.7, — | $-13.9, —, —$, $b_4^3 = -400.0$ | — | [228] |
| $CdCl_2$ | 9.50 | 225 | g = 2.008 | 736, — | $-10.6, —, —$, $b_4^3 = 302.0$ | — | [228] |
| Clay (DCV kaolinite) (site I) | 9.42 | 293 | — | $B_2^0 = 1.104$, $B_2^2 = 0.666$ | $-0.025, —, —$ | — | [229] |
| Clay (DCV kaolinite) (site II) | 9.42 | 293 | — | $B_2^0 = 1.057$, $B_2^2 = 0.661$ | $-0.021, —, —$ | — | [229] |
| Clay (DCV kaolinite) (site I) | 9.42 | 120 | — | $B_2^0 = 1.156$, $B_2^2 = 0.721$ | $-0.033, —, —$ | — | [229] |
| Clay (DCV kaolinite) (site II) | 9.42 | 120 | — | $B_2^0 = 1.120$, $B_2^2 = 0.753$ | $-0.020, —, —$ | — | [229] |
| Clay (DCV kaolinite) (site I) | 9.42 | 4 | — | $B_2^0 = 1.198$, $B_2^2 = 0.739$ | Not determined | — | [229] |
| Clay (DCV kaolinite) (site II) | 9.42 | 4 | — | $B_2^0 = 1.150$, $B_2^2 = 0.768$ | Not determined | — | [229] |
| Clay (SC dickite) (site I) | 9.42 | 293 | — | $B_2^0 = 1.084$, $B_2^2 = 0.851$ | $-0.001, —, —$ | — | [229] |
| Clay (SC dickite) (site II) | 9.42 | 293 | — | $B_2^0 = 1.218$, $B_2^0 = 0.993$ | $0.024, —, —$ | — | [229] |
| Clay (SC dickite) (site I) | 9.42 | 120 | — | $B_2^0 = 1.146$, $B_2^0 = 0.901$ | $-0.005, —, —$ | — | [229] |
| Clay (SC dickite) (site II) | 9.42 | 120 | — | $B_2^0 = 0.1.279$, $B_2^0 = 1.031$ | $-0.016, —, —$ | — | [229] |
| $Cs_2NaAlF_6$ (site I) | 33.97, 94.067 | 10 | g = 2.0023 | D = 215.5 MHz | $B_4^0 = -0.98$ MHz, $B_4^3 = -27.4$ MHz | A-matrix ($^{19}$F) = (47.48, 47.64, 106.70 MHz) | [230] |

(continued)

192 | 4 Multifrequency Transition Ion Data Tabulation

**($Fe^{3+}(3d^5)$) listing – contd.)**

| Host | Frequency (GHz)/ band | $T$ (°K) | $\tilde{g}$ | $b_2^0(=D)$, $b_2^2(=3E)$ ($10^{-4}$ cm$^{-1}$) | $b_4^0, b_4^2, b_4^4$ ($10^{-4}$ cm$^{-1}$) | $A_x, A_y, A_z$ ($10^{-4}$ cm$^{-1}$) | References |
|---|---|---|---|---|---|---|---|
| $Cs_2NaAlF_6$ (site II) | 33.97, 94.067 | 10 | $g = 2.0022$ | $D = -244.7$ MHz | $B_4^0 = -1.24$ MHz, $B_4^3 = -36.4$ MHz | $A(^{19}F)$-matrix = (43.65, 45.3, 106.60 MHz), $A_\parallel(^{23}Na) = 1.34$ MHz, $A_\perp(^{23}Na) = -0.87$ MHz; $\|Q'\| = \|2Q''\| = 0.034$ MHz | [230] |
| $Cs_2NaGaF_6$ | Q | 10 | $g = 2.0023$ | $D = 180.0$ MHz | $B_4^0 = -0.95$ MHz, $B_4^3 = -28.3$ MHz | A-matrix ($^{19}F$) = (47.33, 47.08, 106.16 MHz) | [230] |
| $Cs_2NaGaF_6$ | Q | 10 | $g = 2.0023$ | $D = -216.3$ MHz | $B_4^0 = -1.19$ MHz, $B_4^3 = -37.2$ MHz | A-matrix ($^{19}F$) = (43.82, 44.57, 106.08 MHz) $A_\parallel(^{23}Na) = 1.32$ MHz, $A_\perp(^{23}Na) = -0.90$ MHz; $\|Q'\| = \|2Q''\| = 0.022$ MHz | [230] |
| $Cs_2NaYF_6$ | 9.513, 34 010 | 8–12 | $g = 2.0028$ | — | $B_4^0 = -0.90$ MHz, $B_4^3 = -25.36$ MHz | A-matrix ($^{19}F$) = (44.5, 44.5, 103.7 MHz) | [230] |
| $CuAlO_2$ | X | 294 | $g_\parallel = 1.9942$, $g_\perp = 1.9964$ | $B_2^0 = 87.3$ | $B_4^0 = 4.533$, $B_4^{-3} = 104.23$ | — | [154] |

(continued)

## ($Fe^{3+}(3d^5)$ listing – contd.)

| Host | Frequency (GHz)/band | $T$ (°K) | $\tilde{g}$ | $b_2^0 (= D)$, $b_2^2 (= 3E)$ ($10^{-4}$ cm$^{-1}$) | $b_4^0, b_4^2, b_4^4$ ($10^{-4}$ cm$^{-1}$) | $A_x, A_y, A_z$ ($10^{-4}$ cm$^{-1}$) | References |
|---|---|---|---|---|---|---|---|
| $CuAlO_2$ | X | 110 | $g_\| = 1.9995$, $g_\perp = 1.9995$ | $B_2^0 = 91.0$ | $B_4^0 = 4.80$, $B_4^{-3} = 112.4$ | — | [154] |
| $CuAlO_2$ | Q | 123 | $g_\| = 1.9939$, $g_\perp = 1.9988$ | $B_2^0 = 92.0$ | $B_4^0 = 4.91$, $B_4^{-3} = 119.7$ | — | [154] |
| [(Cyclam-acetato) $FeN_3PF_6$] low spin $Fe^{3+}$ | X | 10 | 1.914, 2.264, 2.545 | — | — | — | [188] |
| [(Cyclam-acetato) $Fe(O_3SCF_3)]PF_6$ low spin $Fe^{3+}$ | Q | 10 | 1.63, 2.19, 2.96 | — | — | — | [188] |
| [{(Cyclam) $Fe(N_3)\}_2(\mu\text{-N})]^{2+}$ (site S = 3/2) | 9.64 | 433 | 3.85, 4.05, 2.00 | 90 000, — | — | — | [231] |
| Diferric transferrin (with oxalate, bicarbonate, or nitriloacetate as synergistic anions) | X | 4.2 | $g_{iso} \sim 2$ | $b_2^0 = 5800$, $b_2^2 = 991.8$ | — | — | [232] |
| GaN | 9.107 | 295 | $g_\| = 2.007$, $g_\perp = 2.009$ | $D = -764$ | $a = 80$, $F = -9$ | — | [233, 234] |
| GaN | 9.49, 34 | 5 | $g_\| = 2.006$, $g_\perp = 2.006$ | $D = -768$ | $(a - F) = 63$, $a = 78$ | — | [235] |
| Iron-doped MFI zeolite catalyst | 9.4/35.2/ 95–475 | 110–390/ 50–270/ 5–100 | $g_{eff} = 2.0$ (average of several reported values) | — | — | — | [236] |
| $KGaF_4$ (phase II) | 9.75 | 508 | 2.0035, 2.0035, 2.0023 | $b_2^0 = 1480$, $b_2^1 = 1855$, $b_2^2 = 435$ | $b_4^0 = -18.3$, $b_4^1 = -93$, $b_4^2 = -23$, $b_4^3 = 50$, $b_4^4 = -38$ | — | [159] |
| $KGaF_4$ (phase III) | 9.75 | RT | 2.0035, 2.0035, 2.0023 | $b_2^{-2} = 83$, $b_2^{-1} = 83$, $b_2^0 = 1518$, $b_2^1 = 2051$, $b_2^2 = 459$ | $b_4^{-4} = 10$, $b_4^{-3} = -87$, $b_4^{-2} = -16$, $b_4^{-1} = 33$, $b_4^0 = -12.4$, $b_4^1 = -97$, $b_4^2 = 59$, $b_4^3 = -87$, $b_4^4 = -5$ | — | [159] |

(continued)

## ($Fe^{3+}$($3d^5$) listing – contd.)

| Host | Frequency T (°K) (GHz)/ band | $\tilde{g}$ | $b_2^0 (= D)$, $b_2^2 (= 3E)$ ($10^{-4}$ cm$^{-1}$) | $b_4^0, b_4^2, b_4^4$ ($10^{-4}$ cm$^{-1}$) | $A_x, A_y, A_z$ ($10^{-4}$ cm$^{-1}$) | References |
|---|---|---|---|---|---|---|
| $KH_2PO_4$ | X | RT 2.00 | −32.7, 593.1 MHz | 42.0, 66.0, −41.4 MHz | — | [237] |
| $KTaO_3$ | X | 300 2.000 | 4 400, 1 960 | — | — | [238] |
| $KTaO_3$ | X | 4.2 2.000 | 4 850, 2 770 | — | — | [238] |
| $KTiOPO_4$ (site 1) | Q | — 2.00269, 1.99949, 1.99822 | $B_2^0 = -817.3$, $B_2^2 = -443.1$ | — | — | [239] |
| $KTiOPO_4$ (site 2) | Q | — 2.00322, 2.00189, 1.99995 | $B_2^0 = -739.1$, $B_2^2 = -442.2$ | — | — | [239] |
| $KTiOPO_4$ (site I) | Q | RT 2.0038, 2.0005, 2.0048 | $B_2^0 = -819.3$, $B_2^2 = -439.8$ | 29.4, −252.4, −146.9 | — | [240] |
| $KTiOPO_4$ (site II) | Q | RT 2.0048, 2.0046, 2.0067 | $B_2^0 = -722.5$, $B_2^2 = -552.9$ | 25.8, 299.1, −101.3 | — | [240] |
| $LiCaAlF_6$ | 9.4 | 290 $g_{iso}$ = 2.002170 | $B_2^0 = 40.072$ | $B_4^0 = -5.799$, $B_4^3 = -4.281$ | — | [241] |
| $LiCaAlF_6$ | 9.4 | 300 $g_{iso}$ = 2.002117 | $B_2^0 = 41.36$ | $B_4^0 = -5.87$, $B_4^3 = 10.91$ | — | [242] |
| $LiCaAlF_6$ | 9.4 | 300 2.002 | $B_2^0 = 41.36$ | $B_4^0 = -5.87$ | — | [242] |
| $LiCaAlF_6$ | — | — 2.00217 | $B_2^0 = 40.072$ | $B_4^0 = -5.799$ | $A_\| = 39.25$, $A_\perp = 20.87$ | [241] |
| $LiCoO_2$ | 9.8–406 | 20 1.9876, 1.9933, 2.0036 | $B_2^0 = 664$, $B_2^2 = 645$ (MHz) | $B_4^0 = 23$, $B_4^2 \approx 0$, $B_4^4 = 91$ (MHz) | — | [243] |
| $LiNbO_3$: 0.1 mol% $Fe_2O_3$ | 9.35 | RT $g_{00} = 2.005$ | $b_2{}^0 = 656$ | $b_4{}^0 = -39$, $b_4{}^3 = 420$ | $c_4{}^3 = 500$ | [244] |
| $LiScGeO_4$ (site I) | X, Q | 300 2.0032, 2.0074, 2.0050 $g_{xy}$ = 0.0002 | 2 296.3, 2 445.5, $b_2^{-2} = 4\,980.2$ | 0.77, 75.49, 64.92 $b_4^{-2} = -196.00$, $b_4^{-4} = -141.09$ | — | [245] |
| $LiScGeO_4$ (site II) | X, Q | 300 2.0051, 2.0072, 2.0054, $g_{xy}$ = 0.0025 | −2 776.2, 2 974.8 $b_2^{-2} = 4\,208.2$ | −7.37, 39.12, −132.60, $b_4^{-2} = 83.66$, $b_4^{-4} = 61.32$ | — | [245] |

(continued)

## ($Fe^{3+}$($3d^5$) listing – contd.)

| Host | Frequency (GHz)/ band | T (°K) | $\tilde{g}$ | $b_2^0(=D)$, $b_2^2(=3E)$ ($10^{-4}$ cm$^{-1}$) | $b_4^0, b_4^2, b_4^4$ ($10^{-4}$ cm$^{-1}$) | $A_x, A_y, A_z$ ($10^{-4}$ cm$^{-1}$) | References |
|---|---|---|---|---|---|---|---|
| LiSrAlF$_6$ | 9.4 | 300 | $g_{iso}$ = 2.002172 | $B_2^0 = -322.40$ | $B_4^0 = -5.723$, $B_4^3 = 4.517$ | — | [242] |
| LiSrA1F$_6$ | 9.4 | 300 | 2.002 | $B_2^0 = -332.104$ | $B_4^0 = -5.72$ | — | [242] |
| LiTaO$_3$ | X | RT | $g_\parallel$ = 2.001, $g_\perp$ = 1.998 | 9 600 MHz, — | 5 MHz, —, —, $b_4^3$ = 4 180 MHz | — | [246] |
| [NaAl$_3$(OH)$_4$(PO$_4$)$_2$ · 2H$_2$O] | 9.33 | — | 2.004 | 31 500, 1 200 | — | — | [247] |
| NaZr$_2$(PO$_4$)$_3$ single crystal | 9.4 | 300 | 2.00593, 2.00593, 2.00512 | −595.957, −10.641 | −14.432, 3.696, −5.220 | — | [248] |
| Na$_2$[Al$_2$Si$_3$O$_{10}$] · 2H$_2$O | 36.772 | RT | 2.006, 2.0069, 2.006 | 1 086.4, 532.3 | — | — | [249] |
| ([NH$_2$CH$_2$COOH]$_3$H$_2$SO$_4$) (site A) | X | 4.2 | 2.0023 | — | 4 070, 330 | — | [250] |
| ([NH$_2$CH$_2$COOH]$_3$H$_2$SO$_4$) (site B) | X | 4.2 | g = 2.0023 | — | 4 880, 380 | — | [250] |
| ([NH$_2$CH$_2$COOH]$_3$H$_2$SO$_4$) (site C) | X | 4.2 | g = 2.0023 | — | 11 630, 400 | — | [250] |
| (NH$_4$)$_2$AlF$_5$ · H$_2$O | 9.21 | 300 | — | 668, −168 | a = −54 G, F = 30 G | — | [251] |
| PbWO$_4$ | X | 4 | 1.721, 1.905, 1.940 | $B_2^0 = -290$, $B_2^2 = -276$, (MHz) | 1.3, −2.8, 0.4 (MHz) | — | [252] |
| SiO$_4$ (amethyst) | X | 215 | g = 2.0036 | 28.63, 4.83 GHz | — | — | [253] |
| Sn$_{1-x}$Fe$_x$O$_2$ intersitial (LS1) | 9.5 | 5 | 3.4, 3.4, — | — | — | — | [254] |
| Sn$_{1-x}$Fe$_x$O$_2$ substitutional (HS 1) | 9.5 | 5 | 2.0, 2.0, 2.0 | D = 645 G, E = −185 G | — | — | [254] |
| Sn$_{1-x}$Fe$_x$O$_2$ substitutional (HS 2) | 9.5 | 5 | 2.0, 2.0, 2.0 | D = 700 G, E = −190 G | — | — | [254] |

*(continued)*

## ($Fe^{3+}(3d^5)$) listing – contd.

| Host | Frequency (GHz)/ band | T (°K) | $\tilde{g}$ | $b_2^0 (= D)$, $b_2^2 (= 3E)$ ($10^{-4}$ cm$^{-1}$) | $b_4^0, b_4^2, b_4^4$ ($10^{-4}$ cm$^{-1}$) | $A_x, A_y, A_z$ ($10^{-4}$ cm$^{-1}$) | References |
|---|---|---|---|---|---|---|---|
| $Sn_{1-x}Fe_xO_2$ intersitial (LS 2) | 9.5 | 5 | 4.1, 4.1, — | — | — | — | [254] |
| $Sn_{1-x}Fe_xO_2$ intersitial (LS 3) | 9.5 | 5 | 5.2, 5.2, — | — | — | — | [254] |
| $Sn_{1-x}Fe_xO_2$ substitutional (HS 3) | 9.5 | 5 | 2.0, 2.0, 2.0 | D = 610 G, E = −165 G | — | — | [254] |
| $Sn_{1-x}Fe_xO_2$ intersitial (LS 4) | 9.5 | 5 | 5.7, 5.7, — | — | — | — | [254] |
| $Sn_{1-x}Fe_xO_2$ nanoparticles (FM 1) | 9.5 | 5 | 3.3, 3.3, 3.3 | — | — | — | [254] |
| $Sn_{1-x}Fe_xO_2$ nanoparticles (FM 2) | 9.5 | 5 | 1.6, 1.6, 1.6 | — | — | — | [254] |
| $Sn_{1-x}Fe_xO_2$ nanoparticles (FM 3) | 9.5 | 5 | 2.35, 2.35, 2.35 | — | — | — | [254] |
| $SrF_2$ (nonirradiated) | 36.7 | 4.2 | 2.000, 2.000, 2.005 | $B_2^0, B_2^2$ (MHz) = 2 206.5, 1 420.98 | $B_4^0, B_4^2$, $B_4^4$(MHz) = 4.9, −24.1, −24.0 | $A_\parallel =$ 25.02, $A_\perp$ = 17.01 | [255] |
| $SrF_2$ (X-irradiated) | 36.7 | 4.2 | 2.003, 2.003, 2.005 | $B_2^0, B_2^2$ (MHz) = 2 489.72, −909.63 | $B_4^0, B_4^2$, $B_4^4$(MHz) = −0.9, 3.6, −1.4 | $A_\parallel =$ 18.01, $A_\perp$ = 14.01 | [255] |
| $Sr(NO_3)_2$ | X | 77 | 1.9989 | $\|D\|$ = 338, $\|E\|$ = 10 | — | — | [256] |
| $TiInS_2$ | 9.8 | RT | g = 2.0 | 800, 566.7 (G) | Negligibly small | — | [257] |
| $TiO_2$ | X | RT | 2.000, 2.000, 2.000 | D = 6 578, E = 699 | — | — | [181] |
| $Tl_2MgF_4$ (center I) | X | 300 | 2.0064, 2.0064, 1.9993 | −1 016.6, — | 31.7, —, 101 | — | [258] |
| $Tl_2ZnF_4$ (center I) | X | 300 | 2.0035, 2.0035, 2.0018 | −900.4, — | 33.6, —, 107 | — | [258] |

(continued)

## ($Fe^{3+}(3d^5)$ listing – contd.)

| Host | Frequency (GHz)/ band | $T$ (°K) | $\tilde{g}$ | $b_2^0(=D)$, $b_2^2(=3E)$ ($10^{-4}$ cm$^{-1}$) | $b_4^0, b_4^2, b_4^4$ ($10^{-4}$ cm$^{-1}$) | $A_x, A_y, A_z$ ($10^{-4}$ cm$^{-1}$) | References |
|---|---|---|---|---|---|---|---|
| $Tl_2ZnF_4$ (center II) | X | 300 | 1.998, 2.004, 2.0032 | −991, 225 | 32, 24, −89 | — | [258] |
| $Tl_2ZnF_4$ (center IV) | X | 300 | 2.0004, 2.0051, 1.9966 | −781.4, 610.0 | 33.5, 25, 111 | — | [258] |
| [{trans-(Cyclam)($N_3$)Fe}$_2$($\mu$-N)]$^{2+}$ (site S = 1/2) | 9.64 | 433 | 2.04, 2.06, 2.2 | — | — | — | [231] |
| $VO_2$ (single crystal) | 35.15 | 150 | $g_\parallel = 2.035$, $g_\perp = 2.027$ | 4.27, −2.82 | 0.3, 1.2, 0.6 | — | [259] |
| $YCaAlO_4$ | X | 295 | $g_\parallel = 1.991$, $g_\perp = 2.021$ | −34.7 GHz, — | — | — | [260] |
| $YCaAlO_4$ | X | 77 | $g_\parallel = 1.980$, $g_\perp = 2.026$ | −35.4 GHz, — | — | — | [260] |
| $YCaAlO_4$ | X | 4.2 | $g_\parallel = 2.00$, $g_\perp = 2.00$ | −36.0 GHz, — | −7.00, −9.63, 6.9 | — | [260] |
| ZnO | X, Q | 300 | $g = 2.006$ | −595.0, — | −12.3, —, —, $b_4^3 = -390.0$ | — | [261] |
| ZnO (site $V_O$) | X, Q | 300 | $g = 2.006$ | 490.0, 2 035.0 | −14.0, —, —, $b_4^3 = -390.0$ | — | [261] |
| ZnO (site $V_{Zn}(I)$) | X, Q | 300 | $g = 2.006$ | −1 282.0, 1 347.0 | −14.0, —, —, $b_4^3 = 390.0$ | — | [261] |
| ZnO (site $V_{Zn}(II)$) | X, Q | 300 | $g = 2.006$ | −796.0, −1 085.0 | −11.0, —, —, $b_4^3 = -390.0$ | — | [261] |
| $ZrSiO_4$ (site I) | 9.5 | 77 | $g_\parallel = g_\perp = 2.003$ | $B_2^0 = 4\,408$ MHz, $B_2^2 = 0.585$ MHz | $B_4^4 = -9.40$ MHz | — | [262] |
| $ZrSiO_4$ (site II) | 9.5 | 77 | $g_\parallel = g_\perp = 2.003$ | $B_2^0 = 2\,301$ MHz, $B_2^2 = 0.247$ MHz | $B_4^4 = -2.42$ MHz | — | [262] |

## Mn$^{2+}$. Data tabulation of SHPs
## Mn$^{2+}$ (3d$^5$)

| Host | Frequency (GHz)/band | T (K) | $g_x, g_y, g_z$ | $b_2^0 (=D), b_2^2 (=3E)$ (10$^{-4}$ cm$^{-1}$) | $b_4^0, b_4^2, b_4^4$ (10$^{-4}$ cm$^{-1}$) | $A_x, A_y, A_z$ (10$^{-4}$ cm$^{-1}$) | References |
|---|---|---|---|---|---|---|---|
| AlF$_3$ · 3H$_2$O | X, Q | 300 | 2.0 | 839.8, 100 MHz | a = 20 MHz | A = 274.3 MHz | [263] |
| BaAl$_{12}$O$_{19}$ | X | — | 1.981 | — | — | 84 G | [264] |
| Diglycine Ba(barium) chloride monohydrate | X | 295 | g = 1.9922 | 237, 80 | a = 5 | A = 89, B = 88 | [265] |
| BaTiO$_3$ (polycrystalline) | 9.15 | 15 | $g_{iso}$ = 2.002 | D = 224.5 | — | |A| = 86.4 | [266] |
| BiVO$_4$ single crystal | X | RT | 2.00953, 2.00339, 2.00305 | $B_2^0$ = 29.194, $B_2^2$ = 81.57, $B_2^1$ = 975.22 | $B_0^0$ = 0.132, $B_4^2$ = 0.362, $B_4^4$ = 0.636, $B_4^1$ = 0.042, $B_4^4$ = −0.175 | $\tilde{A}(^{55}Mn)$ = 28.2, 0, 236.4 (MHz) | [267] |
| BiVO$_4$ | X | 3.8–300 | 2.00305, 2.0033, 2.00953 | 814.5, 382.1 | — | — | [223] |
| Coal 101 Argonne Premium (Upper Freeport) | 9.2 | 286 | $g_{iso}$ = 2.00062 | $c_0^2$ = −6.45 mT | $c_0^4$ = −0.07 mT | $A_{iso}$ = −94.16 (G) | [268] |
| Coal 101 Argonne Premium (Upper Freeport) | 9.3 | 293 | $g_{iso}$ = 2.00078 | — | — | $A_{iso}$ = −94.06 (G) | [268] |
| C$_6$H$_{12}$Cs$_2$ZnO$_{12}$ (diaquacesiumaquabismalonatozincate) | X | RT | 1.8586, 2.0001, 2.0641 | $D_{xx}$ = 5.48, $D_{yy}$ = 18.89, $D_{zz}$ = −24.38 (mT) | — | 8.30, 9.04, 9.20 (mT) | [269] |
| Ca$_{0.999}$Ba$_{0.001}$F$_2$ | X | — | $g_{iso}$ = 2.001 | $B_2^0$ = −20.59, $B_2^2$ = −63.10 (G) | $B_4^0$ = −0.016, $B_4^2$ = −0.016, $B_4^4$ = 0.028(G) | $A_{iso}$ = −102.7 (G) | [270] |
| CaO (900 °C annealing temperature) | X | 295 | 1.997, 1.997, 1.997 | 23.35, 6.67 | — | 91.06, 91.06, 91.06 | [271] |

*(continued)*

## ($Mn^{2+}(3d^5)$) listing – contd.)

| Host | Frequency (GHz)/ band | T (K) | $g_x, g_y, g_z$ | $b_2^0(=D), b_2^2(=3E)$ ($10^{-4}$ cm$^{-1}$) | $b_4^0, b_4^2, b_4^4$ ($10^{-4}$ cm$^{-1}$) | $A_x, A_y, A_z$ ($10^{-4}$ cm$^{-1}$) | References |
|---|---|---|---|---|---|---|---|
| CaCo$_3$ | 9.450 | 295 | $g_\| = 2.00123$, $g_\perp = 2.00131$ | $-76.0$, — | $-2.82$, —, —, $B_4^3 = -1.12$ | $A_\| = -88.23$, $A_\perp = -87.60$ | [272] |
| CaCO$_3$ 350–550 °C (annealing temperature) | X | 295 | 1.990, 1.990, 1.995 | 90.06, 23.35 | — | 91.06, 99.07, 92.73 | [271] |
| CaCO$_3$ (sea water mussel – prismatic layer) | X | 420 | 1.998, 1.998, 2.000 | $-90$, — | — | 81, 81, 83 | [273] |
| CaCO$_3$ (sea water mussel – nacreous layer) | X | 420 | 1.992, 1.997, 1.999 | $-144$, — | — | 75, 78, 79 | [273] |
| CaGa$_2$S$_4$ (set B) | X | RT | 2.0141 | $D = 262.8$, $E = -58.5$ | — | $-76.6$ | [274] |
| CaGa$_2$S$_4$ (set C-1) | X | RT | 2.0205 | $D = 287.1$, $E = 36.0$ | — | $-76.0$ | [274] |
| CaGa$_2$S$_4$ (set C-2) | X | RT | 2.0176 | $D = 215.0$, $E = -35.0$ | — | $-76.5$ | [274] |
| Ca$_{9.4}$Fe$_{0.4}$Mn$_{0.2}$(PO$_4$)$_6$(OH)$_2$ hydroxyapatite | X | — | 2.01 | $D = 250$, $E = 25$ (G) | — | 94 (G) | [227] |
| (CH$_3$CHOH-COOH)$_2 \cdot$ 3H$_2$O | 9.5 | 295 | $g = 1.9769$ | 330, 103 | $a = 10$ | $A = 89$, $B = 84$ | [275] |
| C$_{18}$H$_{18}$Br$_2$MnN$_4$ | 190 | 2.5 | 1.99, 1.98, 2.00 | $-6000$, 950 | — | — | [276] |
| C$_{18}$H$_{18}$Cl$_2$MnN$_4$ | 115 | 2.5 | 2.00, 1.98, 2.00 | $-3600$, 730 | — | — | [276] |
| C$_{18}$H$_{18}$I$_2$MnN$_4$ | 115 | 2.5 | 1.97, 2.00, 2.00 | 1150, 200 | — | — | [276] |
| Co(C$_4$H$_3$O$_4$)$_2 \cdot$ 4H$_2$O | X | 296 | 2.029 | $-333$, $-60$ | $-84$ | — | [277] |
| Co(C$_4$H$_3$O$_4$)$_2 \cdot$ 4H$_2$O | X | 396 | 2.030 | $-295$, 39 | $-84$ | — | [277] |

(continued)

**200** | 4 Multifrequency Transition Ion Data Tabulation

($Mn^{2+}(3d^5)$ listing – contd.)

| Host | Frequency (GHz)/band | $T$ (K) | $g_x, g_y, g_z$ | $b_2^0 (= D), b_2^2 (= 3E)$ ($10^{-4}$ cm$^{-1}$) | $b_4^0, b_4^2, b_4^4$ ($10^{-4}$ cm$^{-1}$) | $A_x, A_y, A_z$ ($10^{-4}$ cm$^{-1}$) | References |
|---|---|---|---|---|---|---|---|
| CoKPO$_4 \cdot$ 6H$_2$O | X | 295 | 2.011, 1.998, 1.991 | ($D_x =$ −141.99, $D_y =$ −87.81, $D_z =$ 229.81) | — | −83.14, −82.21, −78.47 | [279, see also 278] |
| Cs$_2$Co(SO$_4$)$_2 \cdot$ 6H$_2$O | X | 290 | $g_x = 2.001$, $g_z = 2.002$ | −260, 51 | — | $A_x = -87$, $A_z = -88$ | [280] |
| Cs$_2$Mg(SO$_4$)$_2 \cdot$ 6H$_2$O | X | 290 | $g_x = 2.002$, $g_z = 2.004$ | −257, 78 | — | $A_x = -87$, $A_z = -87$ | [280] |
| Cs$_2$Mg(SO$_4$)$_2 \cdot$ 6H$_2$O | X | 77 | $g_x = 2.004$, $g_z = 2.005$ | −283, 72 | — | $A_x = -87$, $A_z = -88$ | [280] |
| Cs$_2$NaLaCl$_6$ (site 1) | 9.2 | 12.8 | $g_\| = 2.0045$, $g_\perp = 2.022$ | $b_2^0 = 943.5$ | — | — | [281] |
| Cs$_2$NaLaCl$_6$ (site 1) | 9.2 | 35 | $g_\| = 2.0044$, $g_\perp = 2.021$ | $b_2^0 = 923.1$ | — | — | [281] |
| Cs$_2$NaLaCl$_6$ (site 1) | 9.2 | 63 | $g_\| = 2.0046$, $g_\perp = 2.026$ | $b_2^0 = 898.3$ | — | — | [281] |
| Cs$_2$NaLaCl$_6$ (site 1) | 9.2 | 93 | $g_\| = 2.0043$, $g_\perp = 2.028$ | $b_2^0 = 866.7$ | — | — | [281] |
| Cs$_2$NaLaCl$_6$ (site 1) | 9.2 | 123 | $g_\| = 2.0047$, $g_\perp = 2.025$ | $b_2^0 = 838.1$ | — | — | [281] |
| Cs$_2$NaLaCl$_6$ (site 1) | 9.2 | 153 | $g_\| = 2.0044$, $g_\perp = 2.029$ | $b_2^0 = 834.2$ | — | — | [281] |
| Cs$_2$NaLaCl$_6$ (site 1) | 9.2 | 183 | $g_\| = 2.0042$, $g_\perp = 2.022$ | $b_2^0 = 822.6$ | — | — | [281] |
| Cs$_2$NaLaCl$_6$ (site 1) | 9.2 | 213 | $g_\| = 2.0045$, $g_\perp = 2.024$ | $b_2^0 = 793.8$ | — | — | [281] |
| Cs$_2$NaLaCl$_6$ (site 1) | 9.2 | 243 | $g_\| = 2.0044$, $g_\perp = 2.021$ | $b_2^0 = 779.6$ | — | — | [281] |
| Cs$_2$NaLaCl$_6$ (site 1) | 9.2 | 273 | $g_\| = 2.0041$, $g_\perp = 2.026$ | $b_2^0 = 765.8$ | — | — | [281] |
| Cs$_2$NaLaCl$_6$ (site 1) | 9.2 | 300 | $g_\| = 2.0045$, $g_\perp = 2.021$ | $b_2^0 = 746.8$ | — | — | [281] |
| Cs$_2$NaLaCl$_6$ (site 2) | 9.2 | 12.8 | $g_\| = 2.0041$, $g_\perp = 2.044$ | $b_2^0 = 1372.2$ | — | — | [281] |

(continued)

## ($Mn^{2+}(3d^5)$) listing – contd.)

| Host | Frequency (GHz)/ band | T (K) | $g_x, g_y, g_z$ | $b_2^0 (=D), b_2^2 (=3E)$ $(10^{-4}\ cm^{-1})$ | $b_4^0, b_4^2, b_4^4$ $(10^{-4}\ cm^{-1})$ | $A_x, A_y, A_z$ $(10^{-4}\ cm^{-1})$ | References |
|---|---|---|---|---|---|---|---|
| $Cs_2NaLaCl_6$ (site 2) | 9.2 | 35 | $g_\| = 2.0043$, $g_\perp = 2.044$ | $b_2^0 = 1331.4$ | — | — | [281] |
| $Cs_2NaLaCl_6$ (site 2) | 9.2 | 63 | $g_\| = 2.0044$, $g_\perp = 2.044$ | $b_2^0 = 1297.8$ | — | — | [281] |
| $Cs_2NaLaCl_6$ (site 2) | 9.2 | 93 | $g_\| = 2.0043$, $g_\perp = 2.045$ | $b_2^0 = 1273.9$ | — | — | [281] |
| $Cs_2NaLaCl_6$ (site 2) | 9.2 | 123 | $g_\| = 2.0045$, $g_\perp = 2.047$ | $b_2^0 = 1246.3$ | — | — | [281] |
| $Cs_2NaLaCl_6$ (site 2) | 9.2 | 153 | $g_\| = 2.0043$, $g_\perp = 2.045$ | $b_2^0 = 1207.1$ | — | — | [281] |
| $Cs_2NaLaCl_6$ (site 2) | 9.2 | 183 | $g_\| = 2.0041$, $g_\perp = 2.046$ | $b_2^0 = 1185.2$ | — | — | [281] |
| $Cs_2NaLaCl_6$ (site 2) | 9.2 | 213 | $g_\| = 2.0043$, $g_\perp = 2.047$ | $b_2^0 = 1163.2$ | — | — | [281] |
| $Cs_2NaLaCl_6$ (site 2) | 9.2 | 243 | $g_\| = 2.0042$, $g_\perp = 2.045$ | $b_2^0 = 1140.3$ | — | — | [281] |
| $Cs_2NaLaCl_6$ (site 2) | 9.2 | 273 | $g_\| = 2.0040$, $g_\perp = 2.048$ | $b_2^0 = 1126.0$ | — | — | [281] |
| $Cs_2NaLaCl_6$ (site 2) | 9.2 | 300 | $g_\| = 2.0044$, $g_\perp = 2.045$ | $b_2^0 = 1096.8$ | — | — | [281] |
| $Cs_2Ni(SO_4)_2 \cdot 6H_2O$ | X | 290 | $g_x = 2.001$, $g_z = 1.998$ | −254, 78 | — | $A_x = -87$, $A_z = -88$ | [280] |
| $Cs_2Zn(SO_4)_2 \cdot 6H_2O$ | X | 290 | $g_x = 2.002$, $g_z = 2.004$ | −268, 39 | — | $A_x = -87$, $A_z = -87$ | [280] |
| $Cs_2Zn(SO_4)_2 \cdot 6H_2O$ | X | 77 | $g_x = 2.003$, $g_z = 2.004$ | 296, 24 | — | $A_x = -88$, $A_z = -88$ | [280] |
| $(Et_4N)_2[Mn(II)_2(salmp)_2]$ | X, Q | 2–7 | — | 400, 171 ? | — | — | [282] |
| Freshwater snail shells (shell: PCL-00) | X | 295 | 2.019, 2.010, 1.998 | 109.30, — | | −74.73, −75.67, −77.54 | [283] |
| Freshwater snail shells (shell: PCL-300) | X | 295 | 2.002, 2.005, 1.996 | 110.23, — | | −74.73, −75.67, −78.00 | [283] |

(continued)

## ($Mn^{2+}(3d^5)$ listing – contd.)

| Host | Frequency (GHz)/band | T (K) | $g_x, g_y, g_z$ | $b_2^0 (= D), b_2^2 (= 3E)$ ($10^{-4}$ cm$^{-1}$) | $b_4^0, b_4^2, b_4^4$ ($10^{-4}$ cm$^{-1}$) | $A_x, A_y, A_z$ ($10^{-4}$ cm$^{-1}$) | References |
|---|---|---|---|---|---|---|---|
| Freshwater snail shells (shell: PCL400) | X | 295 | 2.001, 2.002, 1.998 | 109.30, — | — | −77.54, −77.26, 81.74 | [283] |
| Freshwater snail shells (shell: PCL-450) | X | 295 | 2.009, 2.009, 1.998 | −85.94, — | — | 75.67, 75.67, −78.00 | [283] |
| Freshwater snail shells (shell: PCL-500) | X | 295 | 2.078, 2.078, 1.999 | 107.43, — | — | 81.74, 81.74, 83.14 | [283] |
| Freshwater snail shells (shell: PCL-550) | X | 295 | 2.021, 2.021, 1.998 | 108.36, — | — | 82.67, 82.67, 83.61 | [283] |
| Freshwater snail shells (shell: PCL-900) | X | 295 | 2.078, 2.078, 2.078 | −84.08, — | — | 75.20, 75.20, 75.20 | [283] |
| GaAs:Mn | X | 3.8 | 2.802, 0.315, 1.984 ($g_{xz}$ = 0.014) | $D_{xx} = 220$ MHz, $D_{yy} = 18$ MHz, $D_{zz} = -237$ MHz $B_2^0 = -29$ MHz, $B_2^1 = -232$ MHz, $B_2^2 = -191$ MHz | — | $B_4^0 = 1.56$ MHz, $B_4^1 = -6.3$ MHz, $B_4^2 = 9.7$ MHz, $B_4^3 = -40$ MHz, $B_4^4 = 10.3$ MHz | [284] |
| β-$Ga_2O_3$ | X | RT | 2.014, 2.012, 2.001 | 154.4, 138.8 | 0.04, −0.70, 0.30 | — | [285] |
| $H_2Ni(C_4H_2O_4)_2 \cdot 4H_2O$ | X | 373 | g = 2.008 | 206, 210 | — | A = −82 | [286] |
| $H_2Ni(C_4H_2O_4)_2 \cdot 4H_2O$ | X | 295 | 2.030, 2.008, 1.980 | 232, 186 | — | −84, −89, −92 | [286] |
| $H_2Ni(C_4H_2O_4)_2 \cdot 4H_2O$ | X | 103 | g = 2.008 | 251, 201 | — | A = −91 | [286] |
| $KHSO_4$ | X | 77 | g = 2.0002 | 59, 96 | — | A = 66, B = 26 | [287] |
| $KTaO_3$ ($Mn^{2+}$ mol% = 15) nano powder | X | RT | 1.9815 | — | — | — | [288] |

*(continued)*

## ($Mn^{2+}$($3d^5$) listing – contd.)

| Host | Frequency (GHz)/ band | $T$ (K) | $g_x, g_y, g_z$ | $b_2^0(=D), b_2^2(=3E)$ ($10^{-4}$ cm$^{-1}$) | $b_4^0, b_4^2, b_4^4$ ($10^{-4}$ cm$^{-1}$) | $A_x, A_y, A_z$ ($10^{-4}$ cm$^{-1}$) | References |
|---|---|---|---|---|---|---|---|
| $KTaO_3$ ($Mn^{2+}$ mol% = 5) nano powder | X | RT | 2.0022 | $D = 20$ | — | $A = 79$ | [288] |
| $KTaO_3$ ($Mn^{2+}$ mol% = 0.1) nano powder | X | RT | 1.9991 | $D = 170$ | — | $A = 85$ | [288] |
| $K_2C_2O_4 \cdot H_2O_2$ | X | RT | 2.0010, 2.0010, 2.0015 | $b_2^0 = -440$, $b_2^2 = 150$ | $b_4^0 = 0.1959$ | 97.5, 97.5, 98.5 | [289] |
| $K_3H(SO_4)_2$ | X | 295 | $g = 1.989$ | 577.32, 108.36 | $a - F = 10$ | $A_\| = 84.076$, $A_\perp = 85.01$ | [290] |
| $K_3Na(CrO_4)_2$ | X | 291 | 2.0009, 2.0009, 2.0010 | $D_{xx} = 113.9$ G, $D_{zz} = -227.8$ G | — | 90.9 GHz, —, 89.9 GHz | [291] |
| $LaGa_{1-x}Mn_xO_3$ | X | 438 | 2.000, 2.000, 2.001 | $D = -1165$ MHz | $b_4^0 = -2$ MHz | $-225$, $-225$, $-234$ MHz | [189] |
| L-asparagine monohydrate | X | 77 | $g = 2.0912$ | 299, 297 | $a = -28$ | $A = 98$, $B = 91$ | [292] |
| $La_{7/8}Sr_{1/8}MnO_3$ | 9.4 | 180 | 1.997 | $E/D = -0.4$ | — | — | [293] |
| $LiCsSO_4$ | X | 152 | $g_\| =$ undetermined, $g_\perp = 2.005$ | 2.237, $-0.806$ (GHz) | $b_4^0 = 0.14$; $b_4^2 = 0.26$; $b_4^4 = 0.008$(GHz) | $A_{iso} = -0.267$ GHz | [294] |
| $LiCsSO_4$ | X | 90 | $g_\| =$ undetermined, $g_\perp = 2.007$ | 2.238, $-0.844$ (GHz) | $b_4^0 = 0.25$; $b_4^2 = 0.71$; $b_4^4 = 0.20$(GHz) | $A_{iso} = -0.264$ GHz | [294] |
| $LiGa_5O_8$ | X | 300 | $g = 1.997$ | $D = 33.3$ mT | — | $A = 7.6$ mT | [295] |
| $LiGa_5O_8$ | X | 110 | $g = 1.997$ | $D = 32.8$ mT | — | $A = 7.6$ mT | [295] |
| ($LiHC_2O_4 \cdot H_2O$) | X | 295 | 1.9942 | 180, 171 | $a = 7$ | $A = 114$, $B = 103$ | [296] |

(continued)

## $(Mn^{2+}(3d^5))$ listing – contd.

| Host | Frequency (GHz)/band | T (K) | $g_x, g_y, g_z$ | $b_2^0(= D), b_2^2(= 3E)$ $(10^{-4}$ cm$^{-1})$ | $b_4^0, b_4^2, b_4^4$ $(10^{-4}$ cm$^{-1})$ | $A_x, A_y, A_z$ $(10^{-4}$ cm$^{-1})$ | References |
|---|---|---|---|---|---|---|---|
| [(Me$_3$TACN)$_2$Mn(II)$_2$($\mu$-OAc)$_3$]BPh$_4$ (in pure DFM) | 34.2 | 2 | — | −263, −27.43 | — | — | [282] |
| [(Me$_3$TACN)$_2$Mn(II)$_2$($\mu$-OAc)$_3$]BPh$_4$ (in pure CH$_3$Cn) | 34.2 | 2 | — | 180, 132.7 | — | — | [282] |
| Mg(C$_4$H$_3$O$_4$)$_2$·6H$_2$O (site 1) | X | RT | $g_{iso} = 1.995$ | $b_2^0 = 334$ | $a = 4, F = -13$ | $A_{iso} = -96$ | [297] |
| Mg(C$_4$H$_3$O$_4$)$_2$·6H$_2$O (site 2) | X | RT | $g_{iso} = 2.013$ | $b_2^0 = 322$ | $a = 4, F = -13$ | $A_{iso} = -106$ | [297] |
| Mg(C$_4$H$_3$O$_4$)$_2$·6H$_2$O (site 3) | X | RT | $g_{iso} = 2.001$ | $b_2^0 = 245$ | $a = 4, F = -11$ | $A_{iso} = -84$ | [297] |
| Mg(C$_4$H$_3$O$_4$)$_2$·6H$_2$O | X | 300 | $g_\| = 1.997$, $g_\perp = 2.013$ | $b_2^0 = 312$, $b_2^2 = 347$ | $b_4^0 = 1.2$ | $A_\| = 88$, $A_\perp = 88$ | [298] |
| Mg(C$_4$H$_3$O$_4$)$_2$·6H$_2$O | X | 77 | $g_\| = 2.047$, $g_\perp = 2.060$ | $b_2^0 = 414$, $b_2^2 = 190$ | $b_4^0 = 6.16$ | $A_\| = 93$, $A_\perp = 93$ | [298] |
| Mg$_{0.99}$Mn$_{0.01}$Al$_2$O$_4$ | X | 300 | $g = 2.003$ | $D = 287.726$ | — | $A_{Avg} = 75.668$ | [299] |
| Mg$_{0.99}$Mn$_{0.01}$Al$_2$O$_4$ | X | 110 | $g = 2.003$ | $D = 301.738$ | — | $A_{Avg} = 77.536$ | [299] |
| MgSiF$_6$·6H$_2$O | X | 250 | $g_{iso} = 2.0$ | 278, −117 | $b_4^0 = 2.6$ | $A_{iso} = -91$ | [300] |
| Mn$^{2+}$-ATP-Kp2 dithionite reduced nitrogenase iron-protein for K-pneumoniae | 9.43 | 8.0 | $g = 2.002$ | 125, 38 | — | $A = -76.60$ | [301] |
| [Mn(bipy)(N$_3$)$_2$] (bipy) 2,2'-bipyridine) | X | 2–298 | $g_{iso} = 2.000$ | — | — | — | [302] |
| [Mn(L$^1$)]Cl$_2$ | X | 295 | $g_\| = 3.95$, $g_\perp = 1.89$ | — | — | $A_0 = 11$ | [303] |
| [Mn(L$^2$) Cl$_2$] | X | 295 | $g_\| = 3.77$, $g_\perp = 1.91$ | — | — | $A_0 = 103$ | [303] |

(continued)

## ($Mn^{2+}(3d^5)$ listing – contd.)

| Host | Frequency (GHz)/band | T (K) | $g_x, g_y, g_z$ | $b_2^0(=D), b_2^2(=3E)$ ($10^{-4}$ cm$^{-1}$) | $b_4^0, b_4^2, b_4^4$ ($10^{-4}$ cm$^{-1}$) | $A_x, A_y, A_z$ ($10^{-4}$ cm$^{-1}$) | References |
|---|---|---|---|---|---|---|---|
| [Mn($L^3$)Cl$_2$] | X | 295 | $g_\parallel = 3.99$, $g_\perp = 1.93$ | — | — | $A_0 = 108$ | [303] |
| [Mn($L^4$)]Cl$_2$ | X | 295 | $g_\parallel = 3.95$, $g_\perp = 1.89$ | — | — | $A_0 = 109$ | [303] |
| [Mn($L^5$)Cl]Cl | X | 295 | $g_\parallel = 3.65$, $g_\perp = 1.87$ | — | — | $A_0 = 110$ | [303] |
| [Mn($L^6$)]Cl$_2$ | X | 295 | $g_\parallel = 3.93$, $g_\perp = 1.95$ | — | — | $A_0 = 117$ | [303] |
| [Mn($L^1$)](SCN)$_2$ | X | 295 | $g_\parallel = 3.94$, $g_\perp = 1.99$ | — | — | $A_0 = 105$ | [303] |
| [Mn($L^5$)(SCN)](SCN) | X | 295 | $g_\parallel = 3.99$, $g_\perp = 1.99$ | — | — | $A_0 = 118$ | [303] |
| [Mn($L^6$)](SCN)$_2$ | X | 295 | $g_\parallel = 3.98$, $g_\perp = 1.98$ | — | — | $A_0 = 114$ | [303] |
| [Mn($L^1$)]SO$_4$ | X | 295 | $g_\parallel = 3.64$, $g_\perp = 1.87$ | — | — | $A_0 = 110$ | [303] |
| [Mn($L^5$)]SO$_4$ | X | 295 | $g_\parallel = 3.94$, $g_\perp = 1.99$ | — | — | $A_0 = 121$ | [303] |
| [Mn($L^6$)]SO$_4$ | X | 295 | $g_\parallel = 3.91$, $g_\perp = 1.96$ | — | — | $A_0 = 113$ | [303] |

For above [303] data:

($L^1$):2,4,10,12-tetraphenyl-1,5,7,9,13,15-hexaazatricyclo [15,3,1] —octadeca-1,4,7,9,12,14-hexaene[N6]ane

($L^2$):2,4,10,12-tetraphenyl-1,5,7,13-tetraazacyclohexadeca-1,4,9,12-teraene [N4]ane

($L^3$):2,4,9,11-tetraphenyl-1,5,8,12-tetraazacycloteradeca-1,4,8,11-tetraene[N4]ane

(L4):1,4,/,10,13,16-hexaazacyclooctadecane[N6]ane

($L^5$):1,8-diaza-4,5,11-trithia-2,3:6,7-dibenzo[b,h]-cyclopentadeca-9,13-dione[S3N2]ane

($L^6$):9,18-dimethyl-1,7,10,16-tetraza-4,13-dithiacyclooctadecane-2,6,11,16-teraone

| Mn(t-Buterpy)(N$_3$)$_2$] (t-Buterpy, 4,4′,4″-tri-tert-butyl-2,2′:6′,2″-terpyridine) (neat powder) | 285 | 5 | 2.000, 2.000, 2.000 | −2 500, 440 | — | — | [304] |

(continued)

($Mn^{2+}$($3d^5$) listing – contd.)

| Host | Frequency (GHz)/ band | T (K) | $g_x, g_y, g_z$ | $b_2^0 (= D), b_2^2 (= 3E)$ ($10^{-4}$ cm$^{-1}$) | $b_4^0, b_4^2, b_4^4$ ($10^{-4}$ cm$^{-1}$) | $A_x, A_y, A_z$ ($10^{-4}$ cm$^{-1}$) | References |
|---|---|---|---|---|---|---|---|
| [Mn(terpy)(I)$_2$] (terpy, 2,2':6',2''-terpyridin) | 95–285 | 5–30 | 1.98, 1.99, 1.97 | 10 000, 5 700 | — | — | [305] |
| [Mn(terpy)(Br)$_2$] (terpy, 2,2':6',2''-terpyridin) | 95–285 | 5–30 | 1.985, 1.985, 1.985 | 6 050, 4 770 | — | — | [305] |
| [Mn(terpy)(Cl)$_2$] (terpy, 2,2':6',2''-terpyridin) | 95–285 | 5–30 | 1.994, 2.010, 2.025 | −2 600, 2 250 | — | — | [305] |
| [Mn(terpy)(SCN)$_2$] (terpy, 2,2':6',2''-terpyridin) | 95–285 | 5–30 | 1.99, 1.97, 1.97 | −3 000, 1 500 | — | — | [305] |
| NaHOPD (sodium hydrogen orthophosphate dihydrate) (site I) | X | 273 | g = 2.0042 | 238, 228 | a = 13 | A = 86, B = 83 | [306] |
| NaHOPD (sodium hydrogen orthophosphate dihydrate) (site II) | X | 273 | g = 2.0032 | 238, 228 | a = 13 | A = 86, B = 83 | [306] |
| NaNH$_4$SO$_4$ · 4H$_2$O (SASD) | X | RT | g = 1.999 | 271, 213 | (a − F) = 6.5 | $A_\parallel$ = −84, $A_\perp$ = 90 | [307] |
| Ammonium Hydrogen Oxalate hemihydrate | K | RT | 2.0006, 2.0012, 1.9947 | −135,12 (mT) | a=1.0 (mT) | −9.32, −9.22, −9.32 (mT) | [308] |
| NH$_4$Cl$_{0.9}$I$_{0.1}$ | X | 295 | g = 2.0052 | −4.540, — | −0.014, —, — | A = 0.2438 GHz | [309] |
| NH$_4$Cl$_{0.9}$I$_{0.1}$ | X | 125 | g = 1.9986 | −4.637, — | 0.001, —, — | A = −0.2376 GHz | [309] |
| NH$_4$Cl$_{0.9}$I$_{0.1}$ | Q | 295 | g = 2.0103 | −4.572, — | 0.014, —, — | A = −0.2547 GHz | [309] |

(continued)

## ($Mn^{2+}(3d^5)$ listing – contd.)

| Host | Frequency (GHz)/band | T (K) | $g_x, g_y, g_z$ | $b_2^0(=D), b_2^2(=3E)$ ($10^{-4}$ cm$^{-1}$) | $b_4^0, b_4^2, b_4^4$ ($10^{-4}$ cm$^{-1}$) | $A_x, A_y, A_z$ ($10^{-4}$ cm$^{-1}$) | References |
|---|---|---|---|---|---|---|---|
| $NH_4Cl_{0.9}I_{0.1}$ | Q | 77 | g = 2.0115 | −4.780, — | 0.010, —, — | A = −0.2515 GHz | [309] |
| $NH_4Cl_{0.9}I_{0.1}$ | 249.9 | 253 | g = 1.9997 | −4.538, — | 0.009, —, — | A = −0.2489 GHz | [309] |
| $NH_4Cl_{0.9}I_{0.1}$ | 249.9 | 295/253 | g = 1.9997 | −4.563, — | 0.019, —, — | A = −0.2502 GHz | [309] |
| $(NH_4)_2C_2O_4 \cdot H_2O$ | X | 295 | g = 2.0002 | 257, 85 | — | A = 100, B = 79.5 | [310] |
| $(NH_4)_2Mg(SO_4)_2 \cdot 6H_2O$ | X | 7 | — | −246, 270 | — | −90, −82, −94 | [311] |
| $(NH_4)_2Mg(SO_4)_2 \cdot 6H_2O$ | X | 293 | — | −243, 189 | — | −87, −82, −92 | [311] |
| $NH_4$ (ammonium) selenate (single crystal) | X | 295 | 2.1084 | 345, 243 | a = 30.5 | A = 89, B = 92 | [312] |
| $NH_4$ (ammonium) selenate (powder) | X | 295 | 1.9877 | — | — | A = 98 | [312] |
| Ammonium Tartrate (Site I) | X | RT | 1.9225, 1.9554, 2.1258 | 191, 183 | a = 22 | A = 78, B = 75 | [313] |
| Ammonium Tartrate (Site II) | X | RT | 1.9235, 1.9574, 2.0664 | 180, 171 | a = 22 | A = 78, B = 75 | [313] |
| $Ni(C_4H_3O_4)_2 \cdot 6H_2O$ | X | 300 | $g_\parallel$ = 2.021, $g_\perp$ = 2.024 | 216, 219, a = 0.89 | — | $A_\parallel$ = 82.0, $A_\perp$ = 82.0 | [314] |
| $Ni(C_4H_3O_4)_2 \cdot 6H_2O$ | X | 77 | $g_\parallel$ = 2.026, $g_\perp$ = 2.030 | 244, 390, a = 0.99 | — | $A_\parallel$ = 84.9, $A_\perp$ = 84.9 | [314] |
| $PbWO_4$ | 9.4 | 20 | $g_\parallel$ = 2.0004, $g_\perp$ = 2.0002 | $B_2^0$ = −1.042 mT | $B_4^0$ = −0.0024 mT, $B_4^4$ = −0.0117 mT, $B_4^{-4}$ = −0.0062 mT | $A_\parallel$ = −9.540 mT, $A_\perp$ = −9.637 mT, Q = −0.044 mT | [315] |

*(continued)*

## ($Mn^{2+}(3d^5)$) listing – contd.

| Host | Frequency (GHz)/band | T (K) | $g_x, g_y, g_z$ | $b_2^0 (= D), b_2^2 (= 3E)$ ($10^{-4}$ cm$^{-1}$) | $b_4^0, b_4^2, b_4^4$ ($10^{-4}$ cm$^{-1}$) | $A_x, A_y, A_z$ ($10^{-4}$ cm$^{-1}$) | References |
|---|---|---|---|---|---|---|---|
| $Rb_3H(SO_4)_2$ | X | RT | $g_{iso} = 1.997$ | 667, 186 (G) | (a-F)=19(G) | $A_\parallel = -95$, $A_\perp = -94$ (G) | [316] |
| $REF_2Fe^{2+}Be_2Si_2O_2O_{10}$ | X | 118 | 2.0 | D = 732 | — | A = 89 | [317] |
| $SrAl_{12}O_{19}$ | X | 300 | g = 2.003 | D = 327 G | — | A = 83 G | [318] |
| $SrAl_{12}O_{19}$ | X | 110 | g = 1.999 | D = 309 G | — | A = 83 G | [318] |
| $Sr_{1-x}Ba_xF_2$ | X | RT | 2.001 | −54.87, −135.06 (G) | $B_4^0 = 0.016$, $B_4^2 = 0.042$, $B_4^4 = 0.058$ (G) | $A_{iso} = -100.4$ G | [319] |
| TMATC-Zn (tetramethylammoniumtetrachlorozincate) | X | 77 | 1.9834 | D = 349, E = 106 | a = 21 | A = 105, B = 100 | [320] |
| TMATC-Zn (tetramethylammonium tetrachlorozincate) | 9.11 | 77 | 1.9834 | 349, 318 | — | A = 105, B = 100 | [320] |
| $YAl_3(BO_3)_4$ | 36 | 295 | $g_\parallel = 1.9982$, $g_\perp = 1.9924$ | D = −783.7 G | |a| = 0.015 G, |F| = 13.6 G | $|A_\parallel| = 1.1938$, $|A_\perp| = 1.1435$ | [321] |
| $Y_3Al_5O_{12}$ | X | 300 | 2.005, 2.014, 2.028 | — | — | 87, 83, 84 G | [322] |
| $Y_3Al_5O_{12}$ | X | 110 | g = 2.007 | — | — | $A_{Avg} = 89$ G | [322] |
| Zn ammonium phosphate hexahydrate | X | 295 | g = 1.9527 | 175, 174 | — | A = 92, B = 86 | [323] |
| $ZnAl_2O_4$ | X | RT | 2.003 | D = 322 G | — | 82.58 (G) | [324] |
| $ZnAl_2O_4$ | X | 110 | 2.002 | D = 328 G | — | 83.86 (G) | [324] |
| $Zn(BF_4)_2 \cdot 6H_2O$ | — | 77 | g = 2.001 | −140, — | −3.1, —, — | A = −89.6 | [325] |
| $Zn(BF_4)_2 \cdot 6H_2O$ | — | 293 | g = 2.001 | −170, — | −2.6, —, — | A = −89.6 | [325] |
| $Zn(C_3H_3O_4)_2 \cdot (H_2O)_2$ (site I) | X | RT | 1.959, 1.998, 2.011 | D = 322 G | — | −8.73, −8.55, −9.10 (mT) | [326] |

*(continued)*

## ($Mn^{2+}$($3d^5$) listing – contd.)

| Host | Frequency (GHz)/band | T (K) | $g_x, g_y, g_z$ | $b_2^0(=D), b_2^2(=3E)$ ($10^{-4}$ $cm^{-1}$) | $b_4^0, b_4^2, b_4^4$ ($10^{-4}$ $cm^{-1}$) | $A_x, A_y, A_z$ ($10^{-4}$ $cm^{-1}$) | References |
|---|---|---|---|---|---|---|---|
| $Zn(C_3H_3O_4)_2 \cdot (H_2O)_2$ (site II) | X | RT | 2.015, 1.996, 2.004 | $D = 328$ G | — | −8.18, −8.00, −8.58 (mT) | [326] |
| $Zn(C_5H_5NO)_6 \cdot (BF_4)_2$ | X | 300 | $g_\parallel = 2.0005$, $g_\perp = 2.0096$ | 267, −117.9 | $a = 4.7$ | $A_\parallel = -84.1$, $A_\perp = -87.6$ | [327] |
| $[Zn(H_2O)_6] \cdot [Zn(mal)_2(H_2O)_2]$ | X | RT | 1.972, 2.000, 2.023 | $D_{xx} = -34.49$, $D_{yy} = -3.26$, $D_{zz} = 37.74$, $E = 15.6$ (mT) | — | 8.95, 9.48, 9.93 (mT) | [328] |
| $ZnKPO_4 \cdot 6H_2O$ | 9.12 | 295 | 1.9997, 1.9538, 1.9524 | $D_{xx}$, $D_{yy}$, $D_{zz} = 15.49$, 0.22, −15.71 mT | — | $A_{xx}$, $A_{yy}$, $A_{zz} = 11.70$, 10.53, 10.42 mT | [329] |
| [Zn(Mn)(t-Buterpy)($N_3$)$_2$] (t-Buterpy = 4,4′,4″-tri-tert-butyl-2,2′:6′,2″-terpyridine) (magnetically diluted powder) | 115, 230 | 30 | 2.001, 2.001, 2.0005 | $\|D\| = 2\,600$, $E = 430$ | — | 77.5, 77.5, 77.5 (G) | [304] |
| ZnO nanowires (3 atm%) | 9.3 | 4.2 | $g_\parallel = 2.003$ | $D = -231$ | $\|a - F\| = 6$ | $\|A_\parallel\| = 76$ | [330] |
| ZnO nanowires (10 atm%) | 9.3 | 4.2 | $g_\parallel = 2.000$ | $D = -230$ | $\|a - F\| = 7$ | $\|A_\parallel\| = 78$ | [330] |
| ZnS (cubic) | X | RT | $g_{iso} = 2.00225$ | | $a = 7.987$ | $A = -63.88$ | [331] |
| $ZnV_2O_7$ | 9.61/249.9 | 295/253 | 2.00 | 1 908.2, 360.3 | −6.68, −196.80, −83.39 | — | [226] |
| $ZnV_2O_7$ | 9.61 | 295 | 2.008 | 1 918.2, 363.8 | −76.72, −243.5, −50.03 | $A = B = -78.73$ | [226] |
| $ZnV_2O_7$ | 9.61 | 120 | 2.007 | 2 018.3, 360.3 | −66.71, −60.04, −166.78 | $A = B = -78.73$ | [226] |
| $ZnV_2O_7$ | 9.61 | 77 | 2.007 | 2 028.3, 313.6 | −30.02, 16.68, −133.43 | $A = B = -78.73$ | [226] |
| $ZnV_2O_7$ | 9.61 | 4.2 | 2.09 | 2 051.6, 370.3 | 36.7, 390.3, −93.4 | $A = B = -78.73$ | [226] |

## $Mn^{2+}$ ($3d^5$) – Glasses

| Host | Frequency (GHz or band) | $T$ (K) | $\tilde{g}$ | $b_2^0(=D)$, $b_2^2(=3E)$ ($10^{-4}$ cm$^{-1}$) | $b_4^0$, $b_4^2$, $b_4^4$ ($10^{-4}$ cm$^{-1}$) | $A_x$, $A_y$, $A_z$ ($10^{-4}$ cm$^{-1}$) | References |
|---|---|---|---|---|---|---|---|
| $Ga_{1-x}Mn_xAs$ ($x = 4 \times 10^{-4}$) | X | 2 | $g = 2.00$ | 1.360, — | — | $a = -14.1$ | [332] |
| $GeO_2$-CdO-$CdF_2$-$AlF_3$ glass | X, Q | 77, 300 | $g_{iso} = 2.0$ | — | — | $A_{iso} =$ 81.8–89.3 | [333] |
| $5Na_2O$–$25K_2O$–$69B_2O_3$–1 $MnO_2$ glass | X | RT | $g = 2.019$ | 221, — | — | $A = 76.5$ | [334] |
| $10Na_2O$–$20K_2O$–$69B_2O_3$–1 $MnO_2$ glass | X | RT | $g = 2.019$ | 236, — | — | $A = 78.0$ | [334] |
| $15Na_2O$–$15K_2O$–$69B_2O_3$–1 $MnO_2$ glass | X | RT | $g = 2.027$ | 280, — | — | $A = 83.0$ | [334] |
| $20Na_2O$–$10K_2O$–$69B_2O_3$–1 $MnO_2$ glass | X | RT | $g = 2.027$ | 255, — | — | $A = 84.0$ | [334] |
| $25Na_2O$–$5K_2O$–$69B_2O_3$–1 $MnO_2$ glass | X | RT | $g = 2.027$ | 222, — | — | $A = 79.0$ | [334] |
| $20ZnO \cdot 5Li_2O \cdot 25Na_2O \cdot 50B_2O_3$ glass | X | RT | 4.249, 2.810, 2.023 | $D = 54.87$ (mT) | — | 79.89 | [335] |
| $20ZnO \cdot 10Li_2O \cdot 20Na_2O \cdot 50B_2O_3$ glass | X | RT | 4.249, 2.671, 2.021 | $D = 53.85$ (mT) | — | 80.51 | [335] |
| $20ZnO \cdot 15Li_2O \cdot 15Na_2O \cdot 50B_2O_3$ glass | X | RT | 4.249, 2.721, 2.024 | $D = 55.20$ (mT) | — | 79.69 | [335] |
| $20ZnO \cdot 20Li_2O \cdot 10Na_2O \cdot 50B_2O_3$ glass (site II) | X | RT | 4.249, 2.652, 2.023 | $D = 54.89$ (mT) | — | 79.77 | [335] |
| $20ZnO \cdot 25Li_2O \cdot 5Na_2O \cdot 50B_2O_3$ glass | X | RT | 4.249, 2.677, 2.022 | $D = 55.12$ (mT) | — | 80.31 | [335] |

## 3d⁶ (Co³⁺, Fe²⁺), S = 2
## Co³⁺. Data tabulation of SHPs
### Co³⁺ (3d⁶)

| Host | Frequency (GHz)/band | T (K) | $\tilde{g}$ | $b_2^0(=D), b_2^2(=3E)$ $(10^{-4}\ \text{cm}^{-1})$ | References |
|---|---|---|---|---|---|
| LaCoO₃ | 240 | 40 | $g_\parallel = 3.25, g_\perp = 3.83$ | D = 45 000 | [336] |

## Fe²⁺. Data tabulation of SHPs
### Fe²⁺ (3d⁶)

| Host | Frequency (GHz)/band | T (K) | $\tilde{g}$ | $b_2^0(=D), b_2^2(=3E)$ $(10^{-4}\ \text{cm}^{-1})$ | $b_4^0, b_4^2, b_4^4$ $(10^{-4}\ \text{cm}^{-1})$ | References |
|---|---|---|---|---|---|---|
| Bis(2,2'-bi-2-thiazoline)-bis(isothiocyanato)iron(II) | 610 | 10 | 2.147, 2.166, 2.01 | 124 270, 2 430 | — | [337] |
| CsFe(H₂O)₆PO₄ (site A) | 285, 345, 380 | 10 | 2.36, 2.22, 2.52 | 120 200, 21 230 | — | [338] |
| CsFe(H₂O)₆PO₄ (site B) | 285, 345, 380 | 10 | 2.28, 2.22, 2.52 | 121 500, 13 700 | — | [338] |
| CsFe(D₂O)₆PO₄ (site A) | 285, 345, 380 | 5 | — | −161 100, 33 750 | — | [338] |
| CsFe(D₂O)₆PO₄ (site B) | 285, 345, 380 | 5 | — | 126 020, 40 990 | — | [338] |
| [Fe(H₂O)₆](ClO₄)₂ | 225 | 4.5 | 2.147, 2.166, 2.01 | 111 700, 7 000 | — | [339] |
| [Fe(H₂O)₆]SiF₆ | 321.6 | 20 | 2.099, 2.151, 1.997 | D = 119 500, E = 6 580 | $B_4^0 = 17$, $B_4^4 = 18$ | [340] |
| [Fe(H₂O)₆]SiF₆ | 321.6 | 20 | 2.099, 2.151, 1.997 | D = 119 500, E = 6 580 | $B_4^0 = 17$, $B_4^4 = 18$ | [340] |
| FeSO₄ | 167 | 4.2 | 2.13, 2.20, 2.10 | 102 000, 22 400 | — | [140] |
| Fe(SO₄)·4H₂O | 167 | 4.5 | 2.10, 2.04, 2.11 | 103 200, 22 300 | — | [214] |

(continued)

($Fe^{2+}(3d^6)$ listing contd.)

| Host | Frequency (GHz)/band | T (K) | $\tilde{g}$ | $b_2^0(=D), b_2^2(=3E)$ ($10^{-4}$ cm$^{-1}$) | $b_4^0, b_4^2, b_4^4$ ($10^{-4}$ cm$^{-1}$) | References |
|---|---|---|---|---|---|---|
| $[Fe_2(\mu\text{-}OH)_3(tmtacn)_2]^{2+}$ | 189.84, 216, 279.69, 324.65 | 20 | 2.00, 2.00, 2.00 | 10 800, 0 | — | [341] |
| $Mg_2SiO_4$:Fe | 65–850 | 4.2–15 | $g_\| = 4.30$, $g_\perp = 2.0$ | $D = -627.39$ GHz, $E = 48.14$ GHz | — | [342] |
| $Mg_2SiO_4$ doped with chromium (iron as trace element) | 65–850 | 4.2–15 | $g_\| = 4.28$, $g_\perp = 2.0$ | $D = -777.45$ GHz, $E = 56.01$ GHz | — | [342] |
| $(NH_4)_2[Fe(H_2O)_6](SO_4)_2$ | 225 | 4.5 | 2.226, 2.31, 1.93 | 149 400, 37 780 | — | [339] |
| $[PPh_4]_2[Fe(SPh)_4]$ | 189.38–432.6 | 20 | 2.08, 2.08, 2.00 | 58 400, 14 200 | — | [343] |
| $[Zn(H_2O)_6]SiF_6$ | 321.6 | 20 | 2.25, 2.22, 2.23 | $D = 134\,200$, $E = 500$ | — | [340] |
| $ZnSiF_6 \cdot 6H_2O$ (site a) | 170, 222.4, 333.2 | 5 | $g_\| = 1.984$, $g_\perp = 2.014$ | $D = -635$, $E = 10$ (G) | $B_4^0 = 3.8$ (G) | [344] |
| $ZnSiF_6 \cdot 6H_2O$ (site a) | 170, 222.4, 333.2 | 10 | $g_\| = 1.992$ | $D = -615$, $E = 0$ (G) | $B_4^0 = 3.4$ (G) | [344] |
| $ZnSiF_6 \cdot 6H_2O$ (site a) | 170, 222.4, 333.2 | 17 | $g_\perp = 2.012$ | $D = -860$, $E = 5$ (G) | $B_4^0 = 4.4$ (G) | [344] |
| $ZnSiF_6 \cdot 6H_2O$ (site a) | 170, 222.4, 333.2 | 20 | $g_\| = 1.996$ | $D = -880$, $E = 0$ (G) | $B_4^0 = 6.1$ (G) | [344] |
| $ZnSiF_6 \cdot 6H_2O$ (site a) | 170, 222.4, 333.2 | 35 | $g_\perp = 1.989$ | $D = -798$, $E = 2$ (G) | $B_4^0 = -5.5$ (G) | [344] |
| $ZnSiF_6 \cdot 6H_2O$ (site b) | 170, 222.4, 333.2 | 5 | $g_\| = 2.029$, $g_\perp = 1.986$ | $D = -1920$, $E = 25$ (G) | $B_4^0 = 8.4$ (G) | [344] |
| $ZnSiF_6 \cdot 6H_2O$ (site b) | 170, 222.4, 333.2 | 10 | $g_\perp = 1.992$ | $D = -1995$, $E = 25$ (G) | $B_4^0 = 8.3$ (G) | [344] |
| $ZnSiF_6 \cdot 6H_2O$ (site b) | 170, 222.4, 333.2 | 17 | $g_\| = 2.022$ | $D = -1840$, $E = 0$ (G) | $B_4^0 = 10.5$ (G) | [344] |
| $ZnSiF_6 \cdot 6H_2O$ (site b) | 170, 222.4, 333.2 | 20 | $g_\perp = 1.992$ | $D = -1970$, $E = 25$ (G) | $B_4^0 = 10.2$ (G) | [344] |
| $ZnSiF_6 \cdot 6H_2O$ (site b) | 170, 222.4, 333.2 | 35 | $g_\| = 1.994$ | $D = -765$, $E = 0$ (G) | $B_4^0 B_4^0 = 5.3$ (G) | [344] |

## 3d$^7$ (Co$^{2+}$), $S = 3/2$
## Co$^{2+}$. Data tabulation of SHPs
## Co$^{2+}$ (3d$^7$)

| Host | Frequency (GHz)/ band | T (K) | $\tilde{g}$ | ($b_2^0 = D$), ($b_2^2 = 3E$) ($10^{-4}$ cm$^{-1}$) | $A_x, A_y, A_z$ ($10^{-4}$ cm$^{-1}$) | References |
|---|---|---|---|---|---|---|
| CdNH$_4$PO$_4$ · 6H$_2$O (CAPH) | X | 11.5 | $g_\| = 8.04$, $g_\perp = 3.25$ | — | — | [345] |
| CdPS$_3$ | X | 5 | $g_\| = 4.94$, $g_\perp = 3.99$ | — | $A_\| = 183$, $A_\perp = 97$ (mT) | [346] |
| CdS | 96 | 4.2–35 | $g_\| = 2.269$, $g_\perp = 2.286$ | $b_2^0 = 6400$ | — | [347] |
| CoCl$_2$(PPh$_3$)$_2$ | 190 | 20 | 2.166, 2.170, 2.240 | −147 600, −11 410 | — | [140] |
| Co(PPh$_3$)$_2$Cl$_2$ (Ph = phenyl) | 150–700 | 4.7/20 | 2.166, 2.170, 2.240 | −147 600, 34 230 | — | [348] |
| Co$_2$(OH)(AsO$_4$) | X | 4.2 | $g_1 = 6.22$, $g_2 = 4.21$, $g_3 = 2.87$ | — | $A_1 = 240$, $A_2 = 140$, $A_3 < 20$ | [349] |
| 2, 9, 16, 23-(Diethoxymalonyl)-tetrakis 3, 10, 17, 24-chloro phthalocyanine Pc2 (in powder form) | X | 295 | $g_\| = 1.935$, $g_\perp = 2.365$ | — | 113.97, 80.34 | [350] |
| LiMgPO$_4$ | X | 4.2 | $g_1 = 6.16$, $g_2 = 4.14$, $g_3 = 2.53$ | — | $A_1 = 246$, $A_2 = 89$, $A_3 < 20$ | [351] |
| LiNbO$_3$ | X | 4.3 | $g_\| = 2.671$, $g_\perp = 5.052$ | — | $A_\| = 40$, $A_\perp = 154$ | [352] |
| Li$_2$B$_4$O$_7$ | X | 4 | 5.791, 0.760, 0.327 | — | 567.460, 319.533, 243.183 | [353] |
| Matallophthalo-cyanine | 9.80 | 295 | $g_\| = 2.35$, $g_\perp = 2.05$ | — | $A_\| = 130.78$, $A_\perp = 4.67$ | [354] |

(continued)

## ($Co^{2+}$ ($3d^7$) listing contd.)

| Host | Frequency (GHz)/band | T (K) | $\tilde{g}$ | $(b_2^0 = D)$, $(b_2^2 = 3E)$ ($10^{-4}$ cm$^{-1}$) | $A_x, A_y, A_z$ ($10^{-4}$ cm$^{-1}$) | References |
|---|---|---|---|---|---|---|
| $Mg_{2.997}Co_{0.003}V_2O_8$ (site I) | 9.25 | 4.2 | 7.038, 2.42, 3.36 | — | 373.6, 72.7, 79.2 | [355] |
| $Mg_{2.997}Co_{0.003}V_2O_8$ (site II) | 9.25 | 4.2 | 3.247, 3.47, 6.095 | — | 34.8, 72.2, 242.6 | [355] |
| $NH_4NiPO_4 \cdot 6H_2O$ | X | 4.2 | 4.9091, 5.1389, 2.6680 | — | 160.17, 178.76, 44.37 | [356] |
| Porphyrazine (in powder form) | X | 295 | $g_\parallel = 1.997$, $g_\perp = 2.300$ | — | 118.64, 63.52 | [350] |
| $SrLaGa_3O_7$ (site 1) | 9.5 | 8 | $g_\parallel = 2.91$, $g_\perp = 2.91$ | 87, 0 (mT) | $A_\parallel = 10$, $A_\perp = 1.5$ (mT) | [357] |
| $SrLaGa_3O_7$ (site 2) | 9.5 | 8 | $g_\parallel = 3.35$, $g_\perp = 3.35$ | 108, 0 (mT) | $A_\parallel = 17$, $A_\perp = 4.5$ (mT) | [357] |
| $SrLaGa_3O_7$ | X | 4.2–12.4 | $g_\parallel = 2.26$, $g_\perp = 4.7$ | — | — | [358] |
| $Tp^{t-Bu}CoN_3$ | 377.4 | 4.5 | 2.48, 2.02, 2.31 | 74 570, 15 750 | — | [214] |
| 2, 9, 17, 23-Tetra-(1, 1, 2-(tricarbethoxyethyl) phthalocyanine Pc1. (in powder form) | X | 295 | $g_\parallel = 2.005$, $g_\perp = 2.360$ | — | 113.97, 71.93 | [350] |
| 2, 9, 17, 23-Tetra-(1, 1, 2-(tricarbethoxyethyl) phthalocyanine Pc1 (in chloroform) | X | 295 | $g_\parallel = 2.050$, $g_\perp = 2.380$ | — | 115.84, 78.47 | [350] |
| Co(II)-MBP 1.6 mM (metal binding protein) VanX gene (Van=Vancomycin) | 9.48 | 8 | 2.37, 2.37, 2.03 | $E/D = 0.14$ | — | [359] |
| Co(II)-VanX precipitated | 9.48 | 8 | 2.235, 2.235, 2.112 | $E/D = 0.277$ | — | [359] |

(continued)

## ($Co^{2+}$ ($3d^7$)) listing contd.)

| Host | Frequency (GHz)/band | T (K) | $\tilde{g}$ | ($b_2^0 = D$), ($b_2^2 = 3E$) ($10^{-4}$ cm$^{-1}$) | $A_x, A_y, A_z$ ($10^{-4}$ cm$^{-1}$) | References |
|---|---|---|---|---|---|---|
| Co(II) VanX with bound phosphinate | 9.48 | 8 | 2.245, 2.245, 2.165 | $E/D = 0.233$ | — | [359] |
| Co(II)-MBP-VanX streated with phosphinate | 9.48 | 8 | 2.252, 2.177 | $E/D = 0.245$ | — | [359] |
| $YAlO_3$:$Co^{2+}$ ($\alpha$) | X | 12 | 5.42, 5.02, 1.13 | — | 288, 164, 82 | [360] |
| $YAlO_3$:$Co^{2+}$ ($\beta$) | X | 12 | 6.67, 3.70, 1.81 | — | 178, 84, −8.7 | [360] |
| $Zn_{1-x}Co_xO$ ($x = 1\%$) | X | 10–200 | $g_\parallel = 4.50$, $g_\perp = 2.20$ | $D = 30$ G | $A = 0$ G | [361] |
| $Zn_{1-x}Co_xO$ ($x = 3\%$) | X | 10–200 | $g_\parallel = 4.50$, $g_\perp = 2.20$ | $D = 10$ G | $A = 0$ G | [361] |
| $ZnNH_4PO_4 \cdot 6H_2O$ (ZAPH) | X | 11.5 | $g_\parallel = 2.79$, $g_\perp = 5.72$ | — | — | [345] |
| ZnO nanowires (site A) | 9.3 | 4.2 | $g_\parallel = 2.247$, $g_\perp = 2.276$ | — | $|A_\parallel| = 16.2$, $|A_\perp| = 1.6$ | [330] |
| ZnO nanowires (site B) | 9.3 | 4.2 | $g_\parallel = 2.245$, $g_\perp = 2.276$ | — | $|A_\parallel| = 16.8$, $|A_\perp| = 1.6$ | [330] |

## $3d^8$ ($Ni^{2+}$), $S = 1$

$Ni^{2+}$. Data tabulation of SHPs

$Ni^{2+}$ ($3d^8$)

| Host | Frequency (GHz)/band | T (K) | $\tilde{g}$ | ($b_2^0 = D$), ($b_2^2 = 3E$) ($10^{-4}$ cm$^{-1}$) | References |
|---|---|---|---|---|---|
| Hydrotris(3,5-dimethylpyrazolyl)-borate · Ni($k^3$-$BH_4$) | 150–700 | 5–40 | 2.170, 2.161, 2.133 | $D = 19\,100$, $E = 2850$ | [362] |
| Hydrotris(3,5-dimethylpyrazolyl)-borate · Ni($k^3$-$BD_4$) | 150–700 | 5–40 | 2.174, 2.153, 2.150 | $D = 22\,900$, $E = 2900$ | [362] |
| Hydrotris(3,5-dimethylpyrazole)borate *NiCl | 276 | 10 | 2.280, 2.265, 2.254 | 39 300, 3 480 | [363] |

(continued)

## ($Ni^{2+}$($3d^8$) listing contd.)

| Host | Frequency (GHz)/band | $T$ (K) | $\tilde{g}$ | ($b_2^0 = D$), ($b_2^2 = 3E$) ($10^{-4}$ cm$^{-1}$) | References |
|---|---|---|---|---|---|
| Hydrotris(3,5-dimethylpyrazole)borate *NiBr | 527 | 4.2 | 2.232, 2.232, 2.28 | −114 300, −200 | [363] |
| Hydrotris(3,5-dimethylpyrazole)borate *NiI | 527 | 4.2 | 2.16, 2.16, 2.16 | −230 100, −7 400 | [363] |
| LaNi$_6$ | 189 | 10 | 2.06, 2.21, 2.30 | 55 000, 8 000 | [364] |
| LiNbO$_3$ | — | — | $g_\parallel = 2.24$, $g_\perp = 2.20$ | −50 700, — | [365] |
| NaCl | X | 20 | $g_\parallel = 2.86$, $g_\perp = 2.10$ | — | [366] |
| [Ni(Im)$_2$(L-tyr)$_2$] · 4H$_2$O Im=imidazole, tyr=tyrosine | 100–416 | 50 | 2.170, 2.166, 2.193 | $D = -30\,100$, $E = -4\,066$ | [367] |
| Ni(PPh$_3$)$_2$Br$_2$ (HFEPR) | 325.9 | 4.5 | 2.2, 2.2, 2.0 | 45 000, 15 000 | [368] |
| Ni(PPh$_3$)$_2$Br$_2$ (Ph = phenyl) (powder) | 325.9 | 2 | 2.06, 2.00, 2.22 | 53 800, 17 600 | [368] |
| Ni(PPh$_3$)$_2$Br$_2$ (crystal) (Ph = phenyl) | 325.9 | 4.5 | 1.85, 1.85, 2.77 | 133 000, — | [368] |
| Ni(PPh$_3$)$_2$Br$_2$ (powder, field-dep. magnet.) (Ph = phenyl) | 325.9 | 2 | 2.20, 2.20, 2.20 (fix) | 38 000, 0 | [368] |
| Ni(PPh$_3$)$_2$Cl$_2$ (Ph = phenyl) | 435 | 4.5 | 2.20, 2.17, 2.20 | 132 000, 18 500 | [368] |
| Ni(PPh$_3$)$_2$Cl$_2$ (powder) (Ph = phenyl) | 435 | 2 | 1.99, 2.00, 2.40 | 120 300, 17 800 | [368] |
| Ni(PPh$_3$)$_2$Cl$_2$ (crystal) | 435 | 4.5 | 2.03, 2.03, 2.51 | 140 000, — | [368] |
| Ni(PPh$_3$)$_2$Cl$_2$ (powder) (Ph = phenyl) | 435 | 2 | 2.20, 2.20, 2.20 (fix) | 131 000, 0 | [368] |
| Ni(PPh$_3$)$_2$Cl$_2$ (Ph = phenyl) | 611.2 | 4.2 | 2.200, 2.177, 2.15 | 131 960, 18 480 | [214] |
| Ni(PPh$_3$)$_2$Cl$_2$ (Ph = phenyl) | 10 — 40 | 5.5 | 2.20, 2.20, 2.00 | 133 470, 19 320 | [369] |
| Ni(PPh$_3$)$_2$I$_2$ (powder, mag. sus.) (Ph = phenyl) | 325.9 | 2 | 1.95, 2.00, 2.11 | 279 200, 47 100 | [368] |

(continued)

## (Ni$^{2+}$(3d$^8$) listing contd.)

| Host | Frequency (GHz)/band | T (K) | $\tilde{g}$ | ($b_2^0 = D$), ($b_2^2 = 3E$) ($10^{-4}$ cm$^{-1}$) | References |
|---|---|---|---|---|---|
| Ni(PPh$_3$)$_2$I$_2$ (powder, field-dep. magnet.) (Ph = phenyl) | 325.9 | 2 | 2.00, 2.00, 2.00 (fix) | 256 000, 0 | [368] |
| Ni$_2$CdCl$_6$ · 12H$_2$O (site I) | 9.49 | 295 | $g_\parallel = 2.245$, $g_\perp = 2.257$ | −261.90, — | [226] |
| Ni$_2$CdCl$_6$ · 12H$_2$O (site II) | 9.49 | 295 | $g_\parallel = 2.244$, $g_\perp = 2.207$ | −10 330.50, — | [226] |
| Ni$_2$CdCl$_6$ · 12H$_2$O (site I) | 249.9 | 253 | $g_\parallel = 2.230$, $g_\perp = 2.226$ | — | [226] |
| Ni$_2$CdCl$_6$ · 12H$_2$O (site II) | 249.9 | 253 | $g_\parallel = 2.22$, $g_\perp = 2.219$ | — | [226] |
| Ni$_2$C$_{26}$H$_{22}$N$_4$O$_{10}$C$_{12}$ | X | RT | 2.377, 2.219, 2.071 | $\tilde{D} = 9.7, 4.2$, −13.9 (mT) | [370] |
| Zn(C$_3$H$_4$N$_2$)Cl$_2$ · 4H$_2$O | X | RT | 2.3691, 2.2105, 1.9409 | $D_{11} = 244.07$ mT, $D_{22} = 24.98$ mT, $D_{33} = -269.05$ mT | [371] |

## 3d$^9$ (Cu$^{2+}$, Ni$^+$)
Cu$^{2+}$. Data tabulation of SHPs
Cu$^{2+}$ (3d$^9$)

| Host | Frequency (GHz)/band | T (K) | $\tilde{g}$ | $A_x, A_y, A_z$ ($10^{-4}$ cm$^{-1}$) | References |
|---|---|---|---|---|---|
| Azurin (rapid cooling) 0% glycerol | — | 77 | $g_\parallel = 2.2615$, $g_\perp = 2.0441$ | $A_\parallel = 54.0$, $A_\perp = 4.1$ | [372] |
| Azurin (slow cooling) 0% glycerol | — | 77 | $g_\parallel = 2.2602$, $g_\perp = 2.0439$ | $A_\parallel = 54.0$, $A_\perp = 3.8$ | [372] |
| Azurin (rapid cooling) 40% glycerol | — | 77 | $g_\parallel = 2.2590$, $g_\perp = 2.0441$ | $A_\parallel = 54.0$, $A_\perp = 4.3$ | [372] |
| Azurin (slow cooling) 40% Glycerol | — | 77 | $g_\parallel = 2.2578$, $g_\perp = 2.0441$ | $A_\parallel = 54.0$, $A_\perp = 4.3$ | [372] |
| β-BaB$_2$O$_4$ nano-powder | X | RT | $g_\parallel = 2.368$, $g_\perp = 2.073$ | $A_\parallel = 135$, $A_\perp = 40$ | [373] |
| BaF$_2$ | X, Q | 77 | $g_\parallel = 2.511$, $g_\perp = 2.092$ | $^{63}$Cu: $A_\parallel = 98.3$, $A_\perp = 7$ | [374] |

(continued)

## ($Cu^{2+}$ ($3d^9$)) listing contd.

| Host | Frequency (GHz)/ band | T (K) | $\tilde{g}$ | $A_x, A_y, A_z$ ($10^{-4}$ cm$^{-1}$) | References |
|---|---|---|---|---|---|
| Bis((L-asparaginato)Mg$^{2+}$ | 9.52 | RT | 2.0420, 2.0808, 2.3600 | 99, 108, 140 | [375] |
| Bis(L-asparaginato)zinc(II), C$_8$H$_{14}$N$_4$O$_6$ (single crystal) | 9.5 | 295 | 2.0341, 2.0649, 2.2390 | 51, 75, 169 | [376] |
| Bis(L-asparaginato)zinc(II), C$_8$H$_{14}$N$_4$O$_6$ (powder) | 9.5 | 295 | 2.0344, 2.0653, 2.2388 | 50, 85, 170 | [376] |
| Bis(glycinato)Mg$^{2+}$ monohydrate (site I) | X | RT | 2.1577, 2.2018, 2.3259 | 87, 107, 141 | [377] |
| Bis(glycinato)Mg$^{2+}$ monohydrate (site II) | X | RT | 2.1108, 2.1622, 2.2971 | 69, 117, 134 | [377] |
| (7,16-Bis(p-methylbenzoyl)-6,8,15,17-tetramethyldibenzo[b,i][1,4,8,11]tetraazacyclotetradecinato)copper(II) (in toluene) | X | RT | $g_{iso} = 2.087$ | $A_{iso} = 89.9$ | [106] |
| (7,16-Bis(p-methylbenzoyl)-6,8,15,17-tetramethyldibenzo[b,i][1,4,8,11]tetraazacyclotetradecinato)copper(II) (diluted with an isomorphous tetraaza[14]annulene nickel(II) complex) | X | RT | $g_\| = 2.168$, $g_\perp = g_{2,3} = 2.036$, 2.010 | $A_\| = 199.9$ | [106] |
| (7,16-Bis(p-methoxybenzoyl)-6,8,15,17-tetramethyldibenzo[b,i][1,4,8,11]-tetraazacyclotetradecinato)copper(II) (in toluene) | X | RT | $g_{iso} = 2.087$ | $A_{iso} = 89.9$ | [106] |
| (7,16-Bis(p-methoxybenzoyl)-6,8,15,17-tetramethyldibenzo[b,i][1,4,8,11]tetraazacyclotetradecinato)copper(II) (diluted with an isomorphous tetraaza[14]annulene nickel(II) complex) | X | RT | $g_\| = 2.165$, $g_\perp = g_{2,3} = 2.037$, 2.008 | $A_\| = 199.9$ | [106] |

(continued)

## (Cu$^{2+}$(3d$^9$) listing contd.)

| Host | Frequency (GHz)/band | T (K) | $\tilde{g}$ | $A_x, A_y, A_z$ ($10^{-4}$ cm$^{-1}$) | References |
|---|---|---|---|---|---|
| (7,16-Bis(p-chlorobenzoyl)-6,8,15,17-tetramethyl-dibenzo[b,i][1,4,8,11]tetraazacyclotetradecinato)copper(II) (in toluene) | X | RT | $g_{iso} = 2.086$ | $A_{iso} = 89.5$ | [106] |
| (7,16-Bis(p-chlorobenzoyl)-6,8,15,17-tetramethyl-dibenzo[b,i][1,4,8,11]tetraazacyclotetradecinato)copper(II) (diluted with an isomorphous tetraaza[14] annulene nickel(II) complex) | X | RT | $g_\| = 2.168$, $g_\perp = g_{2,3} = 2.036$, 2.010 | $A_\| = 199.9$ | [106] |
| (7,16-Bis(p-nitrobenzoyl)-6,8,15,17-tetramethyl-dibenzo[b,i][1,4,8,11]tetraazacyclotetradecinato)copper(II) (in toluene) | X | RT | $g_{iso} = 2.087$ | $A_{iso} = 90.2$ | [106] |
| (7,16-Bis(p-nitrobenzoyl)-6,8,15,17-tetramethyl-dibenzo[b,i][1,4,8,11]tetraazacyclotetradecinato)copper(II) (diluted with an isomorphous tetraaza[14] annulene nickel(II) complex) | X | RT | $g_\| = 2.168$, $g_\perp = g_{2,3} = 2.036$, 2.010 | $A_\| = 199.0$ | [106] |
| CaC$_3$H$_2$O$_4$ · 2H$_2$O (site I) | X | 295 | 2.0963, 2.1316, 2.4137 | 32, 34, 49 | [378] |
| CaC$_3$H$_2$O$_4$ · 2H$_2$O (site II) | X | 295 | 2.0668, 2.0800, 2.3561 | 34, 36, 51 | [378] |
| CaC$_3$H$_2$O$_4$ · 2H$_2$O (site III) | X | 295 | 2.0438, 2.0623, 2.2821 | 34, 36, 53 | [378] |
| CaC$_3$H$_2$O$_4$ · 2H$_2$O (site IV) | X | 295 | 2.0063, 2.0241, 2.2357 | 35, 37, 54 | [378] |
| CdBa(HCOO)$_4$ · 2H$_2$O single crystal, site II | X | RT | 2.110, 2.111, 2.360 | 4.28, 4.05, 9.05 (mT) | [379] |
| CdBa(HCOO)$_4$ · 2H$_2$O powder | X | RT | 2.106, 2.118, 2.348 | 4.40, 4.20, 9.06 (mT) | [379] |

(continued)

## ($Cu^{2+}(3d^9)$) listing contd.

| Host | Frequency (GHz)/band | T (K) | $\tilde{g}$ | $A_x, A_y, A_z$ ($10^{-4}$ cm$^{-1}$) | References |
|---|---|---|---|---|---|
| $CdBa(HCOO)_4 \cdot 2H_2O$ powder | X | 113 | 2.106, 2.087, 2.461 | 4.20, 2.50, 10.5 (mT) | [379] |
| $Cd(CH_3NH_2{}^+ CH_2COO^-)Br$ | X | RT | 2.108, 2.0005, 2.207 | −64, −23, −185 | [380] |
| $Cd(CH_3NH_2{}^+ CH_2COO^-)Cl_2$ (powder) | 9.35 | RT | 2.056, 2.036, 2.249 | 30, 34, 180 | [381] |
| $Cd(CH_3NH_2{}^+ CH_2COO^-)Cl_2$ (single crystal) | 9.35 | RT | 2.064, 2.041, 2.213 | −27, −32, −189 | [381] |
| $CdC_4H_4O_5 \cdot 5H_2O$ single crystal | X | 120 | $g_\parallel = 2.102$, $g_\perp = 2.228$ | — | [382] |
| $CdC_4H_4O_5 \cdot 5H_2O$ single crystal | X | 300 | $g_\parallel = 2.102$, $g_\perp = 2.236$ | — | [382] |
| $CdC_4H_4O_5 \cdot 5H_2O$ powder | X | 120 | 2.113, 2.161, 2.215 | — | [382] |
| $Cd(HCO_2)_2 \cdot 2H_2O$ (site I) | X | RT | 2.0917, 2.1166, 2.2887 | 140, 151, 239 | [383] |
| $Cd(HCO_2)_2 \cdot 2H_2O$ (site II) | X | RT | 2.0843, 2.1045, 2.2742 | 141, 158, 267 | [384] |
| $Cd(NH_4)_2(SO_4)_2 \cdot 6H_2O$ | X | 113 | 2.142, 2.052, 2.414 | 45, 35, 104 (GHz) | [385] |
| $Cd(NH_4)_2(SO_4)_2 \cdot 6H_2O$ | X | 113 | 2.137, 2.057, 2.419 | 47, 33, 103 (GHz) | [385] |
| $Cd(stpy)_3(NO_3)2 \cdot 1/2stpy$ (stpy = trans-4-styrylpyridene) | X | 300 | 2.066, 2.108, 2.298 | 23.1, 54.4, 107.3 | [386] |
| Chelidamate [Cu(II)] complex, liquid] | 3.8, 9.307 | 295 | $g_{iso} = 2.123$ | $A^{(63Cu)}_{iso}$ = 140 MHz | [387] |
| Chelidamate [Cu(II)] complex, frozen] | 180 | 10 | $g_\parallel = 2.265$, $g_\perp = 2.056$ | $\tilde{A}(A_X, A_Y, A_Z)$ = 100, 10, 480 (MHz) | [387] |
| $(CH_3)_2NH_2Al(SO_4)_2 \cdot 6H_2O$ (site I) | X | RT | 2.046, 2.082, 2.379 | 62.0, 27.5, 105.2 (absolute values) | [388] |

(continued)

## ($Cu^{2+}$ ($3d^9$) listing contd.)

| Host | Frequency (GHz)/band | T (K) | $\tilde{g}$ | $A_x, A_y, A_z$ ($10^{-4}$ cm$^{-1}$) | References |
|---|---|---|---|---|---|
| $(CH_3)_2NH_2Al(SO_4)_2 \cdot 6H_2O$: (site II) | X | RT | 2.041, 2.086, 2.370 | 60.9, 28.5, 104.2 (absolute values) | [388] |
| $[(CH_3)_2NH_2]_5Cd_3Cl_{11}$ (site I) | X | 6–293 | 2.060, 2.054, 2.296 | 17, 15, 114 (G) | [389] |
| $[(CH_3)_2NH_2]_5Cd_3Cl_{11}$ (site IIa) | X | 6–293 | 2.060, 2.054, 2.284 | 17, 15, 113 (G) | [389] |
| $[(CH_3)_2NH_2]_5Cd_3Cl_{11}$ (site IIb) | X | 6–293 | 2.060, 2.054, 2.282 | 17, 15, 114 (G) | [389] |
| $([Co(H_2O)_4(py)_2](sac)_2 \cdot 4H_2O)$ (site I) | X | RT | 2.106, 2.014, 2.391 | 76.3, 36.5, 117.6 (G) | [42] |
| $([Co(H_2O)_4(py)_2](sac)_2 \cdot 4H_2O)$ (site II) | X | RT | 2.056, 2.086, 2.391 | 56.9, 42.3, 111.4 (G) | [42] |
| $CoNH_4(PO_4) \cdot 6H_2O$ (single crystal) | X | 77, 143–300 | 2.063, 2.155, 2.404 | 2.07, 3.49, 11.58 (mT) | [390] |
| $CoNH_4(PO_4) \cdot 6H_2O$ (powder) | X | 77 | 2.129, —, 2.416 | 2.74, —, 10.78 (mT) | [390] |
| $[Co(nicotinamide)_2 (H_2O)_4]$ saccharinate$_2$ | X | RT | 2.093, 2.056, 2.359 | 3.3, 5.0, 12.2 (mT) | [391] |
| $[Co(tbz)_2(NO_3)(H_2O)](NO_3)$ (tbz, thiabendazole complex) | X | 300 | 2.0626, 2.1351, 2.305 | 23.1, 33.5, 147 | [392] |
| $Cu(agpa) \cdot 2H_2O$ | X | 4.2 | $g_\parallel = 2.220$, $g_\perp = 2.047$ | — | [393] |
| Cu-AlSBA-15 (90A) – (10) mesoporous | — | RT | $g_\parallel = 2.369$, $g_\perp = 2.078$ | $A_\parallel = 131$ | [394] |
| Cu-AlSBA-15 mesoporous | — | 77 | $g_\parallel = 2.391$, $g_\perp = 2.068$ | $A_\parallel = 146$ | [394] |
| $[Cu(bpy)_3]_2[Cr(C_2O_4)_3]NO_3 \cdot 9H_2O$ | 9.6 | — | 2.11 | — | [155] |
| $Cu(CH_2ClCOO)_2$ CuClAc ("neat") | 9.5 | 77 | (2.068, 2.068, 2.348) | $A = 54$ | [395] |
| $Cu(CH_2ClCOO)_2$ CuClAc-Y(zeolite) | 9.5 | 77 | (2.064, 2.070, 2.353) | $A = 66$ | [395] |
| $Cu(CH_3COO)_2$ CuAc ("neat") | 9.5 | 77 | (2.055, 2.095, 2.358) | $A = 73$ | [395] |

(continued)

## ($Cu^{2+}$ ($3d^9$) listing contd.)

| Host | Frequency (GHz)/band | T (K) | $\tilde{g}$ | $A_x, A_y, A_z$ ($10^{-4}$ cm$^{-1}$) | References |
|---|---|---|---|---|---|
| Cu(CH$_3$COO)$_2$ CuAc-Y(zeolite) | 9.5 | 77 | (2.055, 2.095, 2.358) | A = 73 | [395] |
| Cu(2-benzoylpyridine)$_2$ (ClO$_4$)$_2$ ([Cu(C$_{12}$H$_9$NO)$_2$ (ClO$_4$)$_2$) | — | 20–200 | 2.069, 2.043, 2.336 | −12, 24, −179 | [396] |
| Cu(2-benzoylpyridine)$_2$ (ClO$_4$)$_2$ ([Cu(C$_{12}$H$_9$NO)$_2$ (ClO$_4$)$_2$) | — | 295 | 2.097, 2.082, 2.275 | −30, 45, −135 | [396] |
| CuCl$_2$ (PVA polyethylene film) | X | 123–393 | $g_\|$ = 2.322, $g_\perp$ = 2.069 | $A_\|$ = 113 | [397] |
| CuF$_2$ (PVA polyethylene film) | X | 123–393 | $g_\|$ = 2.329, $g_\perp$ = 2.064 | $A_\|$ = 114 | [397] |
| Cu(Hagpa) (Ac) | X | 293 | $g_\|$ = 2.240, $g_\perp$ = 2.075 | — | [393] |
| Cu(Hagpa) (Ac) 2H$_2$O | X | 120 | $g_\|$ = 2.242, $g_\perp$ = 2.071 | — | [393] |
| [Cu(Hagpa)Cl] | X | 293 | $g_\|$ = 2.225, $g_\perp$ = 2.060 | — | [393] |
| Cu(Hagpa) NO$_3$ | X | 293 | $g_\|$ = 2.245, $g_\perp$ = 2.055 | — | [393] |
| [Cu(Hagpa)Br] (H$_2$agpa = aminoguanizone of pyruvic acid) | X, Q | 293 | $g_\|$ = 2.219, $g_\perp$ = 2.061 | — | [393] |
| [Cu(H$_2$L)(ClO$_4$)]ClO$_4$ · H$_2$O | X | 4.2 | g = 2.01 | — | [398] |
| CuH$_2$L$^1$Cl$_2$ as polycrystalline (L = 1-phenyl-1-hydroxymethylene bisphosphonate) | — | 293 | $g_\|$ = 2.069, $g_\perp$ = 2.033 | — | [399] |
| CuH$_2$L$^1$Cl$_2$ as solution (L = 1-phenyl-1-hydroxymethylene bisphosphonate) | — | 77 | $g_\|$ = 2.059, $g_\perp$ = 2.018 | — | [399] |
| CuH$_2$L$^2$Cl$_2$ as polycrystalline (L = 1-phenyl-1-hydroxymethylene bisphosphonate) | — | 293 | $g_\|$ = 2.093, $g_\perp$ = 2.500 | — | [399] |
| CuH$_2$L$^2$Cl$_2$ as solution (L = 1-phenyl-1-hydroxymethylene bisphosphonate) | — | 77 | $g_\|$ = 2.103, $g_\perp$ = 2.056 | — | [399] |

(continued)

## ($Cu^{2+}$ ($3d^9$) listing contd.)

| Host | Frequency (GHz)/band | T (K) | $\tilde{g}$ | $A_x, A_y, A_z$ ($10^{-4}$ cm$^{-1}$) | References |
|---|---|---|---|---|---|
| $CuH_2L^3Cl_2$ as polycrystalline ($L = 1$-phenyl-1-hydroxymethylene bisphosphonate) | — | 293 | $g_\| = 2.093$, $g_\perp = 2.030$ | — | [399] |
| $CuH_2L^3Cl_2$ as solution ($L = 1$-phenyl-1-hydroxymethylene bisphosphonate) | — | 77 | $g_\| = 2.093$, $g_\perp = 2.051$ | — | [399] |
| $[CuH_2L^4]Cl_2$ as polycrystalline ($L = 1$-phenyl-1-hydroxymethylene bisphosphonate) | — | 293 | $g_\| = 2.089$, $g_\perp = 2.034$ | — | [399] |
| $[CuH_2L^4]Cl_2$ as solution ($L = 1$-phenyl-1-hydroxymethylene bisphosphonate) | — | 77 | $g_\| = 2.193$, $g_\perp = 2.014$ | — | [399] |
| $CuH_2L^5Cl_2$ as polycrystalline ($L = 1$-phenyl-1-hydroxymethylene bisphosphonate) | — | 293 | $g_\| = 2.093$, $g_\perp = 2.050$ | — | [399] |
| $CuH_2L^5Cl_2$ as solution ($L = 1$-phenyl-1-hydroxymethylene bisphosphonate) | — | 77 | $g_\| = 2.087$, $g_\perp = 2.015$ | — | [399] |
| $CuH_2L^6Cl_2$ as polycrystalline ($L = 1$-phenyl-1-hydroxymethylene bisphosphonate) | — | 293 | $g_\| = 2.103$, $g_\perp = 2.030$ | — | [399] |
| $CuH_2L^6Cl_2$ as solution ($L = 1$-phenyl-1-hydroxymethylene bisphosphonate) | — | 77 | $g_\| = 2.199$, $g_\perp = 2.056$ | — | [399] |
| $CuH_2L^7Cl_2$ as polycrystalline ($L = 1$-phenyl-1-hydroxymethylene bisphosphonate) | — | 293 | $g_\| = 2.210$, $g_\perp = 2.080$ | — | [399] |
| $CuH_2L^7Cl_2$ as solution ($L = 1$-phenyl-1-hydroxymethylene bisphosphonate) | — | 77 | $g_\| = 2.200$, $g_\perp = 2.074$ | — | [399] |

(continued)

## ($Cu^{2+}$ ($3d^9$) listing contd.)

| Host | Frequency (GHz)/band | $T$ (K) | $\tilde{g}$ | $A_x, A_y, A_z$ ($10^{-4}$ cm$^{-1}$) | References |
|---|---|---|---|---|---|
| [$CuH_2L^8$]$Cl_2$ as polycrystalline ($L$ = 1-phenyl-1-hydroxymethylene bisphosphonate) | — | 293 | $g_\| = 2.109$, $g_\perp = 2.040$ | — | [399] |
| [$CuH_2L^8$]$Cl_2$ as solution ($L$ = 1-phenyl-1-hydroxymethylene bisphosphonate) | — | 77 | $g_\| = 2.300$, $g_\perp = 2.130$ | — | [399] |
| $CuH_2L^1Cl_2$ in solutions (DMSO) ($L$ = 1-phenyl-1-hydroxymethylene bisphosphonate) | — | 293 | $g_\| = 2.069$, $g_\perp = 2.033$ | $A_\| = 60$, $A_\perp = 35$ | [399] |
| $CuH_2L^2Cl_2$ in solutions (DMSO) ($L$ = 1-phenyl-1-hydroxymethylene bisphosphonate) | — | 293 | $g_\| = 2.103$, $g_\perp = 2.056$ | $A_\| = 60$, $A_\perp = 35$ | [399] |
| ($L$ = 1-phenyl-1-hydroxymethylene bisphosphonate) | — | 293 | $g_\| = 2.093$, $g_\perp = 2.050$ | $A_\| = 100$, $A_\perp = 30$ | [399] |
| [$CuH_2L^4$]$Cl_2$ in solutions (DMSO) ($L$ = 1-phenyl-1-hydroxymethylene bisphosphonate) | — | 293 | $g_\| = 2.082$, $g_\perp = 2.047$ | $A_\| = 160$, $A_\perp = 80$ | [399] |
| $CuH_2L^5Cl_2$ in solutions (DMSO) ($L$ = 1-phenyl-1-hydroxymethylene bisphosphonate) | — | 293 | $g_\| = 2.076$, $g_\perp = 2.048$ | $A_\| = 150$, $A_\perp = 85$ | [399] |
| $CuH_2L^6Cl_2$ in solutions (DMSO) ($L$ = 1-phenyl-1-hydroxymethylene bisphosphonate) | — | 293 | $g_\| = 2.047$, $g_\perp = 2.033$ | $A_\| = 100$, $A_\perp = 80$ | [399] |
| $CuH_2L^7Cl_2$ in solutions (DMSO) ($L$ = 1-phenyl-1-hydroxymethylene bisphosphonate) | — | 293 | $g_\| = 2.232$, $g_\perp = 2.132$ | $A_\| = 110$, $A_\perp = 95$ | [399] |
| [$CuH_2L^8$]$Cl_2$ in solutions (DMSO) ($L$ = 1-phenyl-1-hydroxymethylene bisphosphonate) | — | 293 | $g_\| = 2.234$, $g_\perp = 2.125$ | $A_\| = 120$, $A_\perp = 90$ | [399] |

(continued)

## ($Cu^{2+}$ ($3d^9$)) listing contd.)

| Host | Frequency (GHz)/band | $T$ (K) | $\tilde{g}$ | $A_x, A_y, A_z$ ($10^{-4}$ cm$^{-1}$) | References |
|---|---|---|---|---|---|
| [Cu(L$^{II}$)Cl]Cl · H$_2$O (powder) | 9.80 | 295 | $g_\parallel = 2.20$, $g_\perp = 2.09$ | — | [400] |
| CuNaY zeolite (frozen solution) | X | 10 | $g_\parallel = 2.29$–2.37, $g_\perp = 2.04$–2.07 | $A_\parallel = 12.6$–17.3, $A_\perp = 0.5$–2.0 (mT) | [401] |
| Cu(NO$_3$)$_2$ (PVA polyethylene film) | X | 123–393 | $g_\parallel = 2.352$, $g_\perp = 2.068$ | $A_\parallel = 113$ | [397] |
| [Cu(OH,Cl$_2$) · 2H$_2$O] | 9.205 | 123–295 | $g_\parallel = 2.26$, $g_\perp = 2.10$ | — | [402] |
| CuSO$_4$ (PVA polyethylene film) | X | 123–393 | $g_\parallel = 2.328$, $g_\perp = 2.061$ | $A_\parallel = 118$ | [397] |
| [Cu(tolf)$_2$(H$_2$O)]$_2$ | 9, 22.3 | 298 | 2.32, 2.06 | $A_\parallel = 152$, $A_\perp = 25$ | [403] |
| [Cu(tolf)$_2$(H$_2$O)]$_2$ | 9, 22.3 | 97 | 2.32, 2.07 | $A_\parallel = 152$, $A_\perp = 25$ | [403] |
| [Cu$_2$(β-ala)$_4$Cl$_2$]$^{2+}$ (ala, alanine) | X | 80 | 2.061, 2.063, 2.352 | 0, 0, 67 | [404] |
| [Cu$_2$(β-ala)$_4$Cl$_2$]$^{2+}$ | X | 297 | 2.06, 2.06, 2.368 | 0, 0, 72 | [404] |
| Cu$_2$(β-ala)$_4$(NO$_3$)]$^{2+}$ | X | 80 | 2.055, 2.073, 2.362 | 20, 20, 74 | [404] |
| Cu$_2$(β-ala)$_4$(NO$_3$)]$^{2+}$ | X | 297 | 2.07, 2.07, 2.36 | 10, 10, 72 | [404] |
| Cu$_2$(Hagpa)$_2$ · SO$_4$ 3H$_2$O | X | 293 | 2.127, 2.077, 2.239 | — | [393] |
| Cu$_2$SO$_4$(Hagpa)$_2$ (H$_2$O)$_2$ | Q | 293 | $g_\parallel = 2.265$, $g_\perp = 2.061$ | — | [393] |
| Cu$_2$SO$_4$(Hagpa)$_2$ (H$_2$O)$_2$ | X | 293 | $g_\parallel = 2.265$, $g_\perp = 2.062$ | — | [393] |
| Cu$_2$SO$_4$(Hagpa)$_2$ (H$_2$O)$_2$ | X | 4.2 | $g_\parallel = 2.277$, $g_\perp = 2.055$ | — | [393] |
| Cis-cyclohexylcyclam | — | RT | $g_{iso} = 2.099$ | $A_{iso} = 95.2$ | [405] |
| Cis-cyclohexylcyclam | | 77 | $g_\parallel = 2.198$, $g_\perp = 2.069$ | $A_\parallel - 209.2$ | [405] |
| Trans-cyclohexylcyclam | — | RT | $g_{iso} = 2.096$ | $A_{iso} = 96.1$ | [405] |
| Trans-cyclohexylcyclam | | 77 | $g_\parallel = 2.196$, $g_\perp = 2.060$ | $A_\parallel = 209.8$ | [405] |
| C$_6$H$_5$O$_7$Na$_3$ · 2H$_2$O | X | 295 | 2.1076, 2.1289, 2.4454 | 40, 62, 78 | [406] |

(continued)

## ($Cu^{2+}(3d^9)$) listing contd.

| Host | Frequency (GHz)/band | T (K) | $\tilde{g}$ | $A_x, A_y, A_z$ ($10^{-4}$ cm$^{-1}$) | References |
|---|---|---|---|---|---|
| $C_6H_{10}CaO_4$ | X | RT | 2.1285, 2.1868, 2.2205 | 121, 140, 142 | [407] |
| $C_{26}H_{32}O_{13.5}N_6S_2Co$ complex I (site I) | X | 295 | 2.164, 2.049, 2.317 | 46.43, 66.70, 100.89 | [408] |
| $C_{26}H_{32}O_{13.5}N_6S_2Co$ complex I (site II) | X | 295 | 2.165, 2.049, 2.317 | 29.15, 74.36, 99.96 | [408] |
| $C_{26}H_{32}O_{13.5}N_6S_2Co$ complex II (site I) | X | 295 | 2.147, 2.029, 2.328 | 25.97, 79.59, 96.22 | [408] |
| $C_{26}H_{32}O_{13.5}N_6S_2Co$ complex II (site II) | X | 295 | 2.172, 2.030, 2.306 | 33.35, 69.32, 95.29 | [408] |
| $C_{26}H_{32}O_{13.5}N_6S_2Zn$ (site I) | X | 295 | 2.192, 2.038, 2.323 | 22.23, 62.87, 100.24 | [408] |
| $C_{26}H_{32}O_{13.5}N_6S_2Zn$ (site II) | X | 295 | 2.165, 2.037, 2.339 | 24.01, 63.43, 99.68 | [408] |
| $C_{28}H_{30}N_4O_{14}S_2Zn$ [Zn(mein)$_2$(H$_2$O)$_4$]·(sac)$_2$ complex I (site I) | X | RT | 2.212, 2.096, 2.261 | 18.3, 79.7, 89.6 (G) | [409] |
| $C_{28}H_{30}N_4O_{14}S_2Zn$ [Zn(mein)$_2$(H$_2$O)$_4$]·(sac)$_2$ complex I (site II) | X | RT | 2.203, 2.018, 2.344 | 36.4, 67.4, 89.8 (G) | [409] |
| $C_{28}H_{30}N_4O_{14}S_2Zn$ [Zn(mein)$_2$(H$_2$O)$_4$]·(sac)$_2$ complex II (site I) | X | RT | 2.174, 2.020, 2.338 | 40.5, 71.9, 85.9 (G) | [409] |
| $C_{28}H_{30}N_4O_{14}S_2Zn$ [Zn(mein)$_2$(H$_2$O)$_4$]·(sac)$_2$ complex II (site II) | X | RT | 2.175, 2.024, 2.323 | 49.9, 66.7, 101.2 (G) | [409] |
| DAMZ [Diaquamalonatozinc(II)] (powder) | 9.4064 | 295 | $g_\parallel = 2.368$, $g_\perp = 2.093$ | $A_\parallel = 105.00$ | [410] |
| Diaqua(2,2'-bipyridine)malonatozinc(II) | X | RT | 2.121, 2.066, 2.424 | 2.09, 3.62, 14.18 (mT) | [411] |
| Diaquamalonatozinc(II) | 9.09 | 295 | 2.077, 2.087, 2.442 | 17.095, 26.904, 137.697 | [410] |
| Diammonium hexaaqua magnesium sulfate | X | 4.2–320 | 2.089, 2.112, 2.437 | 38, 14, 110 | [412] |

(continued)

## ($Cu^{2+}$($3d^9$) listing contd.)

| Host | Frequency (GHz)/band | T (K) | $\tilde{g}$ | $A_x, A_y, A_z$ ($10^{-4}$ cm$^{-1}$) | References |
|---|---|---|---|---|---|
| (7,16-Dibenzoyl-6,8,15,17-tetramethyldibenzo[b,i][1,4,8,11]tetraazacyclotetradecinato)copper(II) (in toluene) | X | RT | $g_{iso} = 2.086$ | $A_{iso} = 90.1$ | [106] |
| (7,16-Dibenzoyl-6,8,15,17-tetramethyldibenzo[b,i][1,4,8,11]tetraazacyclotetradecinato)copper(II) (diluted with an isomorphous tetraaza[14]annulene nickel(II) complex) | X | RT | $g_\| = 2.166$, $g_\perp = g_{2,3} = 2.036$, 2.010 | $A_\| = 199.8$ | [106] |
| 2, 9, 16, 23-(diethoxymalonyl)-tetrakis 3, 10, 17, 24-chloro phthalocyanine Pc2 (in powder form) | X | 295 | $g_\| = 2.085$, $g_\perp = 2.065$ | — | [350] |
| 2, 9, 16, 23-(diethoxymalonyl)-tetrakis 3, 10, 17, 24-chloro phthalocyanine Pc2 (in chloroform) | X | 295 | $g_\| = 2.083$, $g_\perp = 2.070$ | — | [350] |
| GaN | 35 (ODEPR) | 1.7 | $g_\| = \pm 0.20$, $g_\perp = 1.549$ | $^{63}A_\| = 550$ MHz, $^{63}A_\perp = 570$ MHz | [413] |
| Gly–Gly [Cu(H$_{-1}$L)L]$^{1-}$ peptide (Gly = glycene) | X | — | $g_\| = 2.228$, $g_\perp = 2.052$, $g_{iso} = 2.111$ | $A_\| = -160$, $A_\perp = -12$ | [414] |
| Gly–Gly [Cu(H$_{-1}$L)] | X | — | $g_\| = 2.320$, $g_\perp = 2.063$, $g_{iso} = 2.148$ | $A_\| = -155$, $A_\perp = -10$, $A_{iso} = -58$ | [414] |
| Gly–Gly–Gly [Cu(H$_{-1}$L)(H$_{-1}$L)]$^{2-}$ peptide (Gly = glycene) | X | — | $g_\| = 2.209$, $g_\perp = 2.041$ | $A_\| = -196$, $A_\perp = -19$ | [414] |
| Gly–Gly–Gly [Cu(H$_{-1}$L)] peptide (Gly = glycene) | X | — | $g_\| = 2.320$, $g_\perp = 2.063$, $g_{iso} = 2.148$ | $A_\| = -155$, $A_\perp = -10$, $A_{iso} = -58$ | [414] |
| Gly–Gly–Gly–Gly [Cu(H$_{-1}$L)(H$_{-1}$L)]$^{2-}$ peptide (Gly = glycene) | X | — | $g_\| = 2.209$, $g_\perp = 2.041$ | $A_\| = -196$, $A_\perp = -19$ | [414] |

(continued)

## ($Cu^{2+}$ ($3d^9$) listing contd.)

| Host | Frequency (GHz)/band | $T$ (K) | $\tilde{g}$ | $A_x, A_y, A_z$ ($10^{-4}$ cm$^{-1}$) | References |
|---|---|---|---|---|---|
| Gly–Gly–Gly–Gly [Cu(H$_{-1}$L)] peptide (Gly = glycene) | X | — | $g_{\|} = 2.32$, $g_\perp = 2.063$ $g_{iso} = 2.148$ | $A_{\|} = -155$, $A_\perp = -10$ $A_{iso} = -58$ | [414] |
| His–Gly–Gly [Cu(H$_{-2}$L)L]$^{2-}$ peptide (Gly = glycene; His = histidine) | X | — | $g_{\|} = 2.264$, $g_\perp = 2.056$ $g_{iso} = 2.126$ | $A_{\|} = -179$, $A_\perp = -12$ $A_{iso} = -69$ | [414] |
| Gly–His–Gly [Cu(H$_{-1}$L)] peptide (Gly = glycene; His = histidine) | X | — | $g_{\|} = 2.228$, $g_\perp = 2.051$ $g_{iso} = 2.114$ | $A_{\|} = -191$, $A_\perp = -10$ $A_{iso} = -76$ | [414] |
| Gly–Gly–His–Gly [Cu(H$_{-2}$L)]$^-$ peptide (Gly = glycene; His = histidine) | X | — | $g_{\|} = 2.183$, $g_\perp = 2.041$ $g_{iso} = 2.088$ | $A_{\|} = -197$, $A_\perp = -17$ $A_{iso} = -80$ | [414] |
| KHCO$_3$ | 9.4105 | 295 | $g_{\|} = 2.2349$, $g_\perp = 2.0520$ | $A_{\|} = 18.2$, $A_\perp = 3.2$ (mT) | [415] |
| KRbB$_4$O$_7$ | X | RT | $g_{\|} = 2.0505$, $g_\perp = 2.0590$ | $A_{\|} = 180.5$ | [416] |
| KZnClSO$_4 \cdot 3H_2O$ | X | 77 | 2.1535, 2.0331, 2.4247 | −31, 63, −103 | [417] |
| KZnClSO$_4 \cdot 3H_2O$ | X | RT | 2.039, 2.228, 2.037 | 57, 46, 48 G | [418] |
| KZnClSO$_4 \cdot 3H_2O$ (JT distortion) (site I) | X | 300 | 2.229, 2.041, 2.278 | 48, 46, 56 G | [419] |
| KZnClSO$_4 \cdot 3H_2O$ (JT distortion) (site II) | X | 300 | 2.228, 2.037, 2.309 | 48, 46, 57 G | [419] |
| K$_2$C$_2$O$_4 \cdot$ H$_2$O (site I) | — | — | 2.0562, 2.0735, 2.3248 | 1.0, 10.6, 136.8 | [420] |
| K$_2$C$_2$O$_4 \cdot$ H$_2$O (site II) | — | — | 2.0761, 2.0701, 2.3306 | 4.1, 6.5, 142.5 | [420] |
| K$_2$C$_2$O$_4 \cdot$ H$_2$O$_2$ (site I) | — | — | 2.1463, 2.0900, 2.3516 | 14.7, 16.1, 169.2 | [420] |
| K$_2$C$_2$O$_4 \cdot$ H$_2$O$_2$ (site II) | — | — | 2.0826, 2.0732, 2.3333 | 15.3, 15.3, 173.4 | [420] |
| K$_2$[Zn(H$_2$O)$_6$](SO$_4$)$_2$ | X | RT | 2.0399, 2.2384, 2.3149 | 49.33, 46.79, 58.01 | [421] |
| K$_3$H(CO$_3$)$_2$ (site I) | X | RT | 2.015, 2.039, 2.227 | 2.95, 4.95, 18.70 (mT) | [422] |

(continued)

## ($Cu^{2+}$($3d^9$) listing contd.)

| Host | Frequency (GHz)/band | $T$ (K) | $\tilde{g}$ | $A_x, A_y, A_z$ ($10^{-4}$ cm$^{-1}$) | References |
|---|---|---|---|---|---|
| $K_3H(CO_3)_2$ (site II) | X | RT | 2.013, 2.034, 2.226 | 3.2, 4.2, 18.50 (mT) | [422] |
| $K_3H(SO_4)_2$ complex I | X | RT | 2.066, 2.093, 2.435 | 24, 6, 101 (Gs) | [423] |
| $K_3H(SO_4)_2$ complex II | X | RT | 2.105, 2.065, 2.445 | 30, 54, 80 (Gs) | [423] |
| $K_3H(SO_4)_2$ complex I | X | RT | 2.066, 2.093, 2.435 | 24, 6, 101 (G) | [424] |
| $K_3H(SO_4)_2$ complex II | X | RT | 2.105, 2.065, 2.452 | 30, 54, 88 (G) | [424] |
| $La_2CuO_4$ | — | — | $g_\| = 2.11$, $g_\perp = 2.47$ | $A_\| = 125$, $A_\perp = 108$ (G) | [425] |
| $LiCsSO_4$ | X | RT | 2.564, 2.177, 2.035 | — | [294] |
| $LiCsSO_4$ | X | RT | 2.342, 2.186, 2.022 | — | [294] |
| $LiCsSO_4$ | X | RT | 2.508, 2.141, 2.141 | — | [294] |
| $LiCsSO_4$ | X | RT | 2.318, 2.094, 1.917 | — | [294] |
| $LiKSO_4$ (site I) | 9.1 | RT | 2.0930, 2.1421, 2.2900 | 85, 89, 184 | [426] |
| $LiKSO_4$ (site II) | 9.1 | RT | 2.0795, 2.1580, 2.2876 | 93, 95, 189 | [426] |
| $LiRbB_4O_7$ | X | RT | $g_\| = 2.4451$, $g_\perp = 2.0561$ | $A_\| = 158$, $A_\perp = 30$ | [427] |
| $LiRbB_4O_7$ | X | RT | $g_\| = 2.4358$, $g_\perp = 2.0623$ | $A_\| = 158$, $A_\perp = 37$ | [428] |
| Metallo-phthalocyanine | 9.79 | 295 | $g_\| = 2.10$, $g_\perp = 2.059$ | — | [354] |
| $[Mg(H_2O)_6] \cdot [MgC_6H_5O_7(H_2O)]_2 \cdot 2H_2O$ | X | RT | 2.0346, 2.1400, 2.3874 | 57, 76, 99 | [429] |
| $MgRb(SO_4)_2 \cdot 6H_2O$ MRSH (single crystal) | X | 77 | 2.133, 2.137, 2.327 | 0.0093, 1.345, 10.089 | [430] |
| $MgRb(SO_4)_2 \cdot 6H_2O$ MRSH (powder) | X | 77 | $g_\| = 2.283$, $g_\perp = 2.140$ | $A_\| = 10.3133$ | [430] |

(continued)

## ($Cu^{2+}$($3d^9$)) listing contd.)

| Host | Frequency (GHz)/band | T (K) | $\tilde{g}$ | $A_x, A_y, A_z$ ($10^{-4}$ cm$^{-1}$) | References |
|---|---|---|---|---|---|
| MgRb(SO$_4$)$_2$ · 6H$_2$O MRSH (single crystal) | X | 295 | 2.400, 2.173, 2.108 | — | [430] |
| MgRb(SO$_4$)$_2$ · 6H$_2$O MRSH (powder) | X | 295 | $g_\parallel = 2.466$, $g_\perp = 2.204$ | — | [430] |
| NaHC$_2$O$_4$ · H$_2$O | X | RT | 2.0562, 2.0729, 2.3378 | 36.4, 42.5, 153.8 (G) | [431] |
| Na(NH$_2$CH$_2$COOH)$_2$ NO$_3$ (site I) | X | RT | 2.035, 2.064, 2.245 | 2.59, 4.89, 14.11 mT | [432] |
| Na(NH$_2$CH$_2$COOH)$_2$ NO$_3$ (site II) | X | RT | 2.043, 2.058, 2.274 | 3.37, 4.24, 14.92 mT | [432] |
| Na(NH$_2$CH$_2$COOH)$_2$ NO$_3$ (powder) | X | RT | 2.058, 2.058, 2.256 | 5.28, 5.28, 13.68 mT | [432] |
| NaRbB$_4$O$_7$ | X | RT | $g_\parallel = 2.4434$, $g_\perp = 2.4217$ | $A_\parallel = 162$ | [416] |
| Na$_2$ZnSO$_4$ · 4H$_2$O | X | LNT | 2.2356, 2.0267, 2.3472 | 27, 54, 88 | [433] |
| [(NH$_2$CH$_2$COOH)$_2$ · CaCl$_2$ · 4H$_2$O] | 9.1 | RT | 2.0238, 2.1122, 2.2250 | 83, 86, 118 | [434] |
| (NH$_2$CH$_2$COOH)$_3$ · H$_2$SeO$_4$ | 9.7 | RT | 2.0529, 2.0647, 2.2596 | 42.2, 2.5, 151.2 | [435] |
| (NH$_2$CH$_2$COOH)$_3$ · H$_2$SO$_4$ | 9.7 | RT | 2.054, 2.064, 2.261 | 30, 5, 150 | [435] |
| NH$_4$Br | — | 294 | 2.313, 2.144, 2.046 | 162, 0, 13 | [436] |
| NH$_4$Br | X | 300 | — | $A_\parallel = 259$, $A_\perp = 33$ | [437] |
| NH$_4$Br | — | 77 | $g_\parallel = 2.038$, $g_\perp = 2.190$ | $A_\perp = 67$ | [437] |
| NH$_4$Br | X, Q | 300 | $g_\parallel = 2.032$, $g_\perp = 2.19$ | $A_\parallel = 183.0$, $A_\perp = 33.0$ | [438] |
| NH$_4$Br | X, Q | 77 | $g_\parallel = 2.038$, $g_\perp = 2.19$ | $A_\parallel = 259.0$, $A_\perp = 67.0$ | [438] |
| NH$_4$Br (tetragonal) | Q | 6.5 | $g_\parallel = 2.040$, $g_\perp = 2.192$ | $A_\parallel = 244$, $A_\perp = 31$ | [439] |
| NH$_4$Br (tetragonal) | Q | 166 | $g_\parallel = 2.028$, $g_\perp = 2.185$ | $A_\parallel = 215$, $A_\perp = 36$ | [439] |

(continued)

## ($Cu^{2+}$ ($3d^9$) listing contd.)

| Host | Frequency (GHz)/band | T (K) | $\tilde{g}$ | $A_x, A_y, A_z$ ($10^{-4}$ cm$^{-1}$) | References |
|---|---|---|---|---|---|
| $NH_4Br$ (orthorhombic) | Q | 166 | 2.228, 2.016, 2.024 | 83, 85, 152 | [439] |
| $NH_4Br$ (tetragonal) | Q | 300 | $g_\parallel = 2.040$, $g_\perp = 2.197$ | $A_\parallel = 181$, $A_\perp = 32$ | [439] |
| $NH_4H_2PO_4$ (site I) | X | 77 | 2.1071, 2.0224, 2.4363 | 53, 80, 150 | [440] |
| $NH_4H_2PO_4$ (site II) | X | 77 | 2.0884, 2.0264, 2.3894 | 34, 48, 129 | [440] |
| $NH_4H_2PO_4$ (site III) | X | 77 | 2.1265, 2.0369, 2.3471 | 40, 59, 136 | [440] |
| $NH_4H_2PO_4$ (site IV) | X | 77 | 2.1182, 2.0920, 2.2892 | 38, 68, 129 | [440] |
| $(NH_4)_2C_4H_4O_6$ (site I) | X | RT | 2.08, 2.12, 2.40 | 65, 21, 138 | [70] |
| $(NH_4)_2C_4H_4O_6$ (site II) | X | RT | 2.07, 2.11, 2.37 | 64, 28, 129 | [70] |
| $(NH_4)_2C_4H_4O_6$ (site III) | X | RT | 2.03, 2.09, 2.34 | 31, 23, 135 | [70] |
| $(NH_4)_2C_4H_4O_6$ (complex 1) (site I) | X | RT | 2.0777, 2.0428, 2.3610 | ?.06, 12.254, 13.07 (mT) | [71] |
| $(NH_4)_2C_4H_4O_6$ (complex 1) (site II) | X | RT | 2.0808, 2.0600, 2.3864 | 4.94, 13.72, 13.05 (mT) | [71] |
| $(NH_4)_2C_4H_4O_6$ (complex 2) (site I) | X | RT | 2.0947, 2.0403, 2.3940 | 4.98, 12.51, 12.26 (mT) | [71] |
| $(NH_4)_2C_4H_4O_6$ (complex 2) (site II) | X | RT | 2.0940, 2.0364, 2.3626 | 5.35, 11.26, 12.45 (mT) | [71] |
| $(NH_4)_2C_4H_4O_6$ (complex 3) (site I) | X | RT | 2.0926, 2.0600, 2.3373 | 456, 1.83, 13.01 (mT) | [71] |
| $(NH_4)_2C_4H_4O_6$ (complex 3) (site II) | X | RT | 2.0923, 2.0427, 2.3304 | 3.55, 1.98, 13.05 (mT) | [71] |
| $(NH_4)_2C_4H_4O_6$ (site I) | X | RT | 2.0469, 2.1522, 2.4489 | 56, 66, 79 | [441] |
| $(NH_4)_2C_4H_4O_6$ (site II) | X | RT | 2.0292, 2.1477, 2.4186 | 36, 63, 73 | [441] |
| $[(NH_4)HC_2O_4 \cdot 12H_2O]$ (complex I) (site I) | X | 295 | 2.082, 2.068, 2.322 | 4.39, 61.47, 159.74 | [72] |
| $[(NH_4)HC_2O_4 \cdot 12H_2O]$ (complex I) (site II) | X | 295 | 2.087, 2.062, 2.335 | 9.34, 60.25, 143.12 | [72] |

(continued)

## ($Cu^{2+}(3d^9)$ listing contd.)

| Host | Frequency (GHz)/ band | T (K) | $\tilde{g}$ | $A_x, A_y, A_z$ ($10^{-4}$ cm$^{-1}$) | References |
|---|---|---|---|---|---|
| [$(NH_4)HC_2O_4 \cdot 12H_2O$] (complex II) (site I) | X | 295 | 2.076, 2.056, 2.310 | 4.11, 54.74, 147.69 | [72] |
| [$(NH_4)HC_2O_4 \cdot 12H_2O$] (complex II) (site II) | X | 295 | 2.077, 2.057, 2.308 | 4.30, 58.01, 151.34 | [72] |
| $PbTiO_3$ | 9.8 | — | $\|g_\|\| = 2.340$, $\|g_\perp\| = 2.058$ | $\|^{63}A_\|\| = 155$, $\|^{63}A_\perp\| = 5.4$, $\|^{65}A_\|\| = 166$, $\|^{65}A_\perp\| = 5.8$ | [442] |
| Plastocyanin (rapid cooling) 0% glycerol | — | 77 | $g_\| = 2.2442$, $g_\perp = 2.0462$ | $A_\| = 59.5$, $A_\perp = 5.2$ | [372] |
| Plastocyanin (slow cooling) 0% glycerol | — | 77 | $g_\| = 2.2420$, $g_\perp = 2.0461$ | $A_\| = 59.3$, $A_\perp = 5.2$ | [372] |
| Plastocyanin (rapid cooling) 40% glycerol | — | 77 | $g_\| = 2.2398$, $g_\perp = 2.0439$ | $A_\| = 59.5$, $A_\perp = 5.2$ | [372] |
| Plastocyanin (slow cooling) 40% glycerol | — | 77 | $g_\| = 2.2402$, $g_\perp = 2.0438$ | $A_\| = 59.5$, $A_\perp = 5.2$ | [372] |
| Porphyrazine (in powder form) | X | 295 | $g_\| = 2.150$, $g_\perp = 2.060$ | $A_\| = 219.53$, $A_\perp = 23.35$ | [350] |
| Porphyrazine (in chloroform) | X | 295 | $g_\| = 2.160$, $g_\perp = 2.066$ | $A_\| = 196.18$, $A_\perp = 18.68$ | [350] |
| PTIPDA complex (phthalimide-o-aminophenylene-diamine) | X | 295 | $g_\| = 2.2041$, $g_\perp = 2.0263$ | $A_\| = 166$, $A_\perp = 46$ | [443] |
| $Rb_2CO_3$ | X | RT | 2.031, 2.042, 2.221 | 1.7, 6.0, 18.7 (mT) | [422] |
| $Rb_2KH(CO_3)_2$ | X | RT | 2.101, 2.001, 2.262 | 11.6, 1.8, 17.1 (mT) | [422] |
| $SrB_4O_7$ glass | X | RT | $g_\| = 2.3163$, $g_\perp = 2.0427$ | $A_\| = 139$, $A_\perp = 49$ | [122] |
| $SrF_2$ (normal pressure) | X | 85 | $g_\| = 2.491$, $g_\perp = 2.083$ | $A_\| = 360$ MHz, $A_\perp = 26$ MHz | [444] |
| $SrF_2$ (pressure = 550 MPa) | X | 85 | $g_\| = 2.489$, $g_\perp = 2.083$ | $A_\| = 348$ MHz, $A_\perp = 27$ MHz | [444] |

(continued)

## ($Cu^{2+}(3d^9)$ listing contd.)

| Host | Frequency (GHz)/band | T (K) | $\tilde{g}$ | $A_x, A_y, A_z$ ($10^{-4}$ cm$^{-1}$) | References |
|---|---|---|---|---|---|
| (6,8,15,17-Tetramethyldibenzo[b,i][1,4,8,11] tetraazacyclotetradecinato)copper(II) *(in toluene)* | X | RT | $g_{iso} = 2.087$ | $A_{iso} = 90.2$ | [106] |
| (6,8,15,17-Tetramethyldibenzo[b,i][1,4,8,11] tetraazacyclotetradecinato)copper(II) (diluted with an isomorphous tetraaza[14]annulene nickel(II) complex) | X | RT | $g_\| = 2.168$, $g_\perp = g_{2,3} = 2.040$, 2.017 | $A_\| = 199.9$ | [106] |
| TiO$_2$ (rutile) ($^{65}$Cu) | 9.48 | 18 | 2.10699, 2.09281, 2.34518 | 55.35, 82.34, −261.98 (MHz) | [445] |
| TiO$_2$ (rutile) ($^{63}$Cu) | 9.48 | 18 | 2.10697, 2.09280, 2.34516 | 59.20, 88.21, −280.83 (MHz) | [445] |
| 2,9,17,23-tetra-(1,1,2-(tricarbethoxyethyl) phthalocyanine Pc1 (in powder form) | X | 295 | $g_\| = 2.085$, $g_\perp = 2.055$ | — | [350] |
| 2,9,17,23-tetra-(1,1,2-(tricarbethoxyethyl) phthalocyanine Pc1 (in chloroform) | X | 295 | $g_\| = 2.080$, $g_\perp = 2.020$ | $A_\| = 84.08$, $A_\perp = 70.06$ | [350] |
| XY$_3$Z$_6$B$_3$Si$_6$O$_{27}$(OH)$_4$ (tourmaline) | 9.3 | 293 | 2.054, 2.092, 2.374 | 27.8, 59.3, 133.2 | [446] |
| XY$_3$Z$_6$B$_3$Si$_6$O$_{27}$(OH)$_4$ (tourmaline) | 9.3 | 77 | $g_\| = 2.426$, $g_\perp - 2.106$ | $A_\| = 121$, $A_\perp - 29$ | [446] |
| Zn[CH$_2$NH$_2$COOH]SO$_4$ · 7H$_2$O | 9.8 | 295 | 2.095, 2.062, 2.427 | 28.025, 46.709, 107.430 | [447] |
| ZnC$_5$H$_7$NO$_4$ · 2H$_2$O (site I) | — | — | 2.0170, 2.0768, 2.2334 | 74, 99, 134 | [448] |
| ZnC$_5$H$_7$NO$_4$ · 2H$_2$O (site II) | — | — | 2.0180, 2.0550, 2.1633 | 100, 100, 115 | [448] |

*(continued)*

## ($Cu^{2+}$ ($3d^9$) listing contd.)

| Host | Frequency (GHz)/band | T (K) | $\tilde{g}$ | $A_x, A_y, A_z$ ($10^{-4}$ cm$^{-1}$) | References |
|---|---|---|---|---|---|
| $ZnGa_2O_4$ (site I) | X | 110 | $g_\parallel = 2.355$, $g_\perp = 2.077$ | $A_\parallel = 116$ Oe, $A_\perp = 12$ Oe | [449] |
| $ZnGa_2O_4$ (site II) | X | 110 | $g_\parallel = 2.018$, $g_\perp = 2.246$ | $A_\parallel = 75$ Oe, $A_\perp = 44$ Oe | [449] |
| $[Zn(H_2O)_6][Zn(malonato)_2(H_2O)_2]$ single crystal | X | 300 | 2.034, 2.159, 2.388 | 3.39, 4.89, 13.72 (mT) | [450] |
| $[Zn(H_2O)_6][Zn(malonato)_2(H_2O)_2]$ powder | X | 300, 77 | $g_\parallel = 2.367$, $g_\perp = 2.088$ | $A_\parallel = 11.47$, $A_\perp = 2.63$ (mT) | [450] |
| $ZnKPO_4 \cdot 6H_2O$ | 9.095 | 77 | 2.188, 2.032, 2.372 | 46.71, 60.72, 74.73 G | [451] |
| $[Zn(methylmelonato)(H_2O)]_n$ | X | 300 | 2.076, 2.100, 2.379 | 2.40, 3.22, 13.47 (mT) | [452] |
| $Zn(NH_4)_2PO_4 \cdot 6H_2O$ | X | 77 | 2.462, 2.149, 2.098 | 107, 19, 30 | [453] |
| $Zn(NH_4)_2PO_4 \cdot 6H_2O$ | X | 143 | 2.409, 2.145, 2.064 | 99, 21, 31 | [453] |
| $ZnNa_2(SO_4)_2 \cdot 6H_2O$ crystal | X | 295 | 2.061, 2.275, 2.332 | — | [454] |
| $ZnNa_2(SO_4)_2 \cdot 6H_2O$ powder | X | 295 | 2.095, 2.095, 2.268 | — | [454] |
| $ZnNa_2(SO_4)_2 \cdot 6H_2O$ crystal | X | 123 | 2.039, 2.232, 2.394 | 52.69, 39.24, 74.17 | [454] |
| $ZnNa_2(SO_4)_2 \cdot 6H_2O$ powder | X | 77 | 2.017, 2.110, 2.363 | 64.46, 16.82, 71.93 | [454] |
| $[Zn(nic)_2(H_2O)_4](sac)_2$ | 9.8 | 195 | 2.085, 2.062, 2.375 | 2.803, 3.737, 10.743 | [455] |
| $ZnNH_4PO_4 \cdot 6H_2O$ (site I) | 9.12 | 77 | 2.1135, 2.2216, 2.3937 | 37, 47, 70 | [456] |
| $ZnNH_4PO_4 \cdot 6H_2O$ (site II) | 9.12 | 77 | 2.0845, 2.1995, 2.3777 | 39, 48, 73 | [456] |
| ZnO nanopowders (site I) | X | RT | $g_\parallel = 2.396$, $g_\perp = 2.081$ | $A_\parallel = 123$, $A_\perp = 12$ | [457] |
| ZnO nanopowders (site II) | X | RT | $g_\parallel = 2.348$, $g_\perp = 2.081$ | $A_\parallel = 139$, $A_\perp = 12$ | [457] |

(continued)

## ($Cu^{2+}$ ($3d^9$) listing contd.)

| Host | Frequency (GHz)/ band | T (K) | $\tilde{g}$ | $A_x, A_y, A_z$ ($10^{-4}$ cm$^{-1}$) | References |
|---|---|---|---|---|---|
| [ZnPd(CN)$_4$ (C$_4$H$_{12}$N$_2$O$_2$)] | X | RT | 2.056, 2.042, 2.250 | 4.20, 5.04, 16.77 (mT) | [458] |
| [Zn(picol)$_2$(H$_2$O)$_2$] · 2H$_2$O (site I) | X | 298 | 2.087, 2.088, 2.310 | 49.978, 58.666, 143.583 | [459] |
| [Zn(picol)$_2$(H$_2$O)$_2$] · 2H$_2$O (site II) | X | 298 | 2.083, 2.095, 2.308 | 48.390, 64.365, 140.687 | [459] |
| [Zn(picol)$_2$(H$_2$O)$_2$] · 2H$_2$O (powder) | X | 298 | $g_\parallel = 2.296$, $g_\perp = 2.072$ | $A_\parallel = 137.043$, $A_\perp = 48.297$ | [459] |
| ZnSeO$_4$ | X | RT | 2.097, 2.095, 2.427 | $A_x^{65} = 22.3$, $A_x^{65} = 11.3$, $A_x^{65} = 138.4$ | [460] |
| ZnSiF$_6$ · 6H$_2$O | 9.24 | 5–300 | $g_\parallel = 2.460$, $g_\perp = 2.105$ | $|A| = 100$, $|B| = 14$, $P = 8$ | [461] |
| ZnSiF$_6$ · 6H$_2$O | 9.24 | 29.1 | $g_\parallel = 2.225$, $g_\perp = 2.197$ | $A = 24$ | [461] |
| Zn(stpy)$_3$(NO$_3$)2 · 1/2stpy (stpy = trans-4-styrylpyridene) | X | 300 | 2.067, 2.111, 2.292 | 22.9, 54.7, 107.5 | [386] |

## $Cu^{2+}$ ($3d^9$) Glasses

| Host | Frequency (GHz)/ band | T (K) | $\tilde{g}$ | $A_x, A_y, A_z$ ($10^{-4}$ cm$^{-1}$) | References |
|---|---|---|---|---|---|
| 10BaO · 10Li$_2$O · 10Na$_2$O · 10K$_2$O · 59B$_2$O$_3$ · 1CuO glass | X | RT | $g_\parallel = 2.253$, $g_\perp = 2.031$ | $A_\parallel = 139.27$, $A_\perp = 41.78$ | [26] |
| 5CaO-15SrO-19.9Na$_2$O-60B$_2$O$_3$-0.1CuO glass | X | RT | $g_\parallel = 2.329$, $g_\perp = 2.066$ | $A_\parallel = 117$ | [462] |
| 10CaO-10SrO-19.9Na$_2$O-60B$_2$O$_3$-0.1CuO glass | X | RT | $g_\parallel = 2.335$, $g_\perp = 2.055$ | $A_\parallel = 123$ | [462] |

(continued)

## ($Cu^{2+}$(3d$^9$) glasses listing contd.)

| Host | Frequency (GHz)/band | T (K) | $\tilde{g}$ | $A_x, A_y, A_z$ ($10^{-4}$ cm$^{-1}$) | References |
|---|---|---|---|---|---|
| 15CaO-5SrO-19.9Na$_2$O-60B$_2$O$_3$-0.1CuO glass | X | RT | $g_\parallel = 2.351$<br>$g_\perp = 2.056$ | $A_\parallel = 130$ | [462] |
| CdB$_4$O$_7$ Glass | X | RT | $g_\parallel = 2.345$<br>$g_\perp = 2.060$ | $A_\parallel = 148$<br>$A_\perp = 30$ | [463] |
| Cd(HCOO)$_2$·2H$_2$O (complex I) | X | 77 | 2.064, 2.092, 2.429 | 12, 32, 120 | [464] |
| Cd(HCOO)$_2$·2H$_2$O (complex II) | X | 77 | 2.067, 2.089, 2.417 | 15, 31, 133 | [464] |
| 0.05CuO–0.7B$_2$O$_3$–0.25Li$_2$O glass | X | 300 | $g_\parallel = 2.223$,<br>$g_\perp = 2.035$ | $A_\parallel = 70.0$,<br>$A_\perp = 49.0$ | [125] |
| 0.05CuO–0.7B$_2$O$_3$–0.25Li$_2$O glass | X | 77 | $g_\parallel = 2.388$,<br>$g_\perp = 2.055$ | $A_\parallel = 121.0$,<br>$A_\perp = 42.0$ | [125] |
| 60B$_2$O$_3$ · 30ZnO · 0.5CuO · 10MgCO$_3$ glass | 9.205 | RT | $g_\parallel = 2.356$,<br>$g_\perp = 2.070$ | $A_\parallel = 112$,<br>$A_\perp = 70$ | [465] |
| 60B$_2$O$_3$ · 30ZnO · 0.5CuO · 10CaCO$_3$ glass | 9.205 | RT | $g_\parallel = 2.366$,<br>$g_\perp = 2.076$ | $A_\parallel = 114$,<br>$A_\perp = 80$ | [465] |
| 60B$_2$O$_3$ · 30ZnO · 0.5CuO · 10SrCO$_3$ glass | 9.205 | RT | $g_\parallel = 2.367$,<br>$g_\perp = 2.066$ | $A_\parallel = 115$,<br>$A_\perp = 82$ | [465] |
| Ba$_2$Zn(HCOO)$_6$ · 4H$_2$O | X | 4.2 | 2.065, 2.092, 2.405 | 11, 4, 130 | [466] |
| Ba$_2$Zn(HCOO)$_6$ · 4H$_2$O | X | 295 | 2.075, 2.153, 2.336 | 11, 30, 98 | [466] |
| (50)Bi$_2$O$_3$ · 20Li$_2$O · 30(ZnO · B$_2$O$_3$) | X | RT | $g_\parallel = 2.314$,<br>$g_\perp = 2.068$ | $A_\parallel = 100$,<br>$A_\perp = 39$ | [467] |
| (55)Bi$_2$O$_3$ · 15Li$_2$O · 30(ZnO · B$_2$O$_3$) | X | RT | $g_\parallel = 2.316$,<br>$g_\perp = 2.121$ | $A_\parallel = 90$,<br>$A_\perp = 63$ | [467] |
| (60)Bi$_2$O$_3$ · 10Li$_2$O · 30(ZnO · B$_2$O$_3$) | X | RT | $g_\parallel = 2.322$,<br>$g_\perp = 2.073$ | $A_\parallel = 99$,<br>$A_\perp = 52$ | [467] |
| (65)Bi$_2$O$_3$ · 5Li$_2$O · 30(ZnO · B$_2$O$_3$) | X | RT | $g_\parallel = 2.314$,<br>$g_\perp = 2.073$ | $A_\parallel = 82$,<br>$A_\perp = 54$ | [467] |
| (70)Bi$_2$O$_3$ · 30(ZnO · B$_2$O$_3$) | X | RT | $g_\parallel = 2.327$,<br>$g_\perp = 2.073$ | $A_\parallel = 74$,<br>$A_\perp = 49$ | [467] |
| 45Bi$_2$O$_3$ · 25ZnO · 15B$_2$O$_3$ · 15Li$_2$O glass | X | RT | $g_\parallel = 2.337$,<br>$g_\perp = 2.127$ | $A_\parallel = 160$,<br>$A_\perp = 49$ | [468] |

(continued)

## ($Cu^{2+}$ ($3d^9$)) glasses listing contd.)

| Host | Frequency (GHz)/ band | T (K) | $\tilde{g}$ | $A_x, A_y, A_z$ ($10^{-4}$ cm$^{-1}$) | References |
|---|---|---|---|---|---|
| $50Bi_2O_3 \cdot 25ZnO \cdot 15B_2O_3 \cdot 10Li_2O$ glass | X | RT | $g_\| = 2.332$, $g_\perp = 2.127$ | $A_\| = 114$, $A_\perp = 60$ | [468] |
| $55Bi_2O_3 \cdot 25ZnO \cdot 15B_2O_3 \cdot 5Li_2O$ glass | X | RT | $g_\| = 2.333$, $g_\perp = 2.079$ | $A_\| = 108$, $A_\perp = 48$ | [468] |
| $60Bi_2O_3 \cdot 25ZnO \cdot 15B_2O_3$ glass | X | RT | $g_\| = 2.312$, $g_\perp = 2.056$ | $A_\| = 99$, $A_\perp = 55$ | [468] |
| $xB_2O_3 \cdot (100-x)TeO_2$ glasses ($0 \leq x \leq 50$; varying $x$) | X | RT | $g_\| = 2.039-2.381$, $g_\perp = 2.028-2.050$ | $A_\| = 176-183$, $A_\perp = 57-60$ | [469] |
| (85 mol%) $CdGeO_3 \cdot$ (15 mol%)$AlF_3$ glass | X, Q | 77, 300 | $g_\| = 2.349$, $g_\perp = 2.05$ | $A_\| = 159$, $A_\perp = 25$ | [333] |
| KBaB glass | 9.205 | 123–433 | $g_\| = 2.259$, $g_\perp = 2.048$ | $A_\| = 140$, $A_\perp = 24$ | [470] |
| $90K_2B_4O_7 + 9PbO + CuO$ Potassium lead tetraborate glass | 9.205 | RT | $g_\| = 2.297$, $g_\perp = 2.032$ | $A_\| = 122$, $A_\perp = 25$ | [471] |
| $30K_2O-70B_2O_3$ glass | X | RT | $g_\| = 2.327$, $g_\perp = 2.070$ | $A_\| = 136.6$, $A_\perp = 32.31$ | [472] |
| LiBaB glass | 9.205 | 123–433 | $g_\| = 2.284$, $g_\perp = 2.053$ | $A_\| = 131$, $A_\perp = 25$ | [470] |
| $90Li_2B_4O_7 + 9PbO + CuO$ Lithium lead tetraborate glass | 9.205 | RT | $g_\| = 2.307$, $g_\perp = 2.041$ | $A_\| = 123$, $A_\perp = 21$ | [471] |
| $5Li_2O-25K_2O-70B_2O_3$ glass | X | RT | $g_\| = 2.328$, $g_\perp = 2.069$ | $A_\| = 137.53$, $A_\perp = 34.15$ | [472] |
| $10Li_2O-20K_2O-70B_2O_3$ glass | X | RT | $g_\| = 2.330$, $g_\perp = 2.069$ | $A_\| = 137.53$, $A_\perp = 34.15$ | [472] |
| $15Li_2O-15K_2O-70B_2O_3$ glass | X | RT | $g_\| = 2.339$, $g_\perp = 2.069$ | $A_\| = 138.45$, $A_\perp = 35.07$ | [472] |
| $20Li_2O-10K_2O-70B_2O_3$ glass | X | RT | $g_\| = 2.339$, $g_\perp = 2.069$ | $A_\| = 138.45$, $A_\perp = 35.07$ | [472] |
| $25Li_2O-5K_2O-70B_2O_3$ glass | X | RT | $g_\| = 2.339$, $g_\perp = 2.069$ | $A_\| = 138.45$, $A_\perp = 35.07$ | [472] |
| $8Li_2O-32Na_2O-50B_2O_3-10Bi_2O_3$ glass | X | 295 | $g_\| = 2.326$ | $A_\| = 135.0$ | [473] |
| $16Li_2O-24Na_2O-50B_2O_3-10Bi_2O_3$ glass | X | 295 | $g_\| = 2.329$ | $A_\| = 130.0$ | [473] |

(continued)

## ($Cu^{2+}(3d^9)$) glasses listing contd.)

| Host | Frequency (GHz)/ band | T (K) | $\tilde{g}$ | $A_x, A_y, A_z$ ($10^{-4}$ cm$^{-1}$) | References |
|---|---|---|---|---|---|
| 24Li$_2$O-16Na$_2$O-50B$_2$O$_3$-10Bi$_2$O$_3$ glass | X | 295 | $g_\| = 2.338$ | $A_\| = 125.0$ | [473] |
| 32Li$_2$O-8Na$_2$O-50B$_2$O$_3$-10Bi$_2$O3 glass | X | 295 | $g_\| = 2.334$ | $A_\| = 135.0$ | [473] |
| 40Li$_2$O-50B$_2$O$_3$-10Bi$_2$O$_3$ glass | X | 295 | $g_\| = 2.329$ | $A_\| = 135.5$ | [473] |
| 5Li$_2$O-25Na$_2$O-69.5B$_2$O$_3$ + 0.5CuO glass | 9.205 | 300 | $g_\| = 2.293$, $g_\perp = 2.040$ | $A_\| = 136.1$, $A_\perp = 22.05$ | [474] |
| 10Li$_2$O-20Na$_2$O-69.5B$_2$O$_3$ + 0.5CuO glass | 9.205 | 300 | $g_\| = 2.293$, $g_\perp = 2.041$ | $A_\| = 136.2$, $A_\perp = 22.05$ | [474] |
| 15Li$_2$O-15Na$_2$O-69.5B$_2$O$_3$ + 0.5CuO glass | 9.205 | 300 | $g_\| = 2.293$, $g_\perp = 2.041$ | $A_\| = 136.3$, $A_\perp = 22.04$ | [474] |
| 20Li$_2$O-10Na$_2$O-69.5B$_2$O$_3$ + 0.5CuO glass | 9.205 | 300 | $g_\| = 2.292$, $g_\perp = 2.040$ | $A_\| = 136.3$, $A_\perp = 22.04$ | [474] |
| 25Li$_2$O-5Na$_2$O-69.5B$_2$O$_3$ + 0.5CuO glass | 9-205 | 300 | $g_\| = 2.292$, $g_\perp = 2.041$ | $A_\| = 136.2$, $A_\perp = 22.04$ | [474] |
| 5Li$_2$O + 15Na$_2$O + 20CdO + 59.5P$_2$O$_5$ + 0.5CuO glass | X | RT | $g_\| = 2.437$, $g_\perp = 2.096$ | $A_\| = 117$, $A_\perp = 26$ | [475] |
| 10Li$_2$O + 10Na$_2$O + 20CdO + 59.5P$_2$O$_5$ + 0.5CuO glass | X | RT | $g_\| = 2.441$, $g_\perp = 2.088$ | $A_\| = 121$, $A_\perp = 25$ | [475] |
| 15Li$_2$O+5Na$_2$O+20CdO+59.5P$_2$O$_5$ + 0.5CuO glass | X | RT | $g_\| = 2.433$, $g_\perp = 2.096$ | $A_\| = 125$, $A_\perp = 32$ | [475] |
| Na$_2$O-P$_2$O$_5$-BaCl$_2$ glass | X | RT | $g_\| = 2.423$, $g_\perp = 2.090$ | $A_\| = 104$, $A_\perp = 7.4$ | [476] |
| K$_2$O-P$_2$O$_5$-BaCl$_2$ glass | X | RT | $g_\| = 2.425$, $g_\perp = 2.092$ | $A_\| = 97$, $A_\perp = 7.4$ | [476] |
| Li$_2$O-P$_2$O$_5$-BaCl$_2$ glass | X | RT | $g_\| = 2.432$, $g_\perp = 2.093$ | $A_\| = 104$, $A_\perp = 8.3$ | [476] |
| Li-Na-P$_2$O$_5$-BaCl$_2$ glass | X | RT | $g_\| = 2.410$, $g_\perp = 2.093$ | $A_\| = 109$, $A_\perp = 6.5$ | [476] |
| Na-K-P$_2$O$_5$-BaCl$_2$ glass | X | RT | $g_\| = 2.406$, $g_\perp = 2.092$ | $A_\| = 103$, $A_\perp = 7.4$ | [476] |
| K-Li-P$_2$O$_5$-BaCl$_2$ glass | X | RT | $g_\| = 2.400$, $g_\perp = 2.090$ | $A_\| = 10$, $A_\perp = 8.3$ | [476] |

(continued)

## ($Cu^{2+}$($3d^9$) glasses listing contd.)

| Host | Frequency (GHz)/band | T (K) | $\tilde{g}$ | $A_x, A_y, A_z$ ($10^{-4}$ cm$^{-1}$) | References |
|---|---|---|---|---|---|
| 10Li$_2$O · 5P$_2$O$_5$ · 84TeO$_2$ · CuO | X | RT | $g_\parallel = 2.363$, $g_\perp = 2.073$ | $A_\parallel = 120$ | [477] |
| 10Li$_2$O · 10P$_2$O$_5$ · 79TeO$_2$ · CuO | X | RT | $g_\parallel = 2.375$, $g_\perp = 2.076$ | $A_\parallel = 116$ | [477] |
| 10Li$_2$O · 15P$_2$O$_5$ 74TeO$_2$ · CuO | X | RT | $g_\parallel = 2.386$, $g_\perp = 2.074$ | $A_\parallel = 115$ | [477] |
| 10Li$_2$O · 20P$_2$O$_5$ · 69TeO$_2$ · CuO | X | RT | $g_\parallel = 2.388$, $g_\perp = 2.076$ | $A_\parallel = 112$ | [477] |
| 10Li$_2$O · 25P$_2$O$_5$ 64TeO$_2$ · CuO | X | RT | $g_\parallel = 2.413$, $g_\perp = 2.079$ | $A_\parallel = 110$ | [477] |
| 10MgO · 10Li$_2$O · 10Na$_2$O · 10K$_2$O · 59B$_2$O$_3$ · 1CuO glass | X | RT | $g_\parallel = 2.265$, $g_\perp = 2.045$ | $A_\parallel = 138.88$, $A_\perp = 27.77$ | [26] |
| 5MgO-25Na$_2$O-69B$_2$O$_3$ | X | 310 | $g_\parallel = 2.352$, $g_\perp = 2.098$ | $A_\parallel = 153$, $A_\perp = 19.5$ | [478] |
| 10MgO-20Na$_2$O-69B$_2$O$_3$ | X | 310 | $g_\parallel = 2.348$, $g_\perp = 2.097$ | $A_\parallel = 155$, $A_\perp = 19.4$ | [478] |
| 12MgO-18Na$_2$O-69B$_2$O$_3$ | X | 310 | $g_\parallel = 2.345$, $g_\perp = 2.095$ | $A_\parallel = 158$, $A_\perp = 19.3$ | [478] |
| 15MgO-15Na$_2$O-69B$_2$O$_3$ | X | 310 | $g_\parallel = 2.338$, $g_\perp = 2.095$ | $A_\parallel = 162$, $A_\perp = 19.6$ | [478] |
| 17MgO-13Na$_2$O-69B$_2$O$_3$ | X | 310 | $g_\parallel = 2.336$, $g_\perp = 2.093$ | $A_\parallel = 162$, $A_\perp = 19.5$ | [478] |
| 40MgO · 9.9PbF$_2$ · 50SiO$_2$: 0.1CuO glass | 9.21 | — | $g_\parallel = 2.330$, $g_\perp = 2.073$ | $A_\parallel = 130$, $A_\perp = 27$ | [479] |
| 40MgO · 9.7PbF$_2$ · 50SiO$_2$: 0.3CuO glass | 9.21 | — | $g_\parallel = 2.332$, $g_\perp = 2.073$ | $A_\parallel = 138$, $A_\perp = 29$ | [479] |
| 40MgO · 9.5PbF$_2$ · 50SiO$_2$: 0.5CuO glass | 9.21 | — | $g_\parallel = 2.345$, $g_\perp = 2.071$ | $A_\parallel = 135$, $A_\perp = 26$ | [479] |
| 40MgO · 9.3PbF$_2$ · 50SiO$_2$: 0.7CuO glass | 9.21 | — | $g_\parallel = 2.335$, $g_\perp = 2.072$ | $A_\parallel = 132$, $A_\perp = 28$ | [479] |
| 40MgO · 9.1PbF$_2$ · 50SiO$_2$: 0.9CuO glass | 9.21 | — | $g_\parallel = 2.340$, $g_\perp = 2.073$ | $A_\parallel = 139$, $A_\perp = 27$ | [479] |
| 40MgO · 9PbF$_2$ · 50SiO$_2$: CuO glass | 9.21 | — | $g_\parallel = 2.335$, $g_\perp = 2.073$ | $A_\parallel = 140$, $A_\perp = 29$ | [479] |

(*continued*)

## ($Cu^{2+}$($3d^9$)) glasses listing contd.)

| Host | Frequency (GHz)/band | T (K) | $\tilde{g}$ | $A_x, A_y, A_z$ ($10^{-4}$ cm$^{-1}$) | References |
|---|---|---|---|---|---|
| NaBaB glass | 9.205 | 123–433 | $g_\| = 2.262$, $g_\perp = 2.049$ | $A_\| = 137$, $A_\perp = 24$ | [470] |
| 30NaF-50B$_2$O$_3$-20Bi$_2$O$_3$ glass | X | RT | $g_{\text{eff}} = 2.112$ | — | [480] |
| 5NaI · 25Na$_2$O · 70B$_2$O$_3$ glass | X | RT | $g_\| = 2.334$, $g_\perp = 2.067$ | $A_\| = 151$ | [481] |
| 10NaI · 20Na$_2$O · 70B$_2$O$_3$ glass | X | RT | $g_\| = 2.334$, $g_\perp = 2.067$ | $A_\| = 151$ | [481] |
| 15NaI · 15Na$_2$O · 70B$_2$O$_3$ glass | X | RT | $g_\| = 2.334$, $g_\perp = 2.067$ | $A_\| = 152$ | [481] |
| 20NaI-10Na$_2$O-70B$_2$O$_3$ glass | X | RT | $g_\| = 2.336$, $g_\perp = 2.067$ | $A_\| = 147$ | [481] |
| 25NaI · 5Na$_2$O · 70B$_2$O$_3$ glass | X | RT | $g_\| = 2.336$, $g_\perp = 2.074$ | $A_\| = 147$ | [481] |
| Nanoporous Vycor glasses impregnation time | | | | | |
| 1. 5 min | X | 4.2, 295 | 1. $g_\| = 2.33$, $g_\perp = 2.067$ | 1. $A_\| = 168$, $A_\perp = 10$ | [482] |
| 2. 15 min | | | 2. $g_\| = 2.346$, $g_\perp = 2.067$ | 2. $A_\| = 176$, $A_\perp = 10$ | |
| 3. 1 h | | | 3. $g_\| = 2.317$, $g_\perp = 2.067$ | 3. $A_\| = 173$, $A_\perp = 10$ | |
| (30-x) (NaPO$_3$)$_6$+30PbO + 40B$_2$O$_3$ +x CuO | X | 93–333 | $g_\| = 2.41$, $g_\perp = 2.07$ | $A_\| = 100$ | [483] |
| 80Na$_2$B$_4$O$_7$-19NaF – CuO (NFNB1Cu glass sample) | 9.205 | 163–396 | $g_\| = 2.313$, $g_\perp = 2.056$ | $A_\| = 153$, $A_\perp = 26$ | [484] |
| 90Na$_2$B$_4$O$_7$ + 9PbO + CuO Sodium lead tetraborate glass | 9.205 | RT | $g_\| = 2.298$, $g_\perp = 2.036$ | $A_\| = 128$, $A_\perp = 25$ | [471] |
| 40Na$_2$O-50B$_2$O$_3$-10Bi$_2$O$_3$ glass | X | 295 | $g_\| = 2.321$ | $A_\| = 130.0$ | [473] |
| 30Na$_2$O-50B$_2$O$_3$-20Bi$_2$O$_3$ glass | X | RT | $g_\| = 2.295$, $g_\perp = 2.08$ | $A_\| = 150$, $A_\perp = 70.8$ | [480] |
| 5Na$_2$O– (25)K$_2$O– 70B$_2$O$_3$ (glass) | X | RT | $g_\| = 2.338$, $g_\perp = 2.046$ | $A_\| = 134.9$, $A_\perp = 23.7$ | [485] |
| 10Na$_2$O– (20)K$_2$O– 70B$_2$O$_3$ (glass) | X | RT | $g_\| = 2.333$, $g_\perp = 2.043$ | $A_\| = 134.8$, $A_\perp = 23.7$ | [485] |

(continued)

## ($Cu^{2+}(3d^9)$) glasses listing contd.)

| Host | Frequency (GHz)/band | T (K) | $\tilde{g}$ | $A_x, A_y, A_z$ ($10^{-4}$ cm$^{-1}$) | References |
|---|---|---|---|---|---|
| 15Na$_2$O– (15)K$_2$O– 70B$_2$O$_3$ (glass) | X | RT | $g_\| = 2.333$, $g_\perp = 2.044$ | $A_\| = 140$, $A_\perp = 23.7$ | [485] |
| 20Na$_2$O– (10)K$_2$O– 70B$_2$O$_3$ (glass) | X | RT | $g_\| = 2.338$, $g_\perp = 2.044$ | $A_\| = 139.9$, $A_\perp = 23.7$ | [485] |
| 25Na$_2$O– (5)K$_2$O– 70B$_2$O$_3$ (glass) | X | RT | $g_\| = 2.338$, $g_\perp = 2.045$ | $A_\| = 134.9$, $A_\perp = 23.5$ | [485] |
| 24Na$_2$O-6NaF-50B$_2$O$_3$-20Bi$_2$O$_3$ glass | X | RT | $g_\| = 2.302$, $g_\perp = 2.08$ | $A_\| = 143.8$, $A_\perp = 70.8$ | [480] |
| 18Na$_2$O-12NaF-50B$_2$O$_3$-20Bi$_2$O$_3$ glass | X | RT | $g_\| = 2.311$, $g_\perp = 2.08$ | $A_\| = 137.5$, $A_\perp = 70.8$ | [480] |
| 12Na$_2$O-18NaF-50B$_2$O$_3$-20Bi$_2$O$_3$ glass | X | RT | $g_\| = 2.32$, $g_\perp = 2.086$ | $A_\| = 125$, $A_\perp = 70.8$ | [480] |
| 6Na$_2$O-24NaF-50B$_2$O$_3$-20Bi$_2$O$_3$ glass | X | RT | $g_\| = 2.316$, $g_\perp = 2.086$ | $A_\| = 125$, $A_\perp = 75$ | [480] |
| $x$Na$_2$O · $(1-x)$SiO$_2$ glasses | | | | | |
| 1) $x = 33$, 1% Cu | X | 77, 300 | 1) $g_\| = 2.35$, $g_\perp = 2.065$ | 1) $A_\| = 135$, $A_\perp = 7$ | [486] |
| 2a) $x = 33$, 10% Cu Line 1 | | | 2a) $g_\| = 2.35$, $g_\perp = 2.075$ | 2a) $A_\| = 135$, $A_\perp = 7$ | |
| 2b) $x = 33$, 10% Cu Line 2 | | | 2b) $g_\| = 2.35$, $g_\perp = 2.150$ | 2b) $A_\| = 135$, $A_\perp = 7$ | |
| 10Na$_2$O + 40ZnO + 50B$_2$O$_3$ glass | X | 295 | $g_\| = 2.337$, $g_\perp = 2.070$ | $A_\| = 140$, $A_\perp = 19$ | [487] |
| 20Na$_2$O + 30ZnO + 50B$_2$O$_3$ glass | X | 295 | $g_\| = 2.332$, $g_\perp = 2.069$ | $A_\| = 143$, $A_\perp = 19$ | [487] |
| 25Na$_2$O + 25ZnO + 50B$_2$O$_3$ glass | X | 295 | $g_\| = 2.328$, $g_\perp = 2.068$ | $A_\| = 143$, $A_\perp = 20$ | [487] |
| 30Na$_2$O + 20ZnO + 50B$_2$O$_3$ glass | X | 295 | $g_\| = 2.317$, $g_\perp = 2.069$ | $A_\| = 155$, $A_\perp = 20$ | [487] |
| 20Na$_2$O + 10ZnO + 50B$_2$O$_3$ glass | X | 295 | $g_\| = 2.325$, $g_\perp = 2.068$ | $A_\| = 147$, $A_\perp = 21$ | [487] |
| 50Na$_2$O + 50B$_2$O$_3$ glass | X | 295 | $g_\|g_\| = 2.328$, $g_\perp = 2.068$ | $A_\| = 140$, $A_\perp = 21$ | [487] |
| 14.7PbO-10CaO-5ZnO-70B$_2$O$_3$-0.3CuO glass | X | RT | $g_\| = 2.333$, $g_\perp = 2.067$ | $A_\| = 138$, $A_\perp = 50$ | [488] |

*(continued)*

## ($Cu^{2+}(3d^9)$) glasses listing contd.)

| Host | Frequency (GHz)/band | T (K) | $\tilde{g}$ | $A_x, A_y, A_z$ ($10^{-4}$ cm$^{-1}$) | References |
|---|---|---|---|---|---|
| $10SrO \cdot 10Li_2O \cdot 10Na_2O \cdot 10K_2O \cdot 59B_2O_3 \cdot 1CuO$ glass | X | RT | $g_\| = 2.254$, $g_\perp = 2.036$ | $A_\| = 138.88$, $A_\perp = 27.77$ | [26] |
| $85TeO_2 \cdot 7.5Ag_2O \cdot 7.5WO_3$ glass | X | RT | $g_\| = 2.352$, $g_\perp = 2.103$ | $A_\| = 124$ | [489] |
| $70TeO_2 \cdot 15Ag_2O \cdot 15WO_3$ glass | X | RT | $g_\| = 2.358$, $g_\perp = 2.083$ | $A_\| = 113$ | [489] |
| $55TeO_2 \cdot 22.5Ag_2O \cdot 22.5WO_3$ glass | X | RT | $g_\| = 2.362$, $g_\perp = 2.088$ | $A_\| = 118$ | [489] |
| $40TeO_2 \cdot 30Ag_2O \cdot 30WO_3$ glass | X | RT | $g_\| = 2.357$, $g_\perp = 2.079$ | $A_\| = 114$ | [489] |
| $10TeO_2 + 60 B_2O_3 + 5TiO_2 + 24 Li_2O: 1 CuO$ glass | X | RT | $g_\| = 2.3636$, $g_\perp = 2.0725$ | $A_\| = 122$ | [490] |
| $10TeO_2 + 60 B_2O_3 + 5TiO_2 + 24 Na_2O: 1 CuO$ glass | X | RT | $g_\| = 2.3386$, $g_\perp = 2.0696$ | $A_\| = 139$ | [490] |
| $10TeO_2 + 60 B_2O_3 + 5TiO_2 + 24K_2O: 1 CuO$ glass | X | RT | $g_\| = 2.3409$, $g_\perp = 2.0680$ | $A_\| = 134$ | [490] |
| $10TeO_2 \cdot 60B_2O_3 \cdot 5TiO_2 \cdot 24Li_2O:1CuO$ glass | X | RT | $g_\| = 2.363$, $g_\perp = 2.072$ | $A_\| = 122$ | [491] |
| $10TeO_2 \cdot 60B_2O_3 \cdot 5TiO_2 \cdot 24Na_2O:1CuO$ glass | X | RT | $g_\| = 2.338$, $g_\perp = 2.069$ | $A_\| = 139$ | [491] |
| $10TeO_2 \cdot 60B_2O_3 \cdot 5TiO_2 \cdot 24K_2O:1CuO$ glass | X | RT | $g_\| = 2.340$, $g_\perp = 2.068$ | $A_\| = 134$ | [491] |
| $35TeO_2 \cdot 35B_2O_3 \cdot 5TiO_2 \cdot 24Li_2O:1CuO$ glass | X | RT | $g_\| = 2.298$, $g_\perp = 2.043$ | $A_\| = 143$ | [491] |
| $35TeO_2 \cdot 35B_2O_3 \cdot 5TiO_2 \cdot 24Na_2O:1CuO$ glass | X | RT | $g_\| = 2.299$, $g_\perp = 2.044$ | $A_\| = 140$ | [491] |
| $35TeO_2 \cdot 35B_2O_3 \cdot 5TiO_2 \cdot 24K_2O:1CuO$ glass | X | RT | $g_\| = 2.293$, $g_\perp = 2.041$ | $A_\| = 140$ | [491] |
| $60TeO_2 \cdot 10B_2O_3 \cdot 5TiO_2 \cdot 24Li_2O:1CuO$ glass | X | RT | $g_\| = 2.364$, $g_\perp = 2.068$ | $A_\| = 122$ | [491] |
| $60TeO_2 \cdot 10B_2O_3 \cdot 5TiO_2 \cdot 24Na_2O:1CuO$ glass | X | RT | $g_\| = 2.349$, $g_\perp = 2.069$ | $A_\| = 122$ | [491] |
| $60TeO_2 \cdot 10B_2O_3 \cdot 5TiO_2 \cdot 24K_2O:1CuO$ glass | X | RT | $g_\| = 2.341$, $g_\perp = 2.067$ | $A_\| = 128$ | [491] |

(continued)

## ($Cu^{2+}(3d^9)$) glasses listing contd.)

| Host | Frequency (GHz)/band | T (K) | $\tilde{g}$ | $A_x, A_y, A_z$ ($10^{-4}$ cm$^{-1}$) | References |
|---|---|---|---|---|---|
| 82.5TeO$_2$ · 10GeO$_2$ · 7.5WO$_3$ glass | X | RT | $g_\| = 2.355$, $g_\perp = 2.077$ | $A_\| = 117$ | [492] |
| 75TeO$_2$ · 10GeO$_2$ · 15WO$_3$ glass | X | RT | $g_\| = 2.386$, $g_\perp = 2.083$ | $A_\| = 99$ | [492] |
| 67.5TeO$_2$ · 10GeO$_2$ · 22.5WO$_3$ glass | X | RT | $g_\| = 2.362$, $g_\perp = 2.088$ | $A_\| = 115$ | [492] |
| 60TeO$_2$ · 10GeO$_2$ · 30WO$_3$ glass | X | RT | $g_\| = 2.354$, $g_\perp = 2.099$ | $A_\| = 112$ | [492] |
| 10TeO$_2$+60B$_2$O$_3$+5TiO$_2$+24Li$_2$O glass | X | RT | $g_\| = 2.3636$, $g_\perp = 2.0725$ | $A_\| = 122$ | [493] |
| 10TeO$_2$+60B$_2$O$_3$+5TiO$_2$+24Na$_2$O glass | X | RT | $g_\| = 2.3386$, $g_\perp = 2.0696$ | $A_\| = 139$ | [493] |
| 10TeO$_2$+60B$_2$O$_3$+5TiO$_2$+24K$_2$O glass | X | RT | $g_\| = 2.3409$, $g_\perp = 2.0680$ | $A_\| = 134$ | [493] |
| TriGlycene sulfate phosphate (TGSP) | X | RT | 2.031, 2.071, 2.318 | 40.68, 22.64, 174.0 | [494] |
| ZnB$_4$O$_7$ Glass | X | RT | $g_\| = 2.355$, $g_\perp = 2.079$ | $A_\| = 148$, $A_\perp = 28$ | [463] |
| 50ZnO+50B$_2$O$_3$:Cu | X | 295 | $g_\| = 2.337$, $g_\perp = 2.070$ | $A_\| = 143$, $A_\perp = 20$ | [487] |
| 19.9ZnO + 5Li$_2$O + 25Na$_2$O + 50B$_2$O$_3$ + 0.1CuO | X | RT | $g_\| = 2.319$, $g_\perp = 2.061$ | $A_\| = 173$, $A_\perp = 32.3$ | [495] |
| 19.9ZnO + 15Li$_2$O + 15Na$_2$O + 50B$_2$O$_3$ + 0.1CuO | X | RT | $g_\| = 2.338$, $g_\perp = 2.060$ | $A_\| = 156$, $A_\perp = 31.0$ | [495] |
| 19.9ZnO + 20Li$_2$O + 10Na$_2$O + 50B$_2$O$_3$ + 0.1CuO | X | RT | $g_\| = 2.325$, $g_\perp = 2.061$ | $A_\| = 167$, $A_\perp = 31.3$ | [495] |
| 19.9ZnO + 25Li$_2$O + 5Na$_2$O + 50B$_2$O$_3$ + 0.1CuO | X | RT | $g_\| = 2.325$, $g_\perp = 2.061$ | $A_\| = 171$, $A_\perp = 31.3$ | [495] |
| 10ZnO · 10Li$_2$O · 10Na$_2$O · 10K$_2$O · 59B$_2$O$_3$ · 1CuO glass | X | RT | $g_\| = 2.266$, $g_\perp = 2.030$ | $A_\| = 145.03$, $A_\perp = 25.32$ | [26] |

(continued)

Ni$^+$. Data tabulation of SHPs
Ni$^+$ (3d$^9$)

| Host | Frequency (GHz)/band | T (K) | $\tilde{g}$ | $A_x, A_y, A_z$ ($10^{-4}$ cm$^{-1}$) | References |
|---|---|---|---|---|---|
| AgGaSe$_2$ crystal | 9.5 and ENDOR | 10 | $g_\parallel = 2.6470$, $g_\perp = 2.2493$ | $A(^{61}$Ni$)$: $A_\parallel = -60.2$, $A_\perp = -88.2$ (MHz); $A(^{77}$Se$)$: 36.1, 38.6, 68.6 (MHz); $A(^{69}$Ga$)$: 19.89, 20.63, 22.27 (MHz); $Q(^{61}$Ni$)$: $Q_{zz} = -11.0$ (MHz); $Q(^{69}$Ga$)$: 0.471, 0.031, $-0.502$ (MHz) | [496] |
| CsCaF$_3$ (center I) | X | LNT | $g_\parallel = 2.982$, $g_\perp = 2.142$ | 23, 27, 54 | [497] |
| CsCaF$_3$ (center II) | X | LNT | $g_\parallel = 2.740$, $g_\perp = 2.116$ | 30, 35, 70 | [497] |
| CsCaF$_3$ (center III) | X | LNT | $g_\parallel = 2.676$, $g_\perp = 2.089$ | 32, 37, 75 | [497] |

(4d$^n$) palladium group
4d$^1$ (Mo$^{5+}$), $S = 1/2$
Mo$^{5+}$. Data tabulation of SHPs
Mo$^{5+}$ (4d$^1$)

| Host | Frequency (GHz)/band | T (K) | $\tilde{g}$ | $A_x, A_y, A_z$ ($10^{-4}$ cm$^{-1}$) | References |
|---|---|---|---|---|---|
| 0.5MoO$_3$–0.2Sb$_2$O$_3$–0.3K$_2$O glass | X | 300 | $g_\parallel = 1.998$, $g_\perp = 1.973$ | $A_\parallel = 171.0$, $A_\perp = 63.0$ | [125] |
| 0.5MoO$_3$–0.2Sb$_2$O$_3$–0.3K$_2$O glass | X | 77 | $g_\parallel = 1.986$, $g_\perp = 1.970$ | $A_\parallel = 130.0$, $A_\perp = 50.0$ | [125] |
| 80MoO$_3$–20B$_2$O$_3$ (glass) | X | 300 | $g_\parallel = 1.940$, $g_\perp = 1.974$ | $A_\parallel = 150.0$, $A_\perp = 35.6$ | [498] |
| 80MoO$_3$–20B$_2$O$_3$ (glass) | X | 77 | $g_\parallel = 1.935$, $g_\perp = 1.975$ | $A_\parallel = 141.9$, $A_\perp = 34.5$ | [498] |

## 4d³ (Mo³⁺), S = 3/2
Mo³⁺. Data tabulation of SHPs
Mo³⁺ (4d³)

| Host | Frequency (GHz)/ band | T (K) | $\tilde{g}$ | $A_x, A_y, A_z$ ($10^{-4}$ cm$^{-1}$) | References |
|---|---|---|---|---|---|
| SrTiO$_3$ | X, K | 4.2–77 | 1.9546 | 32.0 | [499] |
| 0.8MoO$_3$–0.2B$_2$O$_3$ glass | X | 300 | $g_\parallel = 1.940$, $g_\perp = 1.974$ | $A_\parallel = 150.0$, $A_\perp = 35.5$ | [125] |
| 0.8MoO$_3$–0.2B$_2$O$_3$ glass | X | 77 | $g_\parallel = 1.935$, $g_\perp = 1.975$ | $A_\parallel = 141.9$, $A_\perp = 34.5$ | [125] |
| YAlO$_3$ (site I) | 9.22 | 20 | 2.441, 1.750, 5.150 | $^{95,97}$Mo hf splitting: $A[001] = 5.1$ mT | [500] |
| YAlO$_3$ (site II) | 9.22 | 20 | 1.45, 5.25, 1.25 | $^{95,97}$Mo hf splitting: $A[001] = 7.65$ mT | [500] |

## 4d⁷ (Rh²⁺), S = 3/2
Rh²⁺. Data tabulation of SHPs
Rh²⁺ (4d⁷)

| Host | Frequency (GHz)/ band | T (K) | $\tilde{g}$ | $A_x, A_y, A_z$ ($10^{-4}$ cm$^{-1}$) | References |
|---|---|---|---|---|---|
| KTiOPO$_4$ (center 1) | X | 90 | 2.5783, 2.5686, 1.9610 | 31.6, 33.9, 26.3 | [501] |
| KTiOPO$_4$ (center 2) | X | 90 | 2.5955, 2.5446, 1.9633 | 33.1, 29.6, 25.2 | [501] |

## 4d⁸ (Rh⁺), S = 1
Rh⁺. Data tabulation of SHPs
Rh⁺ (4d⁸)

| Host | Frequency (GHz)/ band | T (K) | $\tilde{g}$ | $A_x, A_y, A_z$ ($10^{-4}$ cm$^{-1}$) | References |
|---|---|---|---|---|---|
| NaCl | 15–30 (ENDOR) | 8 | $g_{iso} = 2.45$ | ($^{103}$Rh): $A_\parallel = -21.6$, $A_\perp = -21.6$ | [502] |

## $4d^9$ (Pd$^+$), $S = 1/2$
Pd$^+$. Data tabulation of SHPs
Pd$^+$ ($4d^9$)

| Host | Frequency (GHz)/ band | T (K) | $\tilde{g}$ | $A_x, A_y, A_z$ ($10^{-4}$ cm$^{-1}$) | References |
|---|---|---|---|---|---|
| NaCl | 9.2260 | 12 | $g_\| = 2.85, g_\perp = 2.15$ | $A_\| = 29.8, A_\perp = 5.9$ | [503] |

## ($4f^n$) lanthanide group
$4f^1$ (Ce$^{3+}$, Pr$^{4+}$), $S = 1/2$ (Kramers ion)
Ce$^{3+}$. Data tabulation of SHPs
Ce$^{3+}$ ($4f^1$)

| Host | Frequency (GHz)/ band | T (K) | $\tilde{g}$ | References |
|---|---|---|---|---|
| BaLiF$_3$ (site T$_1$) | 9.685 | 5–20 | $g_\| = 2.866, g_\perp = 1.196$ | [504] |
| BaLiF$_3$ (site T$_2$) | 9.685 | 5–20 | $g_\| = 0.772, g_\perp = 2.465$ | [504] |
| BaLiF$_3$ (site R$_1$) | 9.685 | 5–20 | 2.214, 2.141, 0.889 | [504] |
| BaLiF$_3$ (site R$_2$) | 9.685 | 5–20 | 2.079, 1.580, 1.152 | [504] |
| BaMgF$_4$ (site A) | 9.690 | 10 | 1.580 (1.997, 2.443, 0.299) | [505] |
| BaMgF$_4$ (site B) | 9.690 | 10 | 1.698 (0.580, 0.731, 3.784) | [505] |
| CaWO$_3$ | X | 4.2–50 | $g_\| = 2.91, g_\perp = 1.42$ | [506] |
| CaWO$_4$ single crystal | X | — | $g_\| = 2.915, g_\perp = 1.423$ | [507] |
| CaYAlO$_4$ | X | 4.2–50 | $g_\| = 2.52, g_\perp = 1.54$ | [506] |
| K$_2$YF$_5$ | 9.5, 34 | 10 | 0.25, 0.75, 2.74 | [508] |
| LiCaAlF$_6$ | X | 4.2–70 | $g_\| = 1.725, g_\perp = 0.965$ | [509] |
| LiCaAlF$_6$ (site A) | 9.304 | 4.2 | $g_\| = 1.725, g_\perp = 0.965$ | [510] |
| LiCaAlF$_6$ (site B) | 9.304 | 4.2 | 0.84, 1.18, 1.77 | [510] |
| LiCaAlF$_6$ (site C) | 9.304 | 4.2 | 0.95, 1.27, 1.54 | [510] |
| LiYF$_4$ | X | 4.2–50 | $g_\| = 2.765, g_\perp = 1.473$ | [506] |
| LuF$_3$ | X | 15 | 3.374, 0.60, 0.29 | [511] |
| PbGa$_2$S$_4$ (center 1) | 9.23 | 4–30 | 2.43, 2.98, 0.83 | [512] |
| PbGa$_2$S$_4$ (center 2) | 9.23 | 4–30 | 2.65, 2.79, 0.79 | [512] |
| PbGa$_2$S$_4$ (center 3) | 9.23 | 4–30 | 2.34, 3.05, 0.79 | [512] |

(continued)

## ($Ce^{3+}$ ($4f^1$) listing contd.)

| Host | Frequency (GHz)/band | T (K) | $\tilde{g}$ | References |
|---|---|---|---|---|
| $PbGa_2S_4$ (center 4) | 9.23 | 4–30 | 2.43, 3.00, 0.83 | [512] |
| $PbMoO_4$ single crystal | X | — | $g_\| = 2.684$, $g_\perp = 1.514$ | [507] |
| $PbWO_4$ | 9.43 | — | $g_\| = 2.6728$, $g_\perp = 1.5208$ | [513] |
| $PbWO_4$ single crystal | 9.21 | 4.2–300 | $g_\| = 2.677$, $g_\perp = 1.516$ | [507] |
| $PbWO_4$ | X | 4 | $g_\| = 2.6769$, $g_\perp = 1.5220$ | [514] |
| $SrWO_4$ single crystal | X | — | $g_\| = 2.871$, $g_\perp = 1.452$ | [507] |
| $YAlO_3$ | 9.204 | 12 | 3.162, 0.402, 0.395 | [515] |
| $YAl_3(BO_3)_4$ | X | 16 | $g_\| = 1.972$, $g_\perp = 0.737$ | [516] |
| $YF_3$ | X | 15 | 3.384, 0.48, 0.21 | [511] |
| $Y(NO_3) \cdot 6H_2O$ | ~9.45 | <10 | 2.3010, 1.8169, 0.9745 | [517] |
| $Y_2(SO_4)_3 \cdot 8H_2O$ (site 1) | ~9.45 | 4.3 | 3.3000, 1.2669, 0.6261 | [517] |
| $Y_2(SO_4)_3 \cdot 8H_2O$ (site 2) | ~9.45 | 4.3 | 3.1808, 1.1782, 0.6157 | [517] |

## $Pr^{4+}$. Data tabulation of SHPs
## $Pr^{4+}$ ($4f^1$)

| Host | Frequency (GHz)/band | T (K) | $\tilde{g}$ | $A_x, A_y, A_z$ ($10^{-4}$ cm$^{-1}$) | References |
|---|---|---|---|---|---|
| $BaCeO_3$ | 9.095 | 4.2 | $|g| = 0.741$ | 141Pr: $A_{iso} = 609$ | [518] |
| $BaSnO_3$ | 9.095 | 4.2 | $|g| = 0.583$ | 141Pr: $A_{iso} = 589$ | [518] |
| $BaZrO_3$ | 9.095 | 4.2 | $|g| = 0.643$ | 141Pr: $A_{iso} = 597$ | [518] |

## $4f^3$ ($Nd^{3+}$), $S = 1/2$ (Kramers ion)
## $Nd^3$. Data tabulation of SHPs
## $Nd^{3+}$ ($4f^3$)

| Host | Frequency (GHz)/band | T (K) | $\tilde{g}$ | $A_x, A_y, A_z$ ($10^{-4}$ cm$^{-1}$) | References |
|---|---|---|---|---|---|
| $BaTiO_3$ (site III) | X | 4 | $g_\| = 2.461$, $g_\perp = 2.583$ | — | [519] |
| $CaO–Li_2O–B_2O_3$ | 9.5 | 10 | $g_\| = 3.14$, $g_\perp = 1.39$ | — | [520] |

(continued)

## ($Nd^{3+}(4f^3)$) listing contd.)

| Host | Frequency (GHz)/band | T (K) | $\tilde{g}$ | $A_x, A_y, A_z$ ($10^{-4}$ cm$^{-1}$) | References |
|---|---|---|---|---|---|
| CaWO$_4$ single crystal | X | | $g_\| = 2.035$, $g_\perp = 2.537$ | $^{143}A_\| = 203$, $^{143}A_\perp = 260$, $^{145}A_\| = 126$, $^{145}A_\perp = 161$ | [521] |
| KMgF$_3$ (site I) | X | 4 | $g_\| = 2.887$, $g_\perp = 2.391$ | — | [519] |
| KMgF$_3$ (site II) | X | 4 | $g_\| = 2.722$, $g_\perp = 2.472$ | — | [519] |
| KMgF$_3$ (site IV) | X | 4 | 2.557, 2.371, 2.742 | — | [519] |
| KZnF$_3$ (site I) | X | 4 | $g_\| = 2.763$, $g_\perp = 2.452$ | — | [519] |
| KZnF$_3$ (site III) | X | 4 | $g_\| = 3.546$, $g_\perp = 1.154$ | — | [519] |
| K$_2$O–BaO–Al$_2$O$_3$–P$_2$O$_5$ | 9.5 | 10 | $g_\| = 3.08$, $g_\perp = 1.31$ | — | [520] |
| LiNbO$_3$ (center 1) | X | | $g_\| = 1.443$, $g_\perp = 2.963$ | — | [522] |
| LiNbO$_3$ (center 2) | X | | $g_\| = 1.323$, $g_\perp = 3.136$ | — | [522] |
| LiYF$_4$ | 9.38 | 15 | $g_\| = 1.955$, $g_\perp = 2.530$ | $A_\| = 589$ MHz, $A_\perp = 762$ MHz | [511] |
| LiYF$_4$ | X | 10 | $g_\perp = 2.553$, $g_\| = 1.986$ | $A_\| = 198.4$ ($^{143}$Nd), 123.0 ($^{145}$Nd), $A_\perp = 256$ ($^{143}$Nd), 164 ($^{145}$Nd) | [523] |
| Na$_2$O–Al$_2$O$_3$–B$_2$O$_3$ | 9.5 | 10 | $g_\| = 2.99$, $g_\perp = 1.30$ | — | [520] |
| PbMoO$_4$ single crystal | X | | $g_\| = 1.351$, $g_\perp = 2.592$ | $^{143}A_\| = 128$, $^{143}A_\perp = 269$, $^{145}A_\| = 80$, $^{145}A_\perp = 168$ | [521] |

(continued)

## ($Nd^{3+}(4f^3)$ listing contd.)

| Host | Frequency (GHz)/band | T (K) | $\tilde{g}$ | $A_x, A_y, A_z$ ($10^{-4}$ cm$^{-1}$) | References |
|---|---|---|---|---|---|
| PbWO$_4$ single crystal | 9.21 | 4.2–300 | $g_\| = 1.362$, $g_\perp = 2.594$ | $^{143}A_\| = 129$, $^{143}A_\perp = 264$, $^{145}A_\| = 81$, $^{145}A_\perp = 167$ | [521] |
| PbWO$_4$ single crystal | 9.43 | 10 | $g_\| = 1.3614$, $g_\perp = 2.5941$ | $^{143}A_\| = 129.45$, $^{143}A_\perp - 266.8$, $^{145}A_\| = 80.55$, $^{145}A_\perp = 166.0$ | [524] |
| PbWO$_4$ single crystal | 95 | 10 | $g_\| = 1.3543$, $g_\perp = 2.5943$ | $^{143}A_\| = 129.35$, $^{143}A_\perp = 266.8$, $^{145}A_\| = 80.25$, $^{145}A_\perp = 166.0$ | [524] |
| PbWO$_4$ single crystal | 190 | 10 | $g_\| = 1.3357$, $g_\perp = 2.5935$ | $^{143}A_\| = 126.65$, $^{143}A_\perp = 266.8$, $^{145}A_\| = 78.65$, $^{145}A_\perp = 166.0$ | [524] |
| PbWO$_4$ single crystal | 285 | 10 | $g_\perp = 2.5921$ | | [524] |
| SrTiO$_3$ (site II) | X | 4 | $g_\| = 2.609$, $g_\perp = 2.472$ | — | [519] |
| SrWO$_4$ single crystal | X | — | $g_\| = 1.541$, $g_\perp = 2.571$ | $^{143}A_\| = 151$, $^{145}A_\perp = 269$, $^{145}A_\| = 95$, $^{145}A_\perp = 168$ | [521] |
| YAlO$_3$ | 9.204 | 12 | 1.693, 2.570, 2.820 | ($^{145}$Nd) 298, 258, 192 | [515] |
| YF$_3$ | 9.38 | 15 | 4.29, 1.83, 1.19 | 1290, 552, 360 (MHz) | [511] |
| Y(NO$_3$)·6H$_2$O | 9.345 | 4.2 | 0.9688, 0.9672, 0.9004 | ($^{143}$Nd) 0.8613, 0.4880, 0.2072 (GHz) | [517] |
| Y(NO$_3$)·6H$_2$O | 9.345 | 4.2 | 0.9990, 0.9582, 0.1269 | ($^{145}$Nd) 1.0724, 0.9339, 0.3563 (GHz) | [517] |
| Y(NO$_3$)·6H$_2$O | 9.345 | 4.2 | 0.9990, 0.9582, 0.1269 | ($^{145}$Nd) 1.0724, 0.9339, 0.3563 (GHz) | [517] |

(continued)

## ($Nd^{3+}(4f^3)$ listing contd.)

| Host | Frequency (GHz)/ band | T (K) | $\tilde{g}$ | $A_x, A_y, A_z$ ($10^{-4}$ cm$^{-1}$) | References |
|---|---|---|---|---|---|
| YVO$_4$ | X | 10 | $g_\parallel$ = 1.915, $g_\perp$ = 2.361 | $A_\parallel$ = 112.1 ($^{143}$Nd), 70 ($^{145}$Nd) $A_\perp$ = 256.9 ($^{143}$Nd), 159.3 ($^{145}$Nd) | [523] |
| $^{143}$Nd Y$_2$(SO$_4$)$_3 \cdot$ 8H$_2$O (site I) | ~9.45 | 5.6 | 1.1492, 0.9968, 0.8674 | ($^{143}$Nd) 1.175, 0.916, 0.715 (GHz) | [517] |
| $^{143}$Nd Y$_2$(SO$_4$)$_3 \cdot$ 8H$_2$O (site II) | ~9.45 | 5.6 | 0.9934, 0.8534, 0.8461 | ($^{143}$Nd) 1.005, 0.339, 0.138 (GHz) | [517] |
| $^{145}$Nd Y$_2$(SO$_4$)$_3 \cdot$ 8H$_2$O (site I) | ~9.45 | 5.6 | 1.1284, 1.0298, 0.9510 | ($^{145}$Nd) 1.132, 0.246, 0.214 (GHz) | [517] |
| $^{145}$Nd Y$_2$(SO$_4$)$_3 \cdot$ 8H$_2$O (site II) | ~9.45 | 5.6 | 1.1563, 1.1438, 1.0845 | ($^{145}$Nd) 1.248, 0.841, 0.821 (GHz) | [517] |
| ZnO–Li$_2$O–P$_2$O$_5$ | 9.5 | 10 | $g_\parallel$ = 3.03, $g_\perp$ = 1.20 | — | [520] |

$4f^5$ (Sm$^{3+}$), S = 1/2 (Kramers ion)
Sm$^{3+}$. Data tabulation of SHPs
Sm$^{3+}$($4f^5$)

| Host | Frequency (GHz)/ band | T (K) | $\tilde{g}$ | References |
|---|---|---|---|---|
| BaFCl (monoclinic) | X | 4.2 | 0.903, 0.856, 0.19 | [525] |
| BaFCl (axial) | X | 4.2 | $g_\parallel$ = 1.027, $g_\perp$ = 0.23 | [525] |
| KY$_3$F$_{10}$ | 9.69 | 10 | $g_\parallel$ = 0.714, $g_\perp$ = 0.11 | [526] |

## 4.2 Listing of Spin-Hamiltonian Parameters

$4f^7$ ($Eu^{2+}$, $Gd^{3+}$), $S = 7/2$ (S-state ions)
$Eu^{2+}$. Data tabulation of SHPs
$Eu^{2+}$ ($4f^7$)

| Host | Frequency (GHz)/band | T (K) | $\tilde{g}$ | ($b_2^0 = D$), ($b_2^2 = 3E$) ($10^{-4}$ cm$^{-1}$) | $b_4^0, b_4^2, b_4^4$ ($10^{-4}$ cm$^{-1}$) | $b_6^0, b_6^2, b_6^4, b_6^6$ ($10^{-4}$ cm$^{-1}$) | $A_x, A_y, A_z$ ($10^{-4}$ cm$^{-1}$) | References |
|---|---|---|---|---|---|---|---|---|
| $Ba_{12}F_{19}Cl_5$ (site I) | 36 | 78 | $g_{iso} = 1.987$ | −340, 148 | 0.16, 0.14, −0.22 | — | — | [527] |
| $Ba_{12}F_{19}Cl_5$ (site II) | 36 | 78 | $g_{iso} = 1.982$ | 296, 343 | −0.11, −0.17, 0.33 | — | — | [527] |
| $Ba_{12}F_{19}Cl_5$ (site III) | 36 | 78 | $g_{iso} = 1.990$ | −374, 59 | —, —, — | — | — | [527] |
| $Bi_2Se_3$ | X | 300 | 2.001 | $b_2^0 = -0.293$ (GHz) | $b_4^0 = 0.0052$, $b_4^3 = -0.0033$ (GHz) | $b_6^0 = -0.0066$ (GHz) | — | [528] |
| $Bi_2Se_3$ | X | 4.2 | 1.998 | $b_2^0 = -0.349$ (GHz) | $b_4^0 = 0.011$, $b_4^3 = -0.0031$ (GHz) | $b_6^0 = -0.015$ (GHz) | — | [528] |
| $CaF_2$ | 9.447 | 295 | — | — | — | — | $A^{151} = 34.5$, $A^{153} = 15.3$ | [529] |
| $CaO$ | 9.447 | 295 | — | — | — | — | $A^{151} = 29.63$, $A^{153} = 13.05$ | [529] |
| $5CaO \cdot 3Al_2O_3$ | 9.79 | 295 | $g_\parallel = 1.994$, $g_\perp = 2.013$ | 2.956, 0.059 (GHz) | −0.07, −0.07, −0.01 (GHz) | 0.01, 0.06, 0.05, −0.03 (GHz) | — | [530] |
| $5CaO \cdot 3Al_2O_3$ | 9.79 | 77 | $g_\parallel = 1.994$, $g_\perp = 1.941$ | 2.951, 0.004 (GHz) | −0.02, −0.19, 0.07 (GHz) | 0.02, 0.04, 0.03, 0.01 (GHz) | — | [530] |
| $5CaO \cdot 3Al_2O_3$ | 9.79 | 4.2 | $g_\parallel = 2.002$, $g_\perp = 2.007$ | 3.001, −0.020 (GHz) | 0.01, 0.04, 0.06 (GHz) | 0.01, 0.02, 0.01, −0.00 (GHz) | — | [530] |

(continued)

## ($Eu^{2+}$ ($4f^7$) listing contd.)

| Host | Frequency (GHz)/ band | T (K) | $\tilde{g}$ | ($b_2^0 = D$, $b_2^2 = 3E$) ($10^{-4}$ cm$^{-1}$) | $b_4^0, b_4^2, b_4^4$ ($10^{-4}$ cm$^{-1}$) | $b_6^0, b_6^2, b_6^4, b_6^6$ ($10^{-4}$ cm$^{-1}$) | $A_x, A_y, A_z$ ($10^{-4}$ cm$^{-1}$) | References |
|---|---|---|---|---|---|---|---|---|
| $CaWO_4$ | 9.447 | 295 | — | — | — | — | $A_\parallel^{151} = -34.4$, $A_\perp^{151} = -35.0$, $A_\parallel^{153} = -15.5$, $A_\perp^{153} = -16.0$ | [529] |
| CdS | 9.447 | 295 | — | — | — | — | $A^{151} = 23.03$, $A^{153} = 10.32$ | [529] |
| $Cs_2NaLaCl_6$ | 9.1 | RT | $g_{iso} = 1.9934$ | $b_2^0 = 155.7$ | $b_4^0 = -5.4$, $b_4^3 = 110.0$ | $b_6^0 = 0.3$, $b_6^3 = -31.6$, $b_6^6 = -13.9$ | $^{151}A_{iso} = 31.8$, $^{153}A_{iso} = 14.0$ | [531] |
| $Eu@C_{82}$ fullerine (site I) | X | 4.2 | 1.995, 1.993, 1.9946 | 2918, 74 | 8, 33, −43 | — | — | [532] |
| $Eu@C_{82}$ fullerine (site II) | X | 4.2 | 1.9919, 1.9928, 1.9933 | 2915, 605 | 17, −32, −7 | — | — | [532] |
| $Eu@C_{82}$ fullerine (site III) | X | 4.2 | 1.9925, 1.9921, 1.9939 | 2724, 41 | −1, −3, 21 | — | — | [532] |
| $Eu@C_{74}$ fullerine | X | 4.2 | 1.9938, 1.9891, 1.9883 | 1279, 42.2 | −6, 25, −23 | — | — | [532] |
| KCl | 9.447 | 295 | — | — | — | — | $A^{151} = 32.56$, $A^{153} = 14.38$ | [529] |
| $LaCl_2$ | 9.447 | 295 | — | — | — | — | $A^{151} = 38.0$, $A^{153} = 17$ | [529] |
| $Pb_{1-x}Eu_x$ | 9.54 | 4.2, 300 | — | $D = -61.5$ (at 4.2 K) | $b_4 = 270.5$ MHz | $b_6 = -2.0$ MHz | $A(^{151}Eu) = 92.5$ MHz, $A(^{153}Eu) = 36$ MHz | [533] |

*(continued)*

## 4.2 Listing of Spin-Hamiltonian Parameters

**($Eu^{2+}(4f^7)$ listing contd.)**

| Host | Frequency (GHz)/band | T (K) | $\tilde{g}$ | ($b_2^0 = D$), ($b_2^2 = 3E$) ($10^{-4}$ cm$^{-1}$) | $b_4^0, b_4^2, b_4^4$ ($10^{-4}$ cm$^{-1}$) | $b_6^0, b_6^2, b_6^4, b_6^6$ ($10^{-4}$ cm$^{-1}$) | $A_x, A_y, A_z$ ($10^{-4}$ cm$^{-1}$) | References |
|---|---|---|---|---|---|---|---|---|
| $Pb_{1-x}Eu_xSe$ ($x = 0.013$) | X | 4.2 | $g_{iso} = 1.976$ | — | $b_4 = 0.278$ (GHz) | $b_6 = -0.003$ (GHz) | — | [534] |
| $Pb_{1-x}Eu_xSe$ ($x = 0.013$) | X | 110 | $g_{iso} = 1.975$ | — | $b_4 = 0.280$ (GHz) | $b_6 = 0.001$ (GHz) | — | [534] |
| $Pb_{1-x}Eu_xSe$ ($x = 0.013$) | X | 295 | $g_{iso} = 1.981$ | — | $b_4 = 0.275$ (GHz) | $b_6 = 0.003$ (GHz) | — | [534] |
| $Pb_{1-x}Eu_xS$ ($x \sim 0.016$) | 9.56 | 295 | $g_{iso} = 1.972$ | — | $b_4 = 0.438$ (GHz) | $b_6 = -0.019$ (GHz) | — | [535, 536] |
| $Pb_{1-x}Eu_xS$ ($x \sim 0.016$) | 9.56 | 4.2 | $g_{iso} = 1.975$ | — | $b_4 = 0.448$ (GHz) | $b_6 = -0.011$ (GHz) | — | [535, 536] |
| $Pb_{1-x}Eu_xSe$ (cubic site) | X | 300 | $g_{iso} = 1.990$ | — | $b_4 = 267.4$ MHz | $-3.4$ MHz | — | [537] |
| $PbWO_4$ | 9.447 | 295 | 1.9806, 1.9804, 1.9840 | 163.92, 0.21 | $B_4^0, B_4^4, B_4^{-4} =$ 1.00, 8.60, $-0.30$ | — | $A_\parallel^{151} = 34.2, A_\perp^{151} =$ 33.4, $A_\parallel^{153} = 15.3$, $A_\perp^{153} = 15.4$ | [529] |
| $SrAl_2O_4$ (Eu,Dy phosphor) (site I) | 179.4 | 145 | 1.989 | $|D| = 1400$, $|E/D| = 0.258$ | — | — | — | [538] |
| $SrAl_2O_4$ (Eu,Dy phosphor) (site II) | 179.4 | 145 | 1.989 | $|D| = 1120$, $|E/D| = 0.306$ | — | — | — | [538] |
| $SrAl_2O_4$ (site I) | X,W | | $g_\parallel = 1.9934, g_\perp =$ 1.9938 | —, 12, 93 | — | — | — | [539] |

*(continued)*

## (Eu$^{2+}$(4f$^7$) listing contd.)

| Host | Frequency (GHz)/band | T (K) | $g$ | ($b_2^0 = D$), ($b_2^2 = 3E$) ($10^{-4}$ cm$^{-1}$) | $b_4^0, b_4^2, b_4^4$ ($10^{-4}$ cm$^{-1}$) | $b_6^0, b_6^2, b_6^4, b_6^6$ ($10^{-4}$ cm$^{-1}$) | $A_x, A_y, A_z$ ($10^{-4}$ cm$^{-1}$) | References |
|---|---|---|---|---|---|---|---|---|
| SrAl$_2$O$_4$ (site II) | X,W | | $g_\parallel = 1.9912, g_\perp = 1.9930$ | 2, 84, 19 | — | — | — | [539] |
| SrAl$_2$O$_4$ (site III) | X,W | | $g_\parallel = 1.9917, g_\perp = 1.9921$ | 53, −22, −12 | — | — | — | [539] |
| SrAl$_2$O$_4$ (site IV) | X,W | | $g_\parallel = 1.9952, g_\perp = 1.9930$ | −10, 25, −64 | — | — | — | [539] |
| SrCl$_2$ | 9.447 | 295 | — | — | — | — | $A^{151} = 34.5, A^{153} = 15.5$ | [529] |
| SrCl$_2$ | X | 300 | 1.991 | — | $b_4 = -[14.7 + 0.218P]$; P is pressure (kBar) | $b_6 = 0.4$ | $A = [34.4 - 0.145P]$; P is pressure (kBar) | [540] |
| SrCl$_2$ | X | 77 | 1.991 | — | $b_4 = -16.3$ | $b_6 = 0.4$ | 33.5 | [540] |
| SrCl$_2$ | X | 4.2 | 1.991 | — | $b_4 = -16.8$ | $b_6 = 0.4$ | 33.8 | [540] |
| Sr$_4$Al$_{14}$O$_{25}$ (Eu,Dy phosphor) (site I) | 90 | 100 | 1.984 | $\|D\| = 1020, \|E/D\| = 0.333$ | — | — | — | [538] |
| Sr$_4$Al$_{14}$O$_{25}$ (Eu,Dy phosphor) (site II) | 90 | 100 | 1.984 | $\|D\| = 920, \|E/D\| = 0.333$ | — | — | — | [538] |

## Gd$^{3+}$. Data tabulation of SHPs
## Gd$^{3+}$ (4f$^7$)

| Host | Frequency (GHz)/ band | T (K) | $\tilde{g}$ | $(b_2^0 = D)$, $(b_2^2 = 3E)$ $(10^{-4}$ cm$^{-1})$ | $b_4^0, b_4^2, b_4^4$ $(10^{-4}$ cm$^{-1})$ | $b_6^0, b_6^2, b_6^4, b_6^6$ $(10^{-4}$ cm$^{-1})$ | $A_x, A_y, A_z$ $(10^{-4}$ cm$^{-1})$ | References |
|---|---|---|---|---|---|---|---|---|
| BaF$_2$ | X | 293 | 1.9921 | — | $b_4 = \pm 108.52$, —, — (MHz) | $b_6 = \pm 0.6$ MHz | — | [541] |
| BiVO$_4$ | Q | RT | 1.9812, 1.9764, 1.9753 | $D_{xx} = -195.3$; $D_{yy} = -368.0$; $D_{xx} = 563.3$; | — | — | — | [542] |
| BiVO$_4$ | X | RT | 1.9763, 1.9817, 1.9760 | 2533.5, 772.5 (MHz) | $-32.4, 3.0, -240.0$ | 9.1, 3.8, 12.6, 2.5 | — | [543] |
| BiVO$_4$ | X | 3.8–300 | 1.9753, 1.9764, 1.9812 | 845.0, 259.2 (MHz) | $-13.5, -6.9, -42.6$ | $-0.5, -5.0, 1.4, -1.5$ | — | [223] |
| BiVO$_4$ | 33.896 | RT | — | $-532.2, -1157.7$ | 10.9, 1.8, 73.8 | 0.4, 3.9, 4.0, $-1.0$ | — | [544] |
| [Ca$_{10}$(PO$_4$)$_6$(F)$_2$] (site A) | X | 295 | 1.9916, 1.9903, 1.9898 | $\frac{\tilde{D}}{g_e \mu_B}$ 67.06, $-134.43$ (G) | — | — | — | [545] |
| [Ca$_{10}$(PO$_4$)$_6$(F)$_2$] (site A) | W | 295 | 1.9916, 1.9903, 1.9898 | $\frac{\tilde{D}}{g_e \mu_B} = 67.37$, 67.06, $-134.43$ (G) | — | — | — | [546] |
| [Ca$_{10}$(PO$_4$)$_6$(F)$_2$] (site B) | W | 287 | 1.9916, 1.9910, 1.9861 | $\frac{\tilde{D}}{g_e \mu_B} = 664.3$, 51.7, $-716.0$ (G) | — | — | — | [546] |

(continued)

**$Gd^{3+}(4f^7)$ listing contd.**

| Host | Frequency (GHz)/band | T (K) | $\tilde{g}$ | $(b_2^0 = D)$, $(b_2^2 = 3E)$ $(10^{-4}\,cm^{-1})$ | $b_4^0, b_4^2, b_4^4$ $(10^{-4}\,cm^{-1})$ | $b_6^0, b_6^2, b_6^4, b_6^6$ $(10^{-4}\,cm^{-1})$ | $A_x, A_y, A_z$ $(10^{-4}\,cm^{-1})$ | References |
|---|---|---|---|---|---|---|---|---|
| $CdF_2$ | X | 77 | 1.992 | — | $b_4 = \pm 152.14, —,$ — (MHz) | $b_6 = \pm 0.0$ MHz | — | [541] |
| $CdF_2$ | X | 293 | 1.992 | — | $b_4 = \pm 129.66, —,$ — (MHz) | $b_6 = \pm 0.0$ MHz | — | [541] |
| $CdWO_4$ | 35 | 300 | 1.99, 1.989, 1.99 | 786, −276 | 6.6, −66, −96 | — | — | [547] |
| $CsCaF_3$ (cubic center) | X | RT | $g_{iso} = 1.992$ | — | $b_4^0 = -5.49$ | $b_6^0 = 0.89$ | — | [548] |
| $CsCdF_3$ (cubic center) | X | RT | $g_{iso} = 1.992$ | — | $b_4^0 = -4.82$ | $b_6^0 = 0.86$ | — | [548] |
| $CsLa(WO_4)_2$ | 35 | 300 | 1.991, 1.988, 1.989 | 975, 0 | −29.4, 9, 66 | — | — | [547] |
| $Cs_2CdF_4$ | X | RT | $g_{iso} = 1.9918$ | $b_2^0 = -2330.6$, $b_2^2 = 808.3$ | $b_4^0 = 1.6$, $b_4^2 = -2.8$, $b_4^4 = 0.5$ | $b_6^0 = 4.7, b_6^2 = 22$, $b_6^4 = 20.8, b_6^6 = 36.8$ | — | [549] |
| $Cs_2NaYCl_6$ | 9.088 | 473 | 1.9978 | — | −14.48, —, — | 0.6, —, —, — | — | [550] |
| $Cs_2NaYCl_6$ | 9.088 | 300 | 1.9988 | — | −15.6, —, — | 0.6, —, —, — | — | [550] |
| $Cs_2NaYCl_6$ | 9.088 | 6 | 1.9946 | — | −17.7, —, — | 0.6, —, —, — | — | [550] |
| $EuAl_3(BO_3)_4$ | X | 298 | 1.981 | $b_2^0 = 280.18$ | $b_4^0 = -12.95$ | $b_6^0 = 0.61$ | — | [551] |

(continued)

($Gd^{3+}$ ($4f^7$) listing contd.)

| Host | Frequency (GHz)/ band | T (K) | $\tilde{g}$ | ($b_2^0 = D$), ($b_2^2 = 3E$) ($10^{-4}$ cm$^{-1}$) | $b_4^0, b_4^2, b_4^4$ ($10^{-4}$ cm$^{-1}$) | $b_6^0, b_6^2, b_6^4, b_6^6$ ($10^{-4}$ cm$^{-1}$) | $A_x, A_y, A_z$ ($10^{-4}$ cm$^{-1}$) | References |
|---|---|---|---|---|---|---|---|---|
| Gd macrobicyclic complex | 9.5 | 4 | 2.07, 2.02, 2.05 | $D_{tensor}: D_{xx} = 93$ G, $D_{yy} = 114$ G, $D_{zz} = -236$ G | $B_4^0 = 1.1$ G, $B_4^1 =$ 1.8 G, $B_4^2 = 0.0$ G, $B_4^3 = 9.0$ G, $B_4^4 =$ 3.6 G, $B_4^{-4} =$ 0.0 G, $B_4^{-3} =$ −6.0 G, $B_4^{-2} =$ 0.0 G, $B_4^{-1} =$ −3.5 G | $B_6^0 = -0.025$ G, $B_6^1 =$ −0.450 G, $B_6^2 =$ −0.250 G, $B_6^3 =$ −0.650 G, $B_6^4 =$ 0.000 G, $B_6^5 =$ 0.000 G, $B_6^6 =$ 0.000 G, $B_6^{-6} =$ 0.050 G, $B_6^{-5} =$ 0.000 G, $B_6^{-3} =$ −0.300 G, $B_6^{-2} =$ −0.080 G, $B_6^{-1} =$ 0.460 G | — | [552] |
| $Gd(P_2W_{17}O_{61})_2^{17-}$ | 9.4 | RT | — | $b_2^0 = 1050$ (MHz) | — | — | — | [553] |
| $[GdP_5W_{30}O_{110}]^{12-}$ | 9.4 | RT | — | $b_2^0 = 1150$ (MHz) | — | — | — | [553] |
| $[GdSb_9W_{17}O_{86}]^{16-}$ | 9.4 | RT | — | $b_2^0 = 1250$ (MHz) | — | — | — | [553] |
| $Gd(SiW_{11}O_{39})_2^{13-}$ | 9.4 | RT | — | $b_2^0 = 1050$ (MHz) | — | — | — | [553] |
| $GdW_{10}O_{36}^{9-}$ | 9.4 | RT | — | $b_2^0 = 2400$ (MHz) | — | — | — | [553] |
| $KMgF_3$ (cubic center) | X | RT | $g_{iso} = 1.9916$ | — | $b_4^0 = -11.15$ | $b_6^0 = 1.20$ | — | [548] |

*(continued)*

$(Gd^{3+}(4f^7)$ listing contd.)

| Host | Frequency (GHz)/ band | T (K) | $\tilde{g}$ | $(b_2^0 = D)$, $(b_2^2 = 3E)$ $(10^{-4}\,cm^{-1})$ | $b_4^0, b_4^2, b_4^4$ $(10^{-4}\,cm^{-1})$ | $b_6^0, b_6^2, b_6^4, b_6^6$ $(10^{-4}\,cm^{-1})$ | $A_x, A_y, A_z$ $(10^{-4}\,cm^{-1})$ | References |
|---|---|---|---|---|---|---|---|---|
| $KSc(MoO4)_2$ | X | RT | $g_\parallel = 1.9918, g_\perp = 1.9918$ | 359 730, — | −89 282, —, — | 13 801, —, —, — (MHz) | — | [554] |
| $KY(WO_4)_2$ | X | RT | 1.9890, 1.9923, 1.9930 | −336.66, 671.67 | −18.79, 6025, −7.57 | — | — | [555] |
| $KY(WO_4)_2$ | 35 | 300 | 1.9836, 2.0080, 2.0082 | $B_2^0 = 7.456 \cdot 10^{-28}$ J, $B_2^1 = 5.4 \cdot 10^{-30}$ J, $B_2^2 = -2.408 \cdot 10^{-28}$ J | $B_4^0 = -6.46 \cdot 10^{-31}$ J, $B_4^1 = -9.5 \cdot 10^{-31}$ J, $B_4^2 = 7.3 \cdot 10^{-31}$ J, $B_4^3 = -6.7 \cdot 10^{-31}$ J, $B_4^4 = 2.28 \cdot 10^{-30}$ J | $B_6^0 = -8.92 \cdot 10^{-34}$ J, $B_6^1 = -3 \cdot 10^{-33}$ J, $B_6^2 = -9.1 \cdot 10^{-32}$ J, $B_6^3 = 3.4 \cdot 10^{-32}$ J, $B_6^4 = -6.2 \cdot 10^{-32}$ J, $B_6^5 = 1.1 \cdot 10^{-31}$ J, $B_6^6 = -1.77 \cdot 10^{-31}$ J | — | [556] |
| $KY(WO_4)_2$ | 35 | 300 | 1.99, 1.99, 2 | 1248, −540 | −22.8, −78, −38 | — | — | [547] |
| $KY_3F_{10}$ | X | RT | $g_\parallel = 1.986, g_\perp = 1.987$ | $b_2^0 = 816$ | $b_4^0 = -20, b_4^4 = 46$ | $b_6^0 = 0, b_6^4 = 8$ | — | [557] |
| $KZnF_3$ (cubic center) | X | RT | $g_{iso} = 1.992$ | — | $b_4^0 = -8.95$ | $b_6^0 = 1.13$ | — | [548] |
| $K_2YF_5$ (reference frame 1) | 9.5, 34 | 295 | g = 1.990 | 818, 186, $b_2^1 = 254$ | −17, −14, −19, $b_4^1 = -22, b_4^{-3} = -384$ | — | — | [508] |

(continued)

(Gd$^{3+}$(4f$^7$) listing contd.)

| Host | Frequency (GHz)/band | T (K) | $\tilde{g}$ | ($b_2^0 = D$), ($b_2^2 = 3E$) ($10^{-4}$ cm$^{-1}$) | $b_4^0, b_4^2, b_4^4$ ($10^{-4}$ cm$^{-1}$) | $b_6^0, b_6^2, b_6^4, b_6^6$ ($10^{-4}$ cm$^{-1}$) | $A_x, A_y, A_z$ ($10^{-4}$ cm$^{-1}$) | References |
|---|---|---|---|---|---|---|---|---|
| K$_2$YF$_5$ (reference frame 2) | 9.5, 34 | 295 | $g = 1.990$ | −502, −1134, $b_2^{-2} = -128$ | −11, −41, −65, $b_4^{-2} = -102$, $b_4^{-4} = 29$ | — | — | [508] |
| K$_2$YF$_5$ (reference frame 3) | 9.5, 34 | 295 | $g = 1.990$ | 822, 183, $b_2^1 = 0$ | −18, 24, −8 $b_4^1 = -5$, $b_4^3 = 379$ | — | — | [508] |
| K$_2$Zn(SO$_4$)$_2$ · 6H$_2$O | 9.4 | 293 | — | −731.85, −364.2 | 15.1, 1.98, −10.29 | 0.42, 0.54, 0.32, 0.73 | — | [558] |
| K$_2$Zn(SO$_4$)$_2$ · 6H$_2$O | 9.4 | 123 | — | 774.66, −401.85 | 14.18, 0.45, −11.97 | 0.43, 0.81, 1.68, −2.93 | — | [558] |
| LaNbO$_4$ | X | 8 | 2.001 2.001, 1.984 | −2.001, 0.514 (GHz) | 0.066, 0.068, 0.271 (GHz) | — | — | [559] |
| LaNbO$_4$ | X | 73 | 1.997 2.002, 1.987 | −1.982, 0.548 (GHz) | 0.062, 0.038, 0.245 (GHz) | — | — | [559] |
| LaNbO$_4$ | X | 295 | 1.992 1.994, 1.992 | −1.880, 0.907 (GHz) | 0.061, 0.034, 0.275 (GHz) | — | — | [559] |
| La$_2$Si$_2$O$_7$ | 9.61 | 295 | 1.999 1.998, 1.994 | −2.1999, 0.810 GHz, $b_2^{-1} = 0.003$ GHz | −0.10, 0.158, 0.086 (GHz), $b_4^{-1} = 0.017$ (GHz), $b_4^{-3} = 0.037$ GHz | — | — | [560] |
| La$_2$Si$_2$O$_7$ | 9.61 | 8 | 1.990 —, 1.960 | −2.278, 0.787 (GHz) | −0.012, —, —(GHz) | — | — | [560] |

(continued)

## $(Gd^{3+}(4f^7))$ listing contd.

| Host | Frequency (GHz)/band | T (K) | $\tilde{g}$ | ($b_2^0 = D$), ($b_2^2 = 3E$) ($10^{-4}$ cm$^{-1}$) | $b_4^0, b_4^2, b_4^4$ ($10^{-4}$ cm$^{-1}$) | $b_6^0, b_6^2, b_6^4, b_6^6$ ($10^{-4}$ cm$^{-1}$) | $A_x, A_y, A_z$ ($10^{-4}$ cm$^{-1}$) | References |
|---|---|---|---|---|---|---|---|---|
| La$_2$(WO$_4$)$_3$ | 35 | 300 | 1.999, 1.999, 1.999 | 861, −180 | 1.8, −1.2, −96 | — | — | [547] |
| Colquiriite LiCaAlF$_6$ | 9.3 | 300 | $g_\parallel = 1.9921, g_\perp = 1.9920$ | $B_2^0 = 233.86$ | $B_4^0 = 11059$, $B_4^3 = 7029$, $B_4^{-3} = -1193$ | $B_6^0 = -373$, $B_6^3 = -11, B_6^{-3} = 68$, $B_6^6 = 207$, $B_6^{-6} = -114$ | — | [561] |
| MgO | X | 77 | 1.992 | — | $b_4 = -104.78, -,$ — (MHz) | $b_6 = 6.60$ MHz | — | [541] |
| MgO | X | 293 | 1.992 | — | $b_4 = -100.81, -,$ — (MHz) | $b_6 = 6.37$ MHz | — | [541] |
| MgO with 0.8 mol% Gd | 9.50 | RT | 1.9920 | $B_4 = -0.56$ | $B_6 = 0.00017$ | — | — | [562] |
| PbSe | X | 4.2 | 1.990 | | 8.07 | $b_6 = -1.0$ (MHz) | — | [528] |
| PbTe | X | 77 | 1.991 | — | $b_4 = -107.7, -,$ — (MHz) | $b_6 = 2.68$ MHz | — | [541] |
| PbTe | X | 293 | 1.989 | — | $b_4 = -95.5, -,$ — (MHz) | $b_6 = 2.00$ MHz | — | [541] |
| PbWO$_4$ | X | 26 | $g_\parallel = 1.9916, g_\perp = 1.9913$ | −852.89, — | $B_4^0, B_4^2, B_4^4$ (MHz) = −0.914, −4.91, −4.89 | $B_6^0 = 8.6 \cdot 10^{-4}$ MHz, $B_6^4 = -5 \cdot 10^{-4}$ MHz, $B_6^{-4} = 96 \cdot 10^{-4}$ MHz | $^{155}A_\parallel = 8.45$ MHz, $^{155}A_\perp = 12.21$ MHz, $^{157}A_\parallel = 16.11$ MHz, $^{157}A_\perp = 11.18$ MHz | [563] |

(continued)

## (Gd$^{3+}$(4f$^7$)) listing contd.)

| Host | Frequency (GHz)/ band | T (K) | $\tilde{g}$ | $(b_2^0 = D)$, $(b_2^2 = 3E)$ $(10^{-4}\,cm^{-1})$ | $b_4^0, b_4^2, b_4^4$ $(10^{-4}\,cm^{-1})$ | $b_6^0, b_6^2, b_6^4, b_6^6$ $(10^{-4}\,cm^{-1})$ | $A_x, A_y, A_z$ $(10^{-4}\,cm^{-1})$ | References |
|---|---|---|---|---|---|---|---|---|
| PbWO$_4$ | X | 293 | $g_\parallel = 1.9918$, $g_\perp = 1.9916$ | $-821.67$, — | $B_4^0, B_4^2, B_4^4 =$ (MHz) $-0.855$, $-4.65, -4.76$ | $B_6^0 = 9.0 \cdot 10^{-4}$ MHz, $B_6^4 = 9 \cdot 10^{-4}$ MHz, $B_6^{-4} = 85 \cdot 10^{-4}$ MHz | $^{155}A_\parallel = 8.45$ MHz, $^{155}A_\perp = 12.21$ MHz, $^{157}A_\parallel = 16.11$ MHz, $^{157}A_\perp = 11.18$ MHz | [563] |
| Pb(Zr$_{1-x}$Ti$_x$)O$_3$ (x = 0.001) (site B) | 9.45 | 77 | 4.852, 4.852, 8.552 | 51.38, 7.47 | 56.05, 56.05, 65.39 | — | — | [564] |
| Pb(Zr$_{1-x}$Ti$_x$)O$_3$ (x = 0.001) (site A) | 9.45 | 77 | 6.020, 6.020, 8.134 | 43.91, 18.68 | 27.09, 27.09, 45.77 | — | — | [564] |
| Pb(Zr$_{1-x}$Ti$_x$)O$_3$ (x = 0.005) (site B) | 9.45 | 77 | 4.771, 4.771, 8.552 | 53.25, 14.01 | 56.05, 56.05, 66.33 | — | — | [564] |
| Pb(Zr$_{1-x}$Ti$_x$)O$_3$ (x = 0.005) (site A) | 9.45 | 77 | 6.892, 6.892, 8.634 | 42.04, 19.62 | 35.90, 36.90, 56.05 | — | — | [564] |
| Pb(Zr$_{1-x}$Ti$_x$)O$_3$ (x = 0.010) (site B) | 9.45 | 77 | 4.852, 4.852, 8.850 | 55.12, 14.01 | 51.38, 51.38, 69.60 | — | — | [564] |
| Pb(Zr$_{1-x}$Ti$_x$)O$_3$ (x = 0.010) (site A) | 9.45 | 77 | 6.920, 6.920, 7.034 | 42.04, 17.75 | 35.50, 35.50, 55.12 | — | — | [564] |

(continued)

## ($Gd^{3+}$ ($4f^7$) listing contd.)

| Host | Frequency (GHz)/band | T (K) | $\tilde{g}$ | ($b_2^0 = D$), ($b_2^2 = 3E$) ($10^{-4}$ cm$^{-1}$) | $b_4^0, b_4^2, b_4^4$ ($10^{-4}$ cm$^{-1}$) | $b_6^0, b_6^2, b_6^4, b_6^6$ ($10^{-4}$ cm$^{-1}$) | $A_x, A_y, A_z$ ($10^{-4}$ cm$^{-1}$) | References |
|---|---|---|---|---|---|---|---|---|
| $Pb(Zr_{1-x}Ti_x)O_3$ ($x = 0.015$) (site B) | 9.45 | 77 | 4.811, 4.811, 7.560 | 50.45, 19.62 | 46.71, 46.71, 47.64 | — | — | [564] |
| $Pb(Zr_{1-x}Ti_x)O_3$ ($x = 0.015$) (site A) | 9.45 | 77 | 6.892, 6.892, 8.634 | 44.84, 18.68 | 23.82, 23.82, 42.04 | — | — | [564] |
| $Pr(CH_3COO)_3 \cdot H_2O$ | 9.6 | 295 | $g_\parallel = 1.980, g_\perp = 2.026$ | −1.407, 0.186 (GHz) | −0.014, −0.141, −0.059 GHz | 0.000, 0.132, 0.099, 0.180 GHz | — | [565] |
| $Pr(CH_3COO)_3 \cdot H_2O$ | 9.6 | 110 | $g_\parallel = 1.952, g_\perp = 1.982$ | −1.402, 0.160 (GHz) | −0.033, −0.185, −0.071 (GHz) | 0.025, 0.169, −0.055, −0.039 (GHz) | — | [565] |
| $Pr(CH_3COO)_3 \cdot H_2O$ | 9.6 | 4.2 | $g_\parallel = 1.986, g_\perp = 2.077$ | −1.438, 0.117 (GHz) | −0.020, −0.526, 0.116 (GHz) | −0.009, 0.080, −0.354, −0.748 (GHz) | — | [565] |
| $PrNbO_4$ | X | 8 | 1.997, 2.016, 2.002 | −2.012, 0.660 (GHz) | 0.068, 0.046, 0.308 (GHz) | — | — | [559] |
| $PrNbO_4$ | X | 73 | 1.994, 2.012, 1.989 | −2.003, 0.723 (GHz) | 0.066, 0.041, 0.319 (GHz) | — | — | [559] |
| $PrNbO_4$ | X | 295 | 1.998, 1.995, 1.992 | −1.962, 0.947 (GHz) | 0.066, 0.043, 0.308 (GHz) | — | — | [559] |
| $RbCaF_3$ (cubic center) | X | RT | $g_{iso} = 1.992$ | — | $b_4^0 = -4.92$ | $b_6^0 = 0.83$ | — | [548] |
| $RbCdF_3$ (cubic center) | X | RT | $g_{iso} = 1.992$ | — | $b_4^0 = -4.44$ | $b_6^0 = 0.82$ | — | [548] |

(continued)

## $(Gd^{3+}(4f^7)$ listing contd.)

| Host | Frequency (GHz)/band | T (K) | $\tilde{g}$ | $(b_2^0 = D),$ $(b_2^2 = 3E)$ $(10^{-4}\,cm^{-1})$ | $b_4^0, b_4^2, b_4^4$ $(10^{-4}\,cm^{-1})$ | $b_6^0, b_6^2, b_6^4, b_6^6$ $(10^{-4}\,cm^{-1})$ | $A_x, A_y, A_z$ $(10^{-4}\,cm^{-1})$ | References |
|---|---|---|---|---|---|---|---|---|
| RbEu(SO$_4$)$_2 \cdot$ 4H$_2$O | X | RT | $g_\| = 1.998, g_\perp =$ 1.990 | $b_2^0 = 0.360,$ $b_2^2 = -0.213$ | $b_4^0 = -0.001,$ $b_4^2 = -0.041, b_4^4 =$ $-0.028$ (GHz) | $b_6^0 = -0.002,$ $b_6^2 = 0.032,$ $b_6^4 = 0.030,$ $b_6^6 = -0.005$ (GHz) | — | [566] |
| RbNd(SO$_4$)$_2 \cdot$ 4H$_2$O | X | RT | $g_\| = 1.998, g_\perp =$ 1.988 | $b_2^0 = 0.457,$ $b_2^2 = -0.319$ | $b_4^0 = -0.004,$ $b_4^2 = -0.012, b_4^4 =$ $-0.019$ (GHz) | $b_6^0 = -0.001,$ $b_6^2 = 0.002,$ $b_6^4 = -0.018,$ $b_6^6 = -0.015$ (GHz) | — | [566] |
| RbPb$_2$Cl$_5$ | X | RT | $g = 1.991$ | $b_2^0 = -850$ MHz, $b_2^1 = 25$ MHz, $b_2^2$ $= -300$ MHz, $c_2^1$ $= 160$ MHz | $b_4^0 = -10$ MHz, $b_4^1 = 5$ MHz, $b_4^2 =$ $-50$ MHz, $b_4^3 =$ $-280$ MHz, $b_4^4 =$ $-140$ MHz, $c_4^1 =$ $40$ MHz, $c_4^2 =$ $-30$ MHz, $c_4^3 =$ $80$ MHz, $c_4^4 = 10$ MHz, $F = 37$ MHz | — | | [567] |
| RbPr(SO$_4$)$_2 \cdot$ 4H$_2$O | X | RT | $g_\| = 2.001, g_\perp =$ 1.986 | $b_2^0 = 0.526,$ $b_2^2 = -0.392$ | $b_4^0 = -0.008,$ $b_4^2 = -0.019,$ $b_4^4 = -0.017$ (GHz) | $b_6^0 = 0.001,$ $b_6^2 = -0.027,$ $b_6^4 = -0.053,$ $b_6^6 = -0.015$ (GHz) | — | [566] |
| RbSm(SO$_4$)$_2 \cdot$ 4H$_2$O | X | RT | $g_\| = 1.998, g_\perp =$ 1.987 | $b_2^0 = 0.388,$ $b_2^2 = -0.224$ | $b_4^0 = -0.002,$ $b_4^2 = -0.032,$ $b_4^4 = -0.020$ (GHz) | $b_6^0 = -0.001,$ $b_6^2 = 0.017,$ $b_6^4 = -0.002,$ $b_6^6 = -0.020$ (GHz) | — | [566] |

(continued)

## ($Gd^{3+}$ ($4f^7$) listing contd.)

| Host | Frequency (GHz)/band | T (K) | $\tilde{g}$ | ($b_2^0 = D$), ($b_2^2 = 3E$) ($10^{-4}$ cm$^{-1}$) | $b_4^0, b_4^2, b_4^4$ ($10^{-4}$ cm$^{-1}$) | $b_6^0, b_6^2, b_6^4, b_6^6$ ($10^{-4}$ cm$^{-1}$) | $A_x, A_y, A_z$ ($10^{-4}$ cm$^{-1}$) | References |
|---|---|---|---|---|---|---|---|---|
| c-RbZnF$_3$ (tetrahedral center) | X | RT | $g_{iso} = 1.992$ | — | $b_4^2 = -0.032$, $b_4^4 = 72.8$ | $b_6^0 = 0.98$, $b_6^6 = 23.8$ | — | [548] |
| c-RbZnF$_3$ (cubic center) | X | RT | $g_{iso} = 1.992$ | — | $b_4^0 = -8.03$ | $b_6^0 = 1.07$ | — | [548] |
| c-RbZnF$_3$ (tetragonal center) | X | RT | $g_{iso} = 1.992$ | $b_2^0 = -347$ | $b_4^0 = -4.42$, $b_4^4 = -72$ | $b_6^0 = 0.98$, $b_6^6 = -23.8$ | — | [548] |
| h-RbZnF$_3$ (trigonal center) | X | RT | $g_{iso} = 1.992$ | $b_2^0 = 256.33$ | $b_4^0 = -9.1$, $b_4^3 = 85$ | $b_6^0 = -1.7$, $b_6^3 = -26$, $b_6^6 = -17$ | — | [548] |
| Rb$_2$CdF$_4$ | X | RT | $g_{iso} = 1.9918$ | $b_2^0 = -2373.1$, $b_2^2 = 788.9$ | $b_4^0 = 1.3$, $b_4^2 = -5.7$, $b_4^4 = 3.5$ | $b_6^0 = -4.8$, $b_6^2 = 21$, $b_6^4 = -15.2$, $b_6^6 = 30.4$ | — | [549] |
| Rb$_2$ZnF$_4$ | X | RT | $g_{iso} = 1.9918$ | $b_2^0 = -2520.5$, $b_2^2 = 903.2$ | $b_4^0 = 3.8$, $b_4^2 = -8$, $b_4^4 = 13$ | $b_6^0 = 0.3$, $b_6^2 = -6$, $b_6^4 = -18$, $b_6^6 = -7$ | — | [549] |
| TlCdF$_3$ (center III) | X | 300 | $g = 1.992$ | $-201.5$, — | $b_4^0 = -1.96$, $b_4^4 = -33.9$ | $b_6^0 = 0.7$, $b_6^6 = -23$ | — | [568] |
| TlCdF$_3$ (center IV) | X | 300 | $g = 1.992$ | $-153.6$, — | $b_4^0 = -2.88$, $b_4^4 = -24.0$ | $b_6^0 = 0.76$, $b_6^6 = -12$ | — | [568] |
| Tl$_2$ZnF$_4$ (site A) | X | RT | 1.992 | $-557.6$,- | $-3.8$, —, — | $0.8$, —, $-18.6$, — | — | [569] |

*(continued)*

## 4.2 Listing of Spin-Hamiltonian Parameters

($Gd^{3+}$ ($4f^7$) listing contd.)

| Host | Frequency (GHz)/ band | T (K) | $\tilde{g}$ | ($b_2^0 = D$), ($b_2^2 = 3E$) ($10^{-4}$ cm$^{-1}$) | $b_4^0, b_4^2, b_4^4$ ($10^{-4}$ cm$^{-1}$) | $b_6^0, b_6^2, b_6^4, b_6^6$ ($10^{-4}$ cm$^{-1}$) | $A_x, A_y, A_z$ ($10^{-4}$ cm$^{-1}$) | References |
|---|---|---|---|---|---|---|---|---|
| $Tl_2ZnF_4$ (site B) | X | RT | 1.992 | −288.9,- | 1.1, —, — | 0.4, —, −2.0, — | — | [569] |
| $Tl_2ZnF_4$ (site C) | X | RT | 1.992 | −465.7, −256.7 | 0, 7.1, 27 | — | — | [569] |
| $YAl_3(BO_3)_4$ | X | 4–300 | 1.9860, 1.9860, 1.9893 | D = 386.15 | $b_4^0 = -12.65$ | $b_6^0 = 0.35$ | — | [570] |
| $Y_{0.03}Cd_{0.97}F_{2.03}$ (site I) | X | 295 | 1.988, 1.991, 1.991 | 345, −87, $b_2^1 =$ −287 (MHz) | −25, −17, 30, $b_4^1$ = −3, $b_4^3$ = 554 (MHz) | $b_6^0$ = 2.4 MHz, $b_6^1$ = 7 MHz, $b_6^2$ = 4 MHz, $b_6^3$ = −1 MHz, $b_6^4$ = 5 MHz, $b_6^5$ = 60 MHz, $b_6^6$ = −7 MHz | — | [571] |
| $Y_{0.03}Cd_{0.97}F_{2.03}$ (site II) | X | 295 | 1.992, 1.992, 1.992 | 190, — (MHz)) | 28.35, —, —, $b_4^3$ = −2600 (MHz) | $b_6^0$ = 17 MHz, $b_6^3$ = 300 MHz | — | [571] |
| $YGdTi_2O_7$ | X | RT | 2.0193 | $b_2^0$ = 613.58, $b_2^2$ = 347.89 | $b_4^0 = -11.47$, $b_4^4 = 92.02$ | $b_6^0 = 7.43$ | — | [572] |
| $ZnWO_4$ | Q | RT | 1.9835, 1.9685, 1.9638 | D = 644.88 G, E = 161.49 G | $B_4^0 = -0.23$ G, $B_4^1$ = 0.22 G, $B_4^2$ = 0.46 G, $B_4^3$ = 2.26 G, $B_4^4$ = 1.36 G, $C_4^1$ = −0.17 G, $C_4^2$ = −0.20 G, $C_4^3$ = 2.47 G, $C_4^4$ = 0.36 G | — | — | [573] |
| $ZrSiO_4$ | X | RT | 1.9915, 1.991, 1.9919 | $B_2^0 = -379.92$ (G) | $B_4^0 = 8.07$, $B_4^4 = 71.32$ (G) | $B_6^0 = 0.67$, $B_6^4 = 0.45$ (G) | — | [574] |

## Dy$^{3+}$ (4f$^9$)

| Host | Frequency (GHz)/ band | T (K) | $\tilde{g}$ | $A_x, A_y, A_z$ ($10^{-4}$ cm$^{-1}$) | References |
|---|---|---|---|---|---|
| Bi$_4$Ge$_3$O$_{12}$ | X | 5 | $g_\parallel = 0.4$, $\lvert g_\perp \rvert = 9.349$ | $A_\parallel = -11$ ($^{161}$Dy), 16 ($^{163}$Dy) $\lvert A_\perp \rvert = 259$ ($^{161}$Dy), 370 ($^{163}$Dy) | [575] |
| LiYF$_4$ | X | 2 | $g_\parallel = 1.18$, $g_\perp = 6.8$ | — | [576] |
| PbGa$_2$S$_4$ ($^{163}$Dy) | X | 4.2–10 | $g_\parallel = 15.06$, $g_\perp = 2.47$ | $^{163}$Dy: $A_\parallel = 675$, $A_\perp = 111$, $^{161}$Dy: $A_\parallel = 472$, $A_\perp = 77$ | [577] |
| ZrO$_2$ (Y$_2$O$_3$ stabilized) | X | 10 | $g_\parallel \sim 17.5$, $g_\perp \sim 8$ | — | [578] |

## 4f$^{10}$ (Ho$^{3+}$), S = 1/2
Ho$^{3+}$. Data tabulation of SHPs
### Ho$^{3+}$ (4f$^{10}$)

| Host | Frequency (GHz)/ band | T (K) | $\tilde{g}$ | $A_x, A_y, A_z$ ($10^{-4}$ cm$^{-1}$) | References |
|---|---|---|---|---|---|
| MgSiO$_4$ (foresterite) | 65–535 | 4.2 | —, —, 18.5 | 12.3 GHz | [579] |

## 4f$^{11}$ (Er$^{3+}$), S = 1/2 (Kramers ion)
Er$^{3+}$. Data tabulation of SHPs
### Er$^{3+}$ (4f$^{11}$)

| Host | Frequency (GHz)/ band | T (K) | $\tilde{g}$ | $A_x, A_y, A_z$ ($10^{-4}$ cm$^{-1}$) | References |
|---|---|---|---|---|---|
| α-Al$_2$O$_3$ (center I) | 9.3 | 5 | $g_\parallel = 12.176$, $g_\perp = 4.14$ | — | [580] |
| α-Al$_2$O$_3$ (center II) | 9.3 | 5 | $g_\parallel = 17.2$, $g_\perp = 3.92$ | — | [580] |
| Bi$_2$Te$_3$ | Magnetization measurements | 4.2 | $g_\parallel = 13.71$, $g_\perp = 0.0007$ | — | [581] |

(continued)

## ($Er^{3+}(4f^{11})$ listing contd.)

| Host | Frequency (GHz)/ band | T (K) | $\tilde{g}$ | $A_x, A_y, A_z$ ($10^{-4}$ cm$^{-1}$) | References |
|---|---|---|---|---|---|
| KY(WO$_4$)$_2$ | X | 4.2–300 | 0, 3.378, 13.25 | 0.0, −120.75, −469.0 | [582] |
| LiNbO$_3$ | X | 4 | $g_{xx} = 0.546$, $g_{yy} = 1.356$, $g_{zz} = 15.093$, $g_{xy} = -0.293$, $g_{xz} = -0.700$, $g_{yz} = 0.456$ | — | [583] |
| LiNbO$_3$ | X | 4.2 | $g_\| = 15.13$, $g_\perp = 2.14$ | ($^{167}$Er) $A_{iso} = 77$ G | [584] |
| LiNbO$_3$ (center 1) | X | 4 | $g_\| = 14.44$, $g_\perp = 2.11$ | $^{167}$Er: $A_\| = 7.35$ mT | [522] |
| LiNbO$_3$ (center 2) | X | 4 | $g_\| = 14.44$, $g_\perp = 3.136$ | $^{167}$Er: $A_\| = 7.73$ mT | [522] |
| LiYF$_4$ | X | 15 | $g_\| = 3.130$, $g_\perp = 7.927$ | $A_\| = 325$ MHz, $A_\perp = 816$ MHz | [511] |
| 6H-SiC single crystal (low symmetry 1) | 9.3 | 12 | 12.2, 3.35, 1.5 | | [585] |
| 6H-SiC single crystal (low symmetry 2) | 9.3 | 12 | 10.6, 6.16, 1.26 | | [585] |
| 6H-SiC single crystal (low symmetry 3) | 9.3 | 12 | 9.6, 7.52, 1.45 | | [585] |
| 6H-SiC single crystal (axial center 1) | 9.3 | 12 | $g_\| = 1.359$, $g_\perp = 10.251$ | | [585] |
| 6H-SiC single crystal (axial center 2) | 9.3 | 12 | $g_\| = 1.073$, $g_\perp = 8.284$ | | [585] |
| 6H-SiC single crystal (axial center 3) | 9.3 | 12 | $g_\| = 1.164$, $g_\perp = 8.071$ | | [585] |
| 6H-SiC single crystal (axial center 4) | 9.3 | 12 | $g_\| = 0.776$ | | [585] |
| SrLaAlO$_4$ Kramers doublet (tetragonal site) | 9.2 | 4.2 | $g_\| = 2.7$, $g_\perp = 8.5$ | — | [586] |
| SrLaAlO$_4$ Kramers doublet (orthogonal site) | 9.2 | 4.2 | 9.4, 6.8, 2.3 | — | [586] |

(continued)

## ($Er^{3+}$($4f^{11}$) listing contd.)

| Host | Frequency (GHz)/ band | T (K) | $\tilde{g}$ | $A_x, A_y, A_z$ ($10^{-4}$ cm$^{-1}$) | References |
|---|---|---|---|---|---|
| YAlO$_3$ | X | 12 | 8.925, 8.038, 2.896 | 312, 311, 230 | [360] |
| YAlO$_3$ | 9.204 | 12 | 2.810, 8.162, 9.213 | ($^{167}$Er) 280, 335, 350 | [515] |
| YAl$_3$(BO$_3$)$_4$ | X | 16 | $g_\parallel = 1.348$, $g_\perp = 9.505$ | $A_\parallel = 58.5$, $A_\perp = 333.3$ | [516] |
| ($^{167}$Er) Y(NO$_3$) · 6H$_2$O | ~9.45 | 4.2 | 1.0011, 0.6347, 0.5693 | ($^{167}$Er) 1.3655, 0.7511, 0.5025 (GHz) | [517] |
| Y$_2$SiO$_5$ (site I) | 9.5 | 7 | 0.00, 1.79, 14.83 | 14.54, 106.44, 512.59; $\|Q_x\| = 9.523$, $\|Q_y\| = 0.430$, $\|Q_z\| = 9.957$ | [587] |
| Y$_2$SiO$_5$ (site II) | 9.5 | 7 | 0.55, 1.70, 15.54 | 2.10, 132.76, 518.69; $\|Q_x\| = 18.980$, $\|Q_y\| = 0.901$, $\|Q_z\| = 19.88$ | [587] |
| Y$_2$SiO$_5$ | X | RT | 0.00, 1.79, 14.83 (absolute values) | 43.6, 319.1, 1536.7 (absolute values, in MHz) | [588] |
| Y$_2$(SO$_4$)$_3$ · 8H$_2$O (site 1) | ~9.4 | <10 | 0.9638, 0.9545, 0.8608 | ($^{167}$Er) 1.651, 0.789, 0.166 (GHz) | [517] |
| Y$_2$(SO$_4$)$_3$ · 8H$_2$O (site 2) | ~9.4 | <10 | 1.1552, 1.1023, 1.0140 | ($^{167}$Er) 0.913, 0.487, 0.244 (GHz) | [517] |

$4f^{13}$ (Tm$^{3+}$, Yb$^{3+}$); $S = 1/2$ (Kramers ion)
Tm$^{3+}$. Data tabulation of SHPs
Tm$^{3+}$ ($4f^{13}$)

| Host | Frequency (GHz) band | T (K) | $\tilde{g}$ | References |
|---|---|---|---|---|
| KTm(MoO$_4$)$_2$ (Kramers doublet) | 10–190 | 4.2 | $g_c = 13.87$ | [589] |

## $Yb^{3+}$ Data tabulation of SHPs
## $Yb^{3+}(4f^{13})$

| Host | Frequency (GHz) band | T (K) | $\widetilde{g}$ | $A_x, A_y, A_z$ ($10^{-4}$ cm$^{-1}$) | References |
|---|---|---|---|---|---|
| $CaF_2$ | 9.5 | 30 | 3.433 | $A^{171} = 952$, $A^{173} = 260$ | [590] |
| $KTb_{0.2}Yb_{0.8}(WO_4)_2$ | 9.5 | <10 | 4.512, 0.772, 6.4410 | 2706, 2254, 3911 | [591] |
| $KYb(WO_4)_2$ | 9.5 | <10 | 1.532, 0.820, 7.058 | 1700, 1100, 2900 | [591] |
| $KY_3F_{10}$ | 9.69 | 5 | $g_\| = 5.363$, $g_\perp = 1.306$ | $A_\| = 4280$ (MHz) $A_\perp = 1100$ (MHz) | [526] |
| $LiNbO_3$ ($^{171}$Yb) | X | 8 | 3.3260, 2.4455, 0.0471 | 13.178, 2.952, −2.089 (GHz) | [592] |
| $LiNbO_3$ ($^{170}$Yb) (site I) | X | 8 | 4.6075, 3.5945, 1.3980 | — | [592] |
| $LuF_3$ | — | — | 3.26, 1.96, 3.44 | 2620, 1570, 2760 (MHz) | [511] |
| $LiYF_4$ | X | 15 | $g_\| = 1.330$, $g_\perp = 3.903$ | $A_\| = 1040$ MHz, $A_\perp = 3052$ MHz | [511] |
| $YAl_3(BO_3)_4$ | X | 9 | $g_\| = 3.612$, $g_\perp = 1.702$ | $A_\| = 958$, $A_\perp = 454$ | [516] |
| $Y(NO_3) \cdot 6H_2O$ | ~9.45 | 2.5 | 1.0518, 0.7035, 0.5678 | ($^{173}$Yb) −1.8549, −1.1606, −0.4721 (GHz) | [517] |
| $Y(NO_3) \cdot 6H_2O$ | 9.45 | 2.5 | 1.1048, 0.6770, 0.6264 | ($^{171}$Yb) 1.6077, −1.4470, −0.7289 (GHz) | [517] |
| $YF_3$ | X | 15 | 2.42, 1.76, 5.41 | 1940, 1400, 4330 (MHz) | [511] |
| $Y_2(SO_4)_3 \cdot 8H_2O$ (site I) | ~9.45 | 4.3 | 1.0921, 1.0457, 0.9090 | ($^{173}$Yb) 1.184, 0.847, 0.548 (GHz) | [517] |
| $Y_2(SO_4)_3 8H_2O$ (site II) | ~9.45 | 4.3 | 1.0847, 1.0294, 0.9062 | ($^{173}$Yb) 0.973, 0.591, 0.344 (GHz) | [517] |
| $Y_2(SO_4)_3 \cdot 8H_2O$ (site I) | 9.45 | 4.3 | 0.8822, 0.8210, 0.7669 | ($^{171}$Yb) 1.560, 1.551, 0.202 (GHz) | [517] |
| ($^{171}$Yb) $Y_2(SO_4)_3 \cdot 8H_2O$ (site II) | 9.45 | 4.3 | 1.1765, 1.0294, 0.8213 | ($^{171}$Yb) 0.942, 0.603, 0.314 (GHz) | [517] |

## $5s^25p^1$ ($Sb^{2+}$)
## $Sb^{2+}$. Data tabulation of SHPs
## $Sb^{2+}$ ($5s^25p^1$)

| Host | Frequency (GHz) band | T (K) | $\widetilde{g}$ | $A_x, A_y, A_z$ ($10^{-4}$ cm$^{-1}$) | References |
|---|---|---|---|---|---|
| $Sn_2P_2S_6$ | 9.5 | 30 | 1.810, 1.868, 1.887 | $^{121}$Sb=$^{123}$Sb: 1404, 1687, 1849 (MHz) | [593] |

## References

1. Rizzotto, M., Moreno, V., Signorella, S., Daier, V., and Sala, L.F. (2000) *Polyhedron*, **19**, 417.
2. Mazur, M. and Volko, M. (2002) *Phys. Chem. Glasses*, **43**, 237.
3. Ruzay, E., Reyherzx, H.J., Trokssy, J., and Wohlecke, M. (1998) *J. Phys. Condens. Matter*, **10**, 4297.
4. Carver, G., Bendix, J., and TregennaPiggott, P.L.W. (2002) *Chem. Phys.*, **282**, 245.
5. Brant, A.T., Yang, S., Giles, N.C., and Halliburton, L.E. (2011) *J. Appl. Phys.*, **110**, 053714.
6. Yang, S. and Halliburton, L.E. (2010) *Phys. Rev. B*, **81**, 35204.
7. Yang, S., Brant, A.T., and Halliburton, L.E. (2010) *Phys. Rev. B*, **82**, 35209.
8. Wang, G., Gallagher, H.G., Han, T.P.J., Henderson, B., Yamaga, M., and Yosida, T. (1997) *J. Phys. Condens. Matter*, **9**, 1649.
9. Tennant, W.C. and Claridge, R.F.C. (1999) *J. Magn. Reson.*, **137**, 122.
10. Edwards, G.J., Gilliam, O.R., Bartram, R.H., Watterich, A., Voszka, R., Niklas, J.R., GreulichWeber, S., and Spaeth, J.M. (1995) *J. Phys. Condens. Matter*, **7**, 3013.
11. Geifman, I.N., Nagorniy, P.G., and Rotenfeld, M.V. (1997) *Ferroelectrics*, **192**, 87.
12. Zolnierkiewicz, G., Typek, J., Guskos, N., and Bosacka, M. (2008) *Appl. Magn. Reson.*, **34**, 101.
13. Ikram, M., Ahmed, H., Mendes, P., Mir, F.A., Bashir, A., Paila, A., Rossi, A.M., and Eon, J.G. (2007) *Mod. Phys. Lett. B*, **21**, 1489.
14. Gefman, I.N., Golovina, I.S., and Nagorny, P.G. (1998) *Phys. Solid State*, **40**, 491.
15. Takeuchi, H., Ebisu, H., and Arakawa, M. (2008) *J. Phys. Condens. Matter*, **20**, 055221.
16. Luca, V., Thomson, S., and Howe, R.F. (1997) *J. Chem. Soc., Faraday Trans.*, **93**, 2195.
17. Sharma, S., Kumar, A., Chand, P., Sharma, B.K., and Sarkar, S. (2006) *Spectrochim. Acta, Part A*, **63A**, 556.
18. Magon, C.J., Lima, J.F., Donoso, J.P., Lavayen, V., Benavente, E., Navas, D., and Gonzalez, G. (2012) *J. Magn. Reson.*, **222**, 26.
19. Bunton, P.H., Baker, D.B., Engquist, D.E., Klemm, M., Horn, S., Yang, S., Evans, S.M., and Halliburton, L.E. (2009) *Solid State Commun.*, **149**, 1818.
20. Nascimento, O.R., Magon, C.J., Lima, J.F., Donoso, J.P., Benavente, E., Paez, J., Lavayen, V., Santa Ana, M.A., and Gonzalez, G. (2008) *J. Sol-Gel Sci. Technol.*, **45**, 195–204.
21. Ardelean, I., Cozar, O., Vedeanu, N., Rusu, D., and Andronache, C. (2007) *J. Mater. Sci. — Mater. Electron.*, **18**, 963.
22. Garces, N.Y., Stevens, K.T., Foundos, G.K., and Halliburton, L.E. (2004) *J. Phys. Condens. Matter*, **16**, 7095.
23. Bogomolova, L.D., Jachkin, V.A., and Krasil'nikova, N.A. (1998) *J. Non-Cryst. Solids*, **241**, 13.
24. Ramadevudu, G., Shareefuddin, M.D., Chary, M.N., and Rao, M.L. (2008) *Mod. Phys. Lett. B*, **22**, 1579.
25. Cozar, O., Ardelean, I., Simon, V., David, L., Mih, V., and Vedean, N. (1999) *Appl. Magn. Reson.*, **16**, 529.
26. Hameed, A., Ramadevudu, G., Laksmisrinivasa Rao, S., Shareefuddin, M., and Narasimha Chary, M. (2012) *New J. Glass Ceram.*, **2**, 51.
27. Yerli, Y., Zerentuerk, A., and Oezdogan, K. (2007) *Spectrochim. Acta, Part A*, **68A**, 147.
28. Krambrock, K., Guedes, K.J., and Pinheiro, M.V.B. (2008) *Phys. Chem. Miner.*, **35**, 409.
29. Dwivedi, P., Kripal, R., and Shukla, S. (2010) *Chin. Phys. Lett.*, **27**, 017601.
30. Kalkan, H. and Koksal, F.K. (1998) *Phys. Status Solidi B*, **205**, 651.
31. Padiyan, P.D., Muthukrishnan, C., and Murugesan, R. (2003) *J. Mol. Struct.*, **648**, 1.
32. Kalkan, H. and Koksal, F.K. (1998) *Solid State Commun.*, **105**, 307.
33. Deepa, S., Velavan, K., Sougandi, I., Venkatesan, R., and Rao, P.S. (2005) *Spectrochim. Acta, Part A*, **61A**, 2482.

34. Sougandi, I., Rajendiran, T.M., Venkatesan, R., and Rao, P.S. (2005) *Proc. Indian Acad. Sci.*, **114**, 473.
35. Sougandi, I., Venkatesan, R., and Rao, P.S. (2003) *J. Phys. Chem. Solids*, **64**, 1231.
36. Ravikumar, R., Madhu, N., Reddy, B.J., Reddy, Y.P., and Rao, P.S. (1997) *Phys. Scr.*, **55**, 637.
37. Kripal, R., Mishra, I., Gupta, S.K., and Arora, M. (2009) *Spectrochim. Acta, Part A*, **71**, 1969.
38. Duezguen, F. and Karabulut, B. (2007) *J. Mol. Struct.*, **834–836**, 136–140.
39. Kripal, R. and Mishra, I. (2009) *Spectrochim. Acta, Part A*, **72**, 538.
40. Haque, M.I. and Umar, M. (2010) *Turk. J. Phys.*, **34**, 59.
41. Sougandi, I., Velavan, K., Venkatesan, R., and Rao, P.S. (2004) *Phys. Status Solidi B*, **241**, 3014.
42. Bozkurt, E., Karabulut, B., Kartal, İ., and Bozkurt, Y.S. (2009) *Chem. Phys. Lett.*, **477**, 65.
43. Kripal, R., Maurya, M., and Govind, H. (2007) *Physica B (Amsterdam)*, **329**, 281.
44. Zapart, W., Zapart, M.B., Czaja, P., and Barasinski, A. (2006) *Phase Transitions*, **79**, 557.
45. Rasmussen, S.B., Eriksen, K.M., and Fehrmann, R. (2002) *J. Chem. Soc., Dalton Trans.*, 87.
46. Yarbas, Z., Karabulut, B., and Karabulut, A. (2009) *Physica B*, **404**, 3694.
47. Briyik, R. and Tapramaz, R. (2006) *Z. Naturforsch.*, **61a**, 171.
48. Ravikumar, R.V.S.S.N., Chandrasekhar, A.V., Reddy, Y.P., Komatsu, R., Ikeda, K., Yamauchi, J., and Rao, P.S. (2007) *Mater. Chem. Phys.*, **103**, 5.
49. Raju, B.D.P., Narasimhulu, K.V., Gopal, N.O., and Rao, J.L. (2003) *J. Phys. Chem. Solids*, **64**, 1339.
50. Narasimhulu, K.V., Raju, B.D.P., and Rao, J.L. (2002) in *EPR in the 21st Century* (eds A. Kawamori, J. Yamauchi, and H. Ohta), Elsevier Science, p. 207.
51. Kumar, R.R., Bhatnagar, A.K., and Rao, J.L. (2002) *Mater. Lett.*, **57**, 178.
52. Natarajan, B., Mithira, S., Deepa, S., Ravikumar, R., and Rao, P.S. (2006) *Radiat. Eff. Defects Solids*, **161**, 177.
53. Biyik, R. (2009) *Physica B*, **404**, 3483.
54. Ananth, K.M. and Manahoran, P.T. (1993) *J. Phys. Chem. Solids*, **54**, 835.
55. Kripal, R. and Maurya, M. (2010) *Solid State Commun.*, **150**, 95.
56. Kripal, R. and Maurya, M. (2009) *Physica B*, **404**, 1532.
57. Jain, M. (2004) *Z. Naturforsch.*, **59a**, 488.
58. Jain, M. and Jain, V.K. (2011) *Appl. Magn. Reson.*, **40**, 171.
59. Kripal, R. and Shukla, S. (2011) *Spectrosc. Lett.*, **44**, 235.
60. Zapart, M.B., Zapart, W., Czaja, P., Mila, T., and Solecki, J. (2011) *Ferroelectrics*, **417** (1), 70.
61. Kripal, R. and Shukla, S. (2011) *Phys. Scr.*, **83**, 035702.
62. Ravikumar, R.V.S.S.N., Madhu, N., Chandrasekhar, A.V., Reddy, B.J., Reddy, Y.P., Rao, P.S., Rajendiran, T.M., and Venkatesan, R. (2001) *Spectrochim. Acta, Part A*, **57**, 2789.
63. Sougandi, I., Venkatesan, R., Rajendiran, T.M., and Rao, P.S. (2003) *Phys. Scr.*, **67**, 153.
64. Maurya, B.P., Punnoose, A., Umar, M., and Singh, R.J. (1994) *Solid State Commun.*, **89**, 59.
65. Karabulut, B., Tapramaz, R., and Köksal, F. (2004) *Z. Naturforsch.*, **59a**, 669.
66. Kripal, R. and Shukla, S. (2012) *Phys. Scr.*, **85**, 015706.
67. Karabulut, B., Ilkin, I., and Tapramez, R. (2005) *Z. Naturforsch., A: Phys. Sci.*, **60**, 95.
68. Kripal, R. and Bajpai, M. (2010) *J. Alloys Compd.*, **490**, 5.
69. Kartal, I., Karabulut, B., and Bozkurt, E. (2010) *Z. Naturforsch.*, **65a**, 347.
70. Fidan, M., Tapramaz, R., and Sahin, Y. (2010) *J. Phys. Chem. Solids*, **71**, 818.
71. Tapramaz, R., Karabulut, B., and Köksal, F. (2000) *J. Phys. Chem. Solids*, **61**, 1367.
72. Karabulut, B. and Tufan, A. (2006) *Spectrochim. Acta, Part A*, **65A**, 285.
73. Karabulut, B. and Tufan, A. (2008) *Spectrochim. Acta, Part A*, **69**, 642.
74. Kalkan, H. and Koksal, F.K. (1998) *Phys. Stat. Sol.*, **205**, 651.

75. Rao, J.L., Omkaram, I., and Chakradhar, R.P.S. (2007) *Physica B*, **388**, 318.
76. Singh, V. and Jain, V.K. (2002) *Acta Phys. Pol. A*, **102**, 795.
77. Karabulut, B. and Tufan, A. (2006) *Spectrochim. Acta, Part A*, **65A**, 742.
78. Ramesha, H., Parthipan, K., and Rao, P.S. (2012) *Radiat. Eff. Defects Solids*, **167**, 184.
79. Demirci, T.B., Koseogul, Y., Guner, S., and Ulkuseven, B. (2006) *Cent. Eur. J. Chem.*, **4**, 149.
80. Özcesmeci, I., Güner, S., Okur, A.I., and Gül, A. (2007) *J. Porphyrins Phthalocyanines*, **11**, 531.
81. Vasantha, K., Mary, P.A.A., and Dhanuskodi, S. (2002) *Spectrochim. Acta, Part A*, **58**, 311.
82. Kripal, R. and Singh, D.K. (2006) *J. Magn. Magn. Mater.*, **307**, 308.
83. Parthipan, K., Ramesh, H., and Rao, P.S. (2011) *J. Mol. Struct.*, **992**, 59.
84. Natarajan, B., Deepa, S., Mithira, S., Ravikumar, R.V.S.S.N., and Rao, P.S. (2007) *Phys. Scr.*, **76**, 253.
85. Natarajan, B., Mithira, S., Deepa, S., and Sambasiva Rao, P. (2007) *J. Phys. Chem. Solids*, **68**, 1995.
86. Prabhakaran, G., Parthipan, K., and Rao, P.S. (2012) *Appl. Magn. Reson.*, **42**, 187.
87. Kripal, R., Maurya, M., Bajpai, M., and Govind, H. (2009) *Physica B*, **404**, 3493.
88. Kripal, R., Misra, M.G., Lipinski, I.E., and Rudowicz, C. (2012) *Phys. Scr.*, **86**, 045602.
89. Bozkurt, E., Karabulut, B., and Kartal, İ. (2009) *Spectrochim. Acta, Part A*, **73**, 163.
90. Garbarczyk, J.E., Tykarski, L., Machowski, P., and Wasiucionek, M. (2001) *Solid State Ionics*, **140**, 141.
91. Gahlot, P.S., Seth, V.P., Agarwal, A., Khasa, S., and Chand, P. (2003) *Radiat. Eff. Defects Solids*, **158**, 655.
92. Agarwal, A., Seth, V.P., Gahlot, P.S., Khasa, S., Arora, M., and Gupta, S.K. (2004) *J. Alloys Compd.*, **377**, 225.
93. Agarwal, A., Seth, V.P., Gahlot, P.S., Khasa, S., and Chand, P. (2004) *Mater. Chem. Phys.*, **85**, 215.
94. Gahlot, P.S., Agarwal, A., Seth, V.P., Sanghi, S., Gupta, S.K., and Arora, M. (2005) *Spectrochim. Acta, Part A*, **61**, 1189.
95. Yasoda, B., Chakradhar, R.P.S., Rao, J.L., and Gopal, N.O. (2007) *Mater. Chem. Phys.*, **106**, 33.
96. Kumari, J.L., Kumar, J.S., and Cole, S. (2011) *J. Non Cryst. Solids*, **357**, 3734.
97. Prakash, D., Seth, V.P., Chand, I., and Chand, P. (1996) *J. Non Cryst. Solids*, **204**, 46.
98. Khasa, S., Seth, V.P., Gahlot, P.S., Agrawal, A., and Gupta, S.K. (2006) (Proceedings of the 5th International Conference on Borate Glasses, Crystals, and Melts) *Phys. Chem. Glasses: Eur. J. Glasses, Sci. Technol. B*, **47**, 371.
99. Gahlot, P.S., Seth, V.P., Agarwal, A., Kishore, N., Gupta, S.K., and Arora, M. (2005) *J. Phys. Chem. Solids*, **66**, 766.
100. Kumar, V.R., Chakradhar, R.P.S., Murali, A., Gopal, N.O., and Rao, J.L. (2003) *Int. J. Mod. Phys. B*, **17**, 3033.
101. Sreekanth Chakradhar, R.P., Murali, A., and Lakshmana Rao, J. (2000) *Physica B*, **293**, 108.
102. Chakradhar, R.P.S., Sivaramaiah, G., Rao, J.L., and Gopal, N.O. (2005) *Mod. Phys. Lett. B*, **19**, 643.
103. Anshu, S., Sanghi, A., Agarwal, M., Lather, V., and Bhatnagar, S.K. (2009) *IOP Conf. Series: Mater. Sci. Eng.*, **2**, 012054.
104. Kumar, R.R., Rao, A.S., and Venkata Reddy, B.C. (1995) *Opt. Mater.*, **4**, 723.
105. Sreedhar, B., Indira, P., Bhatnagar, A.K., and Kojima, K. (1994) *J. Non Cryst. Solids*, **167**, 106.
106. Sakata, K., Hashimoto, M., and Kashiwamura, T. (1997) *Synth. React. Inorg. Met.-Org. Chem.*, **27**, 797.
107. Seth, V.P., Gupta, S., and Jindal, A. (1993) *J. Non-Cryst. Solids*, **162**, 263.
108. Subhadra, M. and Kistaiah, P. (2011) *J. Non-Cryst. Solids*, **357**, 3442.
109. Subhadra, M. and Kistaiah, P. (2011) *J. Phys. Chem. A*, **115**, 1009.
110. Agarwal, A., Sheoran, A., Sanghi, S., Bhatnagar, V., Gupta, S.K., and Arora, M. (2010) *Spectrochim. Acta, Part A*, **75**, 964.
111. Anshu, Rani, S., Agarwal, A., Sanghi, S., Kishore, N., and Seth, V.P. (2008) *Indian J. Pure Appl. Phys.*, **46**, 382.

112. Khasa, S., Seth, V.P., Agarwal, A., Krishna, R.M., Gupta, S.K., and Chand, P. (2001) *Mater. Chem. Phys.*, **72**, 366.
113. Krishna, R.M., Andre, J.J., Seth, V.P., Khasa, S., and Gupta, S.K. (1999) *Mater. Res. Bull.*, **34**, 1089.
114. Girdhar, G., Rangacharyulu, M., Ravikumar, R.V.S.S.N., and Rao, P.S. (2008) *Bull. Pure Appl. Sci.*, **27**, 105.
115. Dwivedi, P. and Kripal, R. (2010) *Phys. Scr.*, **82**, 045701.
116. Murali, A., Rao, J.L., and Subbaiah, A.V. (1997) *J. Alloys Compd.*, **257**, 96.
117. Shareefuddin, M.D., Jamal, M., Ramadevudu, G., Rao, M.L., and Chary, M.N. (1999) *J. Non Cryst. Solids*, **255**, 228.
118. Sheoran, A., Agarwal, A., Sanghi, S., Seth, V.P., Gupta, S.K., and Arora, M. (2011) *Physica B*, **406**, 4505.
119. Ravikumara, R.V.S.S.N., Reddya, V.R., Chandrasekhara, A.V., Reddya, B.J., Reddy, Y.P., and Rao, P.S. (2002) *J. Alloys Compd.*, **337**, 272.
120. Srinivasulu, K., Omkaram, I., Obeid, H., Suresh Kumar, A., and Rao, J.L. (2012) *J. Phys. Chem. A*, **116**, 3547.
121. Giridhar, G., Sastry, S.S., and Rangacharyulu, M. (2011) *Physica B*, **406**, 4027.
122. Ravikumar, R.V.S.S.N., Kayalvizhi, K., Chandrasekhar, A.V., Reddy, Y.P., Yamauchi, J., Arunakumari, K., and Rao, P.S. (2008) *Appl. Magn. Reson.*, **33**, 185.
123. Sumalatha, B., Omkaram, I., Rao, T.R., and Raju, C.L. (2011) *J. Mol. Struct.*, **1006**, 96.
124. Agarwal, A., Seth, V.P., Gahlot, P., Goyal, D.R., Arora, M., and Gupta, S.K. (2004) *Spectrochim. Acta, Part A*, **60**, 3161.
125. Das, B.B., Ambika, R., Ageetha, S., and Vimala, P. (2004) *J. Magn. Magn. Mater.*, **272**, e1637.
126. Prakash, P.G. and Rao, J.L. (2005) *Spectrochim. Acta, Part A*, **61**, 2595.
127. Rao, T.R., Krishna, C.R., Thampy, U.S.U., Reddy, C.V., Rao, P.S., and Ravikumar, R.V.S.S.N. (2011) *Physica B*, **406**, 2132.
128. Ravikumara, R.V.S.S.N., Jamalaiaha, B.C., Chandrasekhara, A.V., Reddya, B.J., Reddya, Y.P., and Sambasiva Rao, B.P. (1999) *J. Alloys Compd.*, **84**, 287.
129. Chernei, N.V., Nadolinnyi, V.A., Ivannikova, N.V., Gusev, V.A., Kupriyanov, I.N., Shlegel, V.N., and Vasiliev, Y.V. (2005) *J. Struct. Chem.*, **46**, 431.
130. Nosenko, A.E. and Sel'skii, A.A. (1998) *J. Appl. Spectrosc.*, **65**, 992.
131. Whitmore, M.H., Verdun, H.R., Singel, D.J. (1993) *Phys. Rev. B*, **47**, 11479.
132. Misra, S.K., Misiak, L.E., and Capobianco, J.A. (1994) *J. Phys. Condens. Matter*, **6**, 3955.
133. Misra, S.K., Misiak, L.E., Capobianco, J.A., and Bettinelli, M. (1994) *J. Magn. Reson., Ser. A*, **109**, 216.
134. Chandrasekharan, K. and Murty, V.S. (1996) *Solid State Commun.*, **97**, 709.
135. Tregenna‐Piggott, P.L.W., O'Brien, M.C.M., Pilbrow, J.R., Gudel, H.U., Best, S.P., and Noble, C. (1997) *J. Chem. Phys.*, **107**, 8275.
136. Krzystek, J., Ozarowski, A., and Tesler, J. (2006) *Coord. Chem. Rev.*, **250**, 2308.
137. von Bardeleben, H.J., Launay, J.C., and Mazoyer, V. (1993) *Appl. Phys. Lett.*, **63**, 1140.
138. Tesler, J., Wu, C.C., Chen, K.Y., Hsu, H.F., Smirnov, D., Ozarawski, A., and Krzystek, J. (2009) *J. Inorg. Biochem.*, **103**, 487.
139. Tregenna-Piggott, P.L.W., Best, S.U., Gudel, H.U., Weihe, H., and Wilson, C.C. (1999) *J. Solid State Chem.*, **145**, 460.
140. Krzystek, J., Ozarowski, A., Tofimenko, S., Zvyagin, S.A., and Telser, J. (2005) *Monogr. Ser. Int. Conf. Coord. Chem.*, **7143**.
141. Krzystek, J., Fiedler, A.T., Sokol, J.J., Ozarowski, A., Zvyagin, S.A., Brunold, T.C., Long, J.R., Brunel, L.C., and Telser, J. (2004) *Inorg. Chem.*, **43**, 5645.
142. Tatsukawa, T., Shirai, T., Imaizumi, T., Idehara, T., Ogawa, I., and Kanemaki, T. (1998) *Int. J. Infrared Millimeter Waves*, **19**, 859.
143. Schwartz, D.A., Walter, E.D., McIlwain, S.J., Krymov, V.N., and Singel, D.J. (1999) *Appl. Magn. Reson.*, **16**, 223.
144. Weckhuysen, B.M., Schoonheydt, R.A., Mabbs, F.E., and Collison, D. (1996) *J. Chem. Soc., Faraday Trans.*, **92**, 2431.

145. Kittiauchawal, T. and Limsuwan, P. (2008) *Int. J. Mod. Phys. B*, **22**, 4730.
146. Boettcher, R., Erdem, E., Langhammer, H.T., Mueller, T., and Abicht, H. (2005) *J. Phys. Condens. Matter*, **17**, 2763.
147. Ahmad, I., Marinova, V., Vrielinck, H., Callens, F., and Goovaerts, E. (2011) *J. Appl. Phys.*, **109**, 083506.
148. Yamaga, M., Macfarlane, P.I., Henderson, B., Holliday, K., Takeuchi, H., Yosida, T., and Fukui, M. (1997) *J. Phys. Condens. Matter*, **9**, 569.
149. Hrbanski, R., JaniecMateja, M., and Czapla, Z. (2007) *Phase Transitions*, **80**, 163.
150. Krypal, R., Govind, H., Gupta, S.K., and Arora, M. (2006) *J. Magn. Magn. Mater.*, **307**, 257.
151. Kripal, R., Singh, P., and Govind, H. (2009) *Spectrochim. Acta, Part A*, **74**, 357.
152. Zapart, W., Zapart, M.B., and Czaja, P. (2008) *Phase Transitions*, **81**, 1141.
153. Takeuchi, H., Tanaka, H., and Arakawa, M. (1993) *J. Phys. Condens. Matter*, **5**, 9205.
154. Huang, D., Lin, J., Nilges, M.J., and Pan, Y. (2012) *Phys. Status Solidi B*, **249**, 1559.
155. Novosel, N., Zilic, D., Pajic, D., Juric, M., Peric, B., Zadro, K., Rakvin, B., and Planinic, P. (2008) *Solid State Sci.*, **10**, 1387.
156. Park, Y.J., Yeom, T.H., Min, S., Park, I., and Choh, S.H. (1994) *J. Appl. Phys.*, **75**, 7559.
157. Gopal, N.O., Narasimhulu, K.V., Sunandana, C.S., and Rao, J.L. (2004) *Physica B*, **348**, 335.
158. Hermanowicz, K. (2002) *J. Alloys Compd.*, **341**, 179.
159. Fouejio, D. and Rousseau, J.J. (1996) *J. Phys. Condens. Matter*, **8**, 2663.
160. Kripal, R., Singh, P., and Shukla, S. (2011) *Physica B*, **406**, 324.
161. Ahn, S.W., Choh, S.H., and Kim, J.N. (1995) *J. Phys. Condens. Matter*, **7**, 667.
162. Ahn, S.W., Choh, S.H., and Rudowicz, C. (1997) *Appl. Magn. Reson.*, **12**, 351.
163. Takeuchi, H., Tanaka, H., Mori, M., Ebisu, H., and Arakawa, M. (2002) in *EPR in the 21st Century* (eds A. Kawamori, J. Yamauchi, and H. Ohta), Elsevier Science, p. 213.
164. Ostrowski, A. and Waplak, S. (2008) *Appl. Magn. Reson.*, **34**, 55.
165. Kripal, R. and Pandey, S. (2011) *J. Phys. Chem. Solids*, **72**, 67.
166. Yamaga, M., Henderson, B., Holliday, K., Yosida, T., Fukui, M., and Kindo, K. (1999) *J. Phys. Condens. Matter*, **11**, 10499.
167. Kripal, R. and Mishra, I. (2010) *Mater. Chem. Phys.*, **119**, 230.
168. LoyoMenoyo, M., Keeble, D.J., Furukawa, Y., and Kitamura, K. (2005) *J. Appl. Phys.*, **97**, 123905.
169. Malovichko, G., Grachev, V., Kokanyan, E., and Schirmer, O. (1999) *Phys. Rev. B*, **59**, 9113.
170. Yeom, T.H., Chang, Y.M., and Rudowicz, C. (1993) *Solid State Commun.*, **87**, 245.
171. Galeev, A.A., Khasanova, N.M., Rudowicz, C., Shakurov, G.S., Bykov, A.B., Bulka, G.R., Nizamutdinov, N.M., and Vinokurov, V.M. (2000) *J. Phys. Condens. Matter*, **12**, 4465.
172. Shakurov, G.S. and Tarasov, V.F. (2001) *Appl. Magn. Reson.*, **21**, 597.
173. Kripal, R. and Govind, H. (2008) *Physica B*, **403**, 3345.
174. Baker, J.M., Kuriata, J., O'Connell, A.C., Sadlowski, L. (1995) *J. Phys.: Condens. Matter* **7**, 2321.
175. Ravi, S., Selvakumar, P.N., and Subramian, P. (2006) *Solid State Commun.*, **138**, 129.
176. Kripal, R., Govind, H., Gupta, S.K., and Arora, M. (2007) *Solid State Commun.*, **141**, 416.
177. Yeom, T.H., Lee, S.H., Kim, I.G., Choh, S.H., Kim, T.H., and Ro, J.H. (2004) *J. Korean Phys. Soc.*, **44**, 1513.
178. Misra, S.K., Andronenko, S.I., Rao, S., Bhat, S.V., Komen, C.V., and Punnoose, A. (2009) *J. Appl. Phys.*, **105**, 07C514.
179. Sel'skii, A.A. (2001) *J. Appl. Spectrosc.*, **68**, 486.
180. Pan, Y., Mashkovtsev, R.I., Huang, D., Mao, M., and Shatskiy, A. (2011) *Am. Mineral.*, **96**, 1331.
181. Guler, S., Rameev, B., Khaibulla, R.I., Lopatin, O.N., and Atkas, B. (2010) *J. Magn. Magn. Mater.*, **322**, L13.

182. Ebisu, H., Arakawa, M., and Takeuchi, H. (2008) *J. Phys. Condens. Matter*, **20**, 145202.
183. Yamaga, M., Takeuchi, H., Han, T.P.J., and Henderson, B. (1993) *J. Phys. Condens. Matter*, **5**, 8097.
184. Chang, Y.M., Yeom, T.H., Yeung, Y.Y., and Rudowicz, C. (1993) *J. Phys. Condens. Matter*, **5**, 6221.
185. Misra, S.K., Isber, S., and Chand, P. (2000) *Physica B*, **291**, 105.
186. Anandalakshmi, H., Venkatesan, R., and Sambasiva Rao, P. (2004) *Cryst. Res. Technol.*, **39**, 78.
187. Ravikumar, R.V.S.S.N., Komatsu, R., Ikeda, K., Chandrasekhar, A.V., Reddy, B.J., Reddy, Y.P., and Rao, P.S. (2003) *Solid State Commun.*, **126**, 251.
188. Grapperhaus, C.A., Mienert, B., Bill, E., Weyhermiller, T., and Wieghardt, K. (2000) *Inorg. Chem.*, **39**, 5306.
189. Vazhenin, V.A., Potapov, A.P., Guseva, V.B., and Artyomov, M.Y. (2010) *Phys. Solid State*, **52** (3), 515.
190. Keeble, D.J., Li, Z., and PoindeXter, E.H. (1995) *J. Phys. Condens. Matter*, **7**, 6327.
191. Erdem, E., Drahus, M., Eichel, R.A., Ozarowski, A., van Tol, J., and Brunel, L.C. (2008) *Ferroelectrics*, **363**, 39.
192. von Bardeleben, H.J., Miesner, C., Monge, J., Briat, B., Launay, J.C., and Launay, X. (1996) *Semicond. Sci. Technol.*, **11**, 58.
193. Shakurov, G.S., Avanesov, A.G., and Avanesov, S.A. (2009) *Phys. Solid State*, **51**, 2292.
194. Oliete, P.B., Orera, V.M., and Alonso, P.J. (1996) *J. Phys. Condens. Matter*, **8**, 7179.
195. Oliete, P.B., Orera, V.M., and Alonso, P.J. (1996) *Phys. Rev. B*, **53**, 3047.
196. Avanesov, A.G., Badikov, V.V., and Shakurov, G.S. (2003) *Phys. Solid State*, **45**, 1451.
197. Telser, J., Pardi, L.A., Krzystek, J., and Brunel, L.C. (1998) *Inorg. Chem.*, **37**, 5769.
198. Krzystek, J., Telser, J., Hoffman, B.M., Brunel, L.C., and Licoccia, S. (2001) *J. Am. Chem. Soc.*, **123**, 7890.
199. Barra, A., Caneschi, A., Cornia, A., Gateschi, D., Gorini, L., Sessoli, R., and Sorace, L. (2007) *J. Am. Chem. Soc.*, **129**, 10762.
200. Stamatatos, T.C., FoguetAlbiol, D., Lee, S., Stoumpos, C.C., Raptopoulou, C.P., Terzis, A., Wernsdorfer, W., Hill, S.O., Perlepes, S.P., and Christou, G. (2007) *J. Am. Chem. Soc.*, **129**, 9484.
201. Tregenna Pigott, P.L.W., Weihe, H., and Barra, A.L. (2003) *Inorg. Chem.*, **42**, 8504.
202. Krivokapic, I., Noble, C., Klitgaard, S., Tregenna Pigott, P., Weihe, H., and Barra, A. (2005) *Angew. Chem. Int. Ed.*, **44**, 3613.
203. Krzystek, J., Yeagle, G.J., Park, J.H., Britt, R.D., Meisel, M.W., Brunel, L.C., and Telser, J. (2003) *Inorg. Chem.*, **42**, 4610.
204. Mantel, C., Hassan, A.K., Pecaut, J., Deronzier, A., Collomb, M.N., and Duboc-Toia, C. (2003) *J. Am. Chem. Soc.*, **125**, 12337.
205. Krzystek, J., Pardi, L.A., Brunel, L.C., Goldberg, D.P., Hoffman, B.M., Licoccia, S., and Telser, J. (2002) *Spectrochim. Acta, Part A*, **58**, 1113.
206. Barra, A.L., Gatteschi, D., Sessoli, R., Abbati, G.L., Cornia, A., Fabretti, A.C., and Uyttherhoeven, M.G. (1997) *Angew. Chem., Int. Ed. Engl.*, **36**, 2329.
207. Barra, A.L., Gatteschi, D., Sessoli, R., Abbati, G.L., Corina, A., Fabretti, A.C., and Uyttherhoeven, M.G. (1997) *Angew. Chem., Int. Ed. Engl.*, **36** (21), 2329–2331.
208. Aromi, G., Telser, J., Ozarowski, A., Brunel, L.C., Stoeckli-Evans, H.M., and Krzystek, J. (2005) *Inorg. Chem.*, **44**, 187.
209. Goldberg, D.P., Telser, J., Krzystek, J., Montalban, A.G., Brunel, L.C., Barrett, A.G.M., and Hoffman, B.M. (1997) *J. Am. Chem. Soc.*, **119**, 8723.
210. Misra, S.K. and Regler, B. (2013) *Appl. Magn. Reson.*, **44**, 401.
211. Harvey, J.D., Ziegler, C.J., Telser, J., Ozarowski, A., and Krzystek, J. (2005) *Inorg. Chem.*, **44**, 4451.
212. Goldberg, D.P., Telser, J., Krystek, J., Montalban, A.G., Brunel, L.C., Barrett, A.G.M., and Hoffman, B.M. (1997) *J. Am. Chem. Soc.*, **119**, 8722.
213. Krzystek, J. and Telser, J. (2003) *J. Magn. Reson.*, **162**, 454.

214. J. Krzystek, S. A. Zvyagin, A. Ozarowski, S. Trofimenko, J. Telser, *J. Magn. Reson.* **178** 174 (2006).
215. Azamat, D.V., Dejneka, A., Lancok, J., Trepakov, V.A., Jastrabik, L., and Badalyan, A.G. (2011) *Arch. Condens. Matter*, **117**.
216. Laguta, V.V., Glinchuk, M.D., Bykov, I.P., Zaritskii, M.I., Rosa, J., Jastrabik, L., Trepakov, V., and Syrnikov, P.P. (1996) *Solid State Commun.*, **98**, 1003.
217. Stosser, R. and Scholz, G. (1997) *Appl. Magn. Reson.*, **12**, 167.
218. Stosser, R. and Scholz, G. (1998) *Appl. Magn. Reson.*, **15**, 449.
219. Buzare, J.Y., Silly, G., Klein, J., Scholz, G., Stosser, R., and Nofz, M. (2002) *J. Phys. Condens. Matter*, **14**, 10331.
220. Limusuwan, P., Udomkan, N., and Winotai, P. (2007) *Mod. Phys. Lett. B*, **21**, 225.
221. Narduzzo, A., Ardavan, A., Singleton, J., Pardi, L., Bercu, V., AkutsoSato, A., Akutso, A., Turner, S.S., and Day, P. (2004) *J. Phys. IV France*, **114**, 347.
222. Yeom, T.H., Choh, S.H., Du, M.L., and Jang, M.S. (1996) *Phys. Rev. B*, **53**, 3415.
223. Yeom, T.H., Rudowicz, C., and Choh, S.H. (2002) *J. Korean Phys. Soc.*, **41**, 756.
224. Martin, A., Bravo, D., Dieguez, E., and Lopez, F.J. (1996) *Phys. Rev. B*, **54**, 12915.
225. Claridge, R.F.C., Tennant, W.C., and McGavin, D.G. (1997) *J. Phys. Chem. Solids*, **58**, 813.
226. Misra, S.K., Andronenko, S.I., Earle, K.A., and Freed, J.H. (2001) *Appl. Magn. Reson.*, **21**, 549.
227. Pan-On, W., Meejo, S., Tang, I. (2008) *Mat. Res. Bull.*, **43**, 2137.
228. Nistor, S.V., Goovaerts, E., and Schoemaker, D. (1994) *J. Phys. Condens. Matter*, **6**, 2619.
229. Balan, E., Allard, T., Boizot, B., Morin, G., and Muller, J.P. (1999) *Clays Clay Miner.*, **47**, 605.
230. Loncke, F., De Cooman, H., Khaidukov, N.M., Vrielinck, H., Goovaerts, E., Matthys, P., and Callens, F. (2007) *Phys. Chem. Chem. Phys.*, **9**, 5320.
231. Meyer, K., Bill, E., Mienert, B., Weyhermuller, T., and Wieghardt, K. (1999) *J. Am. Chem. Soc.*, **121**, 4859.
232. Seidel, A., Bill, E., Haggstrom, E., Nordblad, P., and Kilar, F. (1994) *Arch. Biochem. Biophys.*, **308**, 52.
233. Kashiwagi, T., Sonoda, S., Yashiro, H., Akasaka, Y., and Hagiwara, M. (2007) *J. Magn. Magn. Mater.*, **310**, 2152.
234. Kashiwagi, T., Sonoda, S., Yashiro, H., Akasaka, Y., and Hagiwara, M. (2007) *Jpn. J. Appl. Phys.*, **46**, 581.
235. Gehlhoff, W., Azamat, D., Haboeck, U., and Hoffmann, A. (2006) *Physica B*, **376–377**, 486.
236. Ferretti, A.M., Barra, A.L., Forni, L., Oliva, C., Schweiger, A., and Ponti, A. (2004) *J. Phys. Chem. B*, **108**, 1999.
237. Garces, N.Y., Stevens, K.T., Halliburton, L.E., Yan, M., Zaitseva, N.P., and DeYoreo, J.J. (2001) *J. Cryst. Growth*, **225**, 435.
238. Laguta, V.V. (1998) *Phys. Solid State*, **40**, 1989.
239. Ahn, S.W., Choh, S.H., and Choi, B.C. (1995) *J. Phys. Condens. Matter*, **7**, 9615.
240. Ahn, S.W. and Choh, S.H. (1998) *J. Phys. Condens. Matter*, **10**, 341.
241. Abdulsabirov, R.Y., Antonova, I.I., Korableva, S.L., Nizamutdinov, N.M., Stepanov, V.G., and Khasanova, N.M. (1997) *Phys. Solid State*, **39**, 423.
242. Nizamutidinov, N.M., Khasanova, N.M., Antonova, I.I., Abdulsabirov, R.Y., Korableva, S.L., Galeev, A.A., and Stepanov, V.G. (1998) *Appl. Magn. Reson.*, **15**, 145.
243. Jakes, P., Erdem, E., Ozarowski, A., van Tol, J., Buckan, R., Mikhailova, D., Ehrenberg, H., and Eichel, R.A. (2011) *Phys. Chem. Chem. Phys.*, **13**, 9344.
244. Razdobarin, A.G., Basun, S.A., Bursian, V.É., Sochava, L.S., and Evans, D.R. (2010) *Phys. Solid State*, **52**, 706.
245. Galeev, A.A., Khasanova, N.M., Rudowicz, C., Shakurov, G.S., Bulka, G.R., Nizamutdinov, N.M., and Vinokurov, V.M. (2004) *Appl. Magn. Reson.*, **26**, 533.
246. Vazhenin, V.A., Guseva, V.B., Yu Artyomov, M., Route, R.K., Fejer, M.M., and Byer, R.L. (2003) *J. Phys. Condens. Matter*, **15**, 275.

247. VassilikouDova, A.B. (1993) *Appl. Magn. Reson.*, **5**, 25.
248. Bulka, G.R., Vinokurov, V.M., Galeev, A.A., Denisenko, G.A., Khasanova, N.M., Konunnikov, G.V., Nizamutdinov, N.M., Stefanovski, S.V., and Trul, A.Y. (2005) *Crystallogr. Rep.*, **50**, 827.
249. Vinokurov, V.M., Gaite, J.M., Bulka, G.R., Khasanova, N.M., Nizamutdinov, N.M., Galeev, A.A., and Rudowicz, C. (2002) *J. Magn. Reson.*, **155**, 57.
250. Los, S. and Trybula, Z. (2002) *Acta Phys. Pol. A*, **101**, 279–287.
251. Darabon, A., Neamtu, C., Frcaz, S.I., and Tapaszto, L. (2003) *Appl. Magn. Reson.*, **25**, 1.
252. Yeom, T.H. (2004) *J. Korean Phys. Soc.*, **44**, 376.
253. Cortezao, S.U., Pontuschka, W.M., Da Rocha, M.S.F., and Blak, A.R. (2003) *J. Phys. Chem. Solids*, **64**, 1151.
254. Misra, S.K., Andronenko, S.I., Reddy, K.M., Hays, J., Thurber, A., and Punnoose, A. (2007) *J. Appl. Phys.*, **101**, 09H120.
255. Zhiteisev, E.R., Ulanov, V.A., and Zaripov, M.M. (2007) *Phys. Solid State*, **49**, 845.
256. Pandey, S. and Kripal, R. (2011) *J. Magn. Reson.*, **209**, 220.
257. Mikailov, F.A., Rameev, B.Z., Kazan, S., Yildiz, F., Mammadov, T.G., and Aktas, B. (2004) *Phys. Status Solidi C*, **1**, 3567.
258. Arakawa, M., Okamato, A., Ebisu, H., and Takeuchi, H. (2006) *J. Phys. Condens. Matter*, **18**, 3053.
259. Misra, S.K., Andronenko, S.I., and Andronenko, R.R. (1998) *Phys. Rev. B*, **57**, 8203.
260. Misra, S.K. and Andronenko, S.I. (2002) *Phys. Rev. B*, **65**, 104435.
261. Azamat, D.V. and Fanciulli, M. (2007) *Physica B*, **401–402**, 382.
262. Ball, D. and Van Wyk, J.A. (2000) *Phys. Status Solidi B*, **218**, 545.
263. Scholz, G., Stosser, R., Sebastian, S., Kemnitz, E., and Bartoll, J. (1998) *J. Phys. Chem. Solids*, **60**, 153.
264. Singh, V., Chakradhar, R.P.S., Rao, J.L., and Kwak, H.Y. (2011) *J. Mater. Sci.*, **46**, 3928.
265. Kripal, R. and Singh, D.K. (2006) *J. Phys. Soc. Jpn.*, **75**, 114711.
266. Eichel, R. and Boettcher, R. (2007) *Mol. Phys.*, **105**, 2195.
267. Yeom, T.H., Rudowicz, C., Choh, S.H., and McGavin, D.G. (1996) *Phys. Status Solidi B*, **198**, 839.
268. Zhang, Y.P., Buckmaster, H.A., and Kudynska, J. (1995) *Fuel*, **74**, 1307.
269. Geetha, P., Parthipan, K., Sathya, P., and Balaji, S. (2012) *J. Chem. Pharm. Res.*, **4**, 2724.
270. Tanaka, K., Suzuki, H., Kawano, K., and Nakata, R. (1995) *J. Phys. Chem. Solids*, **56**, 703.
271. Vongsavat, V., Winotai, P., and Meejo, S. (2006) *Nucl. Instrum. Methods Phys. Res., Sect. B*, **243**, 167.
272. Piligkos, S., Laursen, I., Morgenstjerne, A., and Weihe, H. (2007) *Mol. Phys.*, **105**, 2025.
273. Narasimhulu, K.V. and Rao, J.L. (2000) *Spectrochim. Acta*, **56**, 1345.
274. Obonai, T., Hidaka, C., Nomura, S., and Takizawa, T. (2010) *Opt. Mater.*, **32**, 1637.
275. Kripal, R. and Singh, D.K. (2006) *J. Magn. Magn. Mater.*, **306**, 112.
276. Duboc, C., Phoeung, T., Zein, S., Pecaut, J., Collomb, M., and Neese, F. (2007) *Inorg. Chem.*, **46**, 4905.
277. Narasimhulu, K.V. and Rao, J.L. (1998) *Physica B (Amsterdam)*, **254** (1–2), 37.
278. Deepa, S., Natarajan, B., Mithira, S., Velavan, K., and Rao, P.S. (2005) *Radiat. Eff. Defects Solids*, **160**, 357.
279. Velavan, K., Venkatesan, R., and Rao, P.S. (2005) *J. Phys. Chem. Solids*, **66**, 876.
280. Jain, V.K., Kapoor, V., and Prakash, V. (1996) *Solid State Commun.*, **97**, 425.
281. Gleason, R.J., Buldu, J.L., Cabrera, E., Quintanar, C., and Munoz, E. (1997) *J. Phys. Chem. Solids*, **58**, 1507.
282. Golombek, A.P. and Hendrich, M.P. (2003) *J. Magn. Reson.*, **165**, 33.
283. Udomkan, N., Limsuwan, P., and Chaimanee, Y. (2006) *Int. J. Mod. Phys. B*, **20**, 1097.
284. Yeom, T.H. (2012) *J. Phys. Soc. Jpn.*, **81**, 104702.
285. Kim, I.G., Yeom, T.H., Lee, S.H., Yu, Y.M., Shin, H.W., and Choh, S.H. (2001) *J. Appl. Phys.*, **89**, 4470.

286. Gopal, N.O., Narasimhulu, K.V., and Rao, J.L. (2002) *J. Phys. Chem. Solids*, **63**, 295.
287. Kripal, R. and Mishra, V. (2005) *Solid State Commun.*, **133**, 23.
288. Golovina, I.S., Shanina, B.D., Geifman, I.N., Andriiko, A.A., and Chernenko, L.V. (2012) *Phys. Solid State*, **54**, 551.
289. Maurya, B.P., Punnoose, A., Ikram, M., and Singh, R.J. (1995) *Polyhedron*, **14**, 2561.
290. Ostrowski, A. and Krupska, A. (2006) *Phase Transitions*, **79**, 569.
291. Jerzak, S. (2003) *J. Phys. Condens. Matter*, **15**, 8725.
292. Kripal, R. and Singh, P. (2007) *Physica B*, **387**, 222.
293. Alejandro, G., Passeggi, M.C.G., Vega, D., Ramos, C.A., Causa, M.T., Tovar, M., and Senis, R. (2003) *Phys. Rev. B*, **68**, 214429.
294. Misra, S.K. and Misiak, L.E. (1993) *Phys. Rev. B*, **48**, 13579.
295. Singh, V., Chakradhar, R.P.S., Rao, J.L., and Kim, D.K. (2009) *J. Lumin.*, **129**, 755.
296. Kripal, R., Govind, H., and Maurya, M. (2007) *J. Magn. Magn. Mater.*, **308**, 243.
297. Joseph, J. and Rao, P. (1996) *Spectrochim. Acta A*, **52**, 607.
298. Krishna, R.M., Seth, V.P., Gupta, S.K., Prakash, D., Chand, I., and Rao, J.L. (1997) *Spectrochim. Acta, Part A*, **53**, 253.
299. Singh, V., Chakradhar, R.P.S., Rao, J.L., and Kim, D. (2007) *J. Solid State Chem.*, **180**, 2067.
300. Kassiba, A., Hrabanski, R., Bonhomme, D., and Hader, A. (1995) *J. Phys. Condens. Matter*, **7**, 3339.
301. Petersen, J., Gessner, C., Fisher, K., Mitchell, C.J., Lowe, D.J., and Lubitz, W. (2005) *Biochem. J.*, **391**, 527.
302. Cortes, R., Drillon, M., Solans, X., Lezama, L., and Rojo, T. (1997) *Inorg. Chem.*, **36**, 677.
303. Kumar, R. and Chandra, S. (2007) *Spectrochim. Acta, Part A*, **67**, 188.
304. Mantel, C., Philouze, C., Collomb, M.N., and Duboc, C. (2004) *Eur. J. Inorg. Chem.*, **2004**, 3880.
305. Mantel, C., Baffert, C., Romero, I., Deronzier, A., Pécaut, J., Collomb, M.N., and Duboc, C. (2004) *Inorg. Chem.*, **43**, 6455.
306. Kripal, R. and Singh, D.K. (2008) *Spectrochim. Acta, Part A*, **69**, 889.
307. Bodziony, T., Lipinski, I.E., Kuriata, J., and Bednarski, W. (2001) *Physica B*, **299**, 70.
308. Ozturk, S.T. (2012) *J. Phys. Soc. Jpn.*, **81**, 114708.
309. Misra, S.K., Andronenko, S.I., Rinaldi, G., Chand, P., Earle, K.A., and Freed, J.H. (2003) *J. Magn. Reson.*, **160**, 131.
310. Kripal, R. and Mishra, V. (2005) *Solid State Commun.*, **134**, 699.
311. Hoffmann, S.K., Augustyniak, M.A., Goslar, J., and Hilczer, W. (1998) *Mol. Phys.*, **9**, 1265.
312. Kripal, R. and Shukla, A.K. (2007) *Spectrochim. Acta, Part A*, **66A**, 453.
313. Kripal, R., Govind, H., Bajpai, M., and Maurya, M. (2008) *Spectrochim. Acta, Part A*, **71**, 1302.
314. Krishna, R.M., Seth, V.P., Bansal, R.S., Chand, I., Gupta, S.K., and Andre, J.J. (1998) *Spectrochim. Acta, Part A*, **54**, 517.
315. Nistor, S.V., Stefan, M., Goovaerts, E., Nikl, M., and Bohacek, P. (2004) *Radiat. Meas.*, **38**, 655.
316. Ostrowski, A. and Bednarski, W. (2009) *J. Phys. Condens. Matter*, **21**, 205401.
317. Malczewski, D., Skrzypek, D., and Molak, A. (2004) *Arch. Mater. Sci.*, **25**, 561.
318. Singh, V., Chakradhar, R.P.S., Rao, J.L., and Kim, D. (2008) *Physica B*, **403**, 120.
319. Tanaka, K., Oomori, Y., Inoue, H., Kawano, K., Nakata, R., and Sumita, M. (1993) *J. Phys. Chem. Solids.*, **54**, 315.
320. Kripal, R. and Maurya, M. (2008) *Mater. Chem. Phys.*, **108**, 257.
321. Vorotynov, A.M., Petrakovskii, G.A., Shiyan, Y.G., Bezmaternykh, L.N., Temerov, V.E., Bovina, A.F., and Aleshkevych, P. (2007) *Phys. Solid State*, **49**, 463.
322. Singh, V., Chakradhar, R.P.S., Rao, J.L., and Kwak, H.Y. (2010) *Appl. Phys. B*, **98**, 407.
323. Kripal, R., Govind, H., Gupta, S.K., and Arora, M. (2007) *Physica B*, **392**, 92.

324. Singh, V., Chakradhar, R.P.S., Rao, J.L., and Kim, D. (2008) *J. Lumin.*, **128**, 394.
325. Neĭlo, G.N., Prokhorov, A.A., and Prokhorov, A.D. (2000) *Phys. Solid State*, **42**, 1134.
326. Natarajan, B., Mithira, S., and Sambasiva Rao, P. (2008) *Solid State Sci.*, **10**, 1916.
327. Benial, A.M.F., Ramakrishnan, V., and Murugesan, R. (1999) *Spectrochim. Acta, Part A*, **55**, 2573.
328. Boobalan, S. and Rao, P.S. (2010) *J. Organomet. Chem.*, **695**, 963.
329. Anandalakshmi, H., Sougandi, I., Velavan, K., Venkatesan, R., and Rao, P.S. (2004) *Spectrochim. Acta, Part A*, **60**, 2661.
330. Ankiewicz, A.O., Carmo, M.C., Sobolev, N.A., Gehlhoff, W., Kaidashev, E.M., Rahm, A., Lorenz, M., and Grundman, M. (2007) *J. Appl. Phys.*, **101**, 024324.
331. Nistor, S.V. and Stefan, M. (2009) *J. Phys. Condens. Matter*, **21**, 145408.
332. Fedorych, O.M., Hankiewicz, E.M., and Wilamowski, Z. (2002) *Phys. Rev. B*, **66**, 045201.
333. Bogomolova, L.D., Krasil'nikova, N.A., Bogdanov, B.L., Khalilev, V.D., and Mitrofanov, V.V. (1995) *J. Non-Cryst. Solids*, **188**, 130.
334. Chakradhar, R.P.S., Ramesh, K.P., Rao, J.L., and Ramakrishna, J. (2003) *J. Phys. Chem. Solids*, **64**, 641.
335. Rao, T.R., Reddy, C.V., Krishna, C.R., Sathish, D.V., Rao, P.S., and Ravikumar, R.V.S.S.N. (2011) *Mater. Res. Bull.*, **46**, ????
336. Hoch, M.J.R., Nellutla, S., van Tol, J., Choi, E.S., Lu, J., Zheng, H., and Mitchell, J.F. (2009) *Phys. Rev. B*, **79**, 214421.
337. Ozarowski, A., Zvyagin, S.A., Reiff, W.M., Telser, J., Brunel, L.C., and Krzystek, J. (2004) *J. Am. Chem. Soc.*, **126**, 6574–6575.
338. Carver, G., Dobe, C., Jensen, T.B., Tregenna Piggott, P.L.W., Janssen, S., Bill, E., McIntyre, G.J., and Barra, A. (2006) *Inorg. Chem.*, **45**, 4695.
339. Telser, J., van Slageren, J., Vongtragool, S., Dressel, M., Reiff, W.M., Zvyagin, S.A., Ozarowski, A., and Krzystek, J. (2005) *Magn. Reson. Chem.*, **43**, 130.
340. Krzystek, J., Smirnov, D., Schlegel, C., van Slageren, J., Telser, J., and Ozarowski, A. (2011) *J. Magn. Reson.*, **213**, 158.
341. Knapp, M.J., Krzystek, J., Brunel, L.C., and Hendrickson, D.N. (1999) *Inorg. Chem.*, **38**, 3321.
342. Shakurov, G.S., Shcherbakova, T.A., and Shustov, V.A. (2011) *Appl. Magn. Reson.*, **40**, 135.
343. Knapp, M.J., Krzystek, J., Brunel, L.C., and Hendrickson, D.N. (2000) *Inorg. Chem.*, **39**, 281.
344. Misra, S.K., Diehl, S., Tipikin, D., and Freed, J.H. (2010) *J. Magn. Reson.*, **205**, 14.
345. Ravikumar, R.V.S.S.N., Chandrasekhar, A.V., Reddy, B.J., Reddy, Y.P., and Yamauchi, J. (2002) *Ferroelectrics*, **274**, 127.
346. Hefni, M.A. (1992) *J. Phys. Soc. Jpn.*, **61**, 2534.
347. Bindilatti, V., Anisimov, A.N., Oliveira, N.F. Jr., Shapira, Y., Goiran, M., Yang, F., Isber, S., Averous, M., and Demianiuk, M. (1994) *Phys. Rev. B*, **50**, 16464.
348. Krzystek, J., Zvyagin, S.A., Ozarowski, A., Fiedler, A.T., Brunold, T.C., and Telser, J. (2004) *J. Am. Chem. Soc.*, **126**, 2148.
349. Rojo, J.M., Mesa, J.L., Lezama, L., Barberis, G.E., and Rojo, T. (1996) *J. Magn. Magn. Mater.*, **157-158**, 493.
350. Guener, S., Sener, M.K., Dincer, H., Koeseoglu, Y., Kazan, S., and Kocak, M.B. (2006) *J. Magn. Magn. Mater.*, **300**, e530.
351. Goni, A., Lezama, L., Barberis, G.E., Pizarro, J.L., Arriortua, M.I., and Rojo, T. (1996) *J. Magn. Magn. Mater.*, **164**, 251.
352. Choi, Y.N., Park, I.W., Kim, S.S., Park, S.S., and Choh, S.H. (1999) *J. Phys. Condens. Matter*, **11**, 4723.
353. Piwowarska, D., Kaczmarek, S.M., Berkowski, M., and Stefaniuk, I. (2006) *J. Cryst. Growth*, **291**, 123.
354. Kalkan, A., Guener, S., and Bayir, Z.A. (2007) *Dyes Pigm.*, **74**, 636.

355. Aleshkevych, P., FinkFinowicki, J., and Szymczak, H. (2007) *Acta Phys. Pol.*, **111**, 105.
356. Goñi, A., Lezama, L.M., Rojo, T., Foglio, M.E., Valdivia, J.A., and Barberis, G.E. (1998) *Phys. Rev. B*, **57**, 246.
357. Fuks, H., Kaczmarek, S.M., Bodziony, T., and Berkowski, M. (2008) *Appl. Magn. Reson.*, **34**, 27.
358. Kaczmarek, S.M. and Boulon, G. (2003) *Opt. Mater.*, **24**, 151.
359. Breece, R.M., Costello, A., Bennet, B., Sigdel, T.K., Matthews, M.L., Tierney, D.L., and Crowder, M.W. (2005) *J. Biol. Chem.*, **280**, 11074.
360. Stefaniuk, I., Mathovskii, A., Rudowicz, C., Suchocki, A., Wilamowski, Z., Lukasiewicz, T., and Galazka, Z. (2006) *J. Phys. Condens. Matter*, **18**, 4751.
361. Raita, O., Popa, A., Toloman, D., Stan, M., Darabont, A., and Giurgiu, L. (2011) *Appl. Magn. Reson.*, **40**, 245.
362. Desrochers, P.J., Sutton, C.A., Abrams, M.L., Ye, S., Neese, F., Tesler, J., Ozarowski, A., and Krzystek, J. (2012) *Inorg. Chem.*, **51**, 2793.
363. Desrochers, P.J., Telser, J., Zvyagin, S.A., Ozarowski, A., Krzystek, J., and Vicici, D.A. (2006) *Inorg. Chem.*, **45**, 8930.
364. Yukawa, Y., Aromi, G., Igarashi, S., Ribas, J., Zvyagin, S.A., and Krzystek, J. (2005) *Angew. Chem. Int. Ed.*, **44**, 1997–2001.
365. Feng, W., Wu, X., and Zheng, W. (2008) *Radiat. Eff. Defects Solids*, **163**, 29.
366. Shengelaya, A., Drulis, H., Macalik, B., and Suszynska, M. (1996) *Z. Phys. B*, **101**, 373.
367. Wojciechowska, A., Daszkiewicz, M., Staszak, Z., TruszZdybek, A., Bienko, A., and Ozarowski, A. (2011) *Inorg. Chem.*, **50**, 11532.
368. Krzystek, J., Park, J.H., Meisel, M.W., Hitchman, M.A., Stratemeier, H., Brunel, L.C., and Telser, J. (2002) *Inorg. Chem.*, **41**, 4478–4487.
369. Vongtragool, S., Gorshunov, B., Dressel, M., Krzystek, J., Eichhorn, D.M., and Telser, J. (2003) *Inorg. Chem.*, **42**, 1788.
370. Srinivasan, R., Sougandi, I., Venkatesan, R., and Rao, P.S. (2003) *Proc. Indian Acad. Sci. (Chem. Sci.)*, **115**, 91.
371. Velavan, K., Rajendiran, T.M., Venatesan, R., and Rao, P.S. (2002) *Solid State Commun.*, **122**, 15.
372. Guzzi, R., Bizzarri, A.R., Sportelli, L., and Cannistraro, S. (1997) *Biophys. Chem.*, **63**, 211.
373. Reddy, C.V., Krishna, C.R., Thampy, U.S.U., Reddy, Y.P., Rao, P.S., and Ravikumar, R.V.S.S.N. (2011) *Phys. Scr.*, **84**, 025602.
374. Hoffmann, S.K., Goslar, J., Lijewski, S., and Ulanov, V.A. (2007) *J. Chem. Phys.*, **127** (12), 124705.
375. Dwivedi, P. and Kripal, R. (2010) *Appl. Magn. Reson.*, **38**, 403.
376. Kripal, R. and Singh, D.K. (2006) *J. Phys. Chem. Solids*, **67**, 2559.
377. Dwivedi, P. and Kripal, R. (2010) *Spectrochim. Acta, Part A*, **75**, 830.
378. Kripal, R. and Singh, D.K. (2007) *Spectrochim. Acta, Part A*, **67A**, 815.
379. Kalkan, H., Atalay, S., and Senel, I. (1998) *Z. Naturforsch.*, **53a**, 945.
380. Padiyan, D.P., Muthukrishnan, C., and Murugesan, R. (2000) *Cryst. Res. Technol.*, **35**, 595.
381. Padiyan, D.P., Muthukrishnan, C., and Murugesan, R. (2000) *J. Magn. Magn. Mater.*, **222**, 251.
382. Rao, J.L., Padmavathi, V., Narasimhulu, K.V., and Nagaraja Naidu, Y. (1995) *Spectrochim. Acta, Part A*, **51**, 2531.
383. Kripal, R. and Misra, S. (2004) *J. Phys. Chem. Solids*, **65**, 939.
384. Kripal, R. and Misra, S. (2004) *J. Phys. Chem. Solids*, **65**, 939.
385. Koksal, F., Kartal, I., and Ucun, F. (1996) *Solid State Commun.*, **98**, 1087.
386. Karunakaran, C., Thomas, K.R.J., Shunmugasundaram, A., and Murugesan, R. (2001) *Spectrochim. Acta, Part A*, **57**, 441.
387. Ramic, E., Eichel, R., Dinse, K., Titz, A., and Schmidt, B. (2006) *J. Phys. Chem. B*, **110**, 20655.
388. Bednarski, W., Waplak, S., and Kirpichnikova, L.F. (1999) *J. Phys. Chem. Solids*, **60**, 1669.

389. Ostrowski, A., Bednarski, W., Waplak, S., and Czapla, Z. (2002) *Acta Phys. Pol. A*, **101**, 893.
390. Sougandi, I., Venkatesan, R., and Sambasiva Rao, P. (2004) *Spectrochim. Acta, Part A*, **60**, 2653.
391. Bozkurt, E., Kartal, I., Karabulut, B., and Uçar, Í. (2008) *Spectrochim. Acta, Part A*, **71**, 794.
392. Mothilal, K.K., Karunakaran, C., Rao, P.S., and Murugesan, R. (2003) *Spectrochim. Acta, Part A*, **59**, 3337.
393. Turta, C.I., Chapurina, L.F., Donoca, I.G., Voronkova, V., Healey, E.R., and Kravtsov, V.C. (2008) *Inorg. Chim. Acta*, **361**, 309.
394. Murali, A., Chang, Z., Ranjit, K.T., Krishna, R.M., Kurshev, V., and Kevan, L. (2002) *J. Phys. Chem. B*, **106**, 6913.
395. Chavan, S., Srinivas, D., and Ratnasamy, E. (2000) *J. Catal.*, **192**, 286.
396. Hoffmann, S.K., Goslar, J., Hilzcer, W., Goher, M.A.S., Luo, B.S., and Mak, T.C.W. (1997) *J. Phys. Chem. Solids*, **58**, 1351.
397. Raju, C.L., Rao, J.L., Gopal, N.O., and Reddy, B.C.V. (2007) *Mater. Chem. Phys.*, **101**, 423.
398. DominguezVera, J.M., Camara, F., Moreno, J.M., Colacio, E., and Stoeckli-Evans, H. (1998) *Inorg. Chem.*, **37**, 3046.
399. Chandra, S., Sangeetika, and Thakur, S. (2004) *Transition Met. Chem.*, **29**, 925.
400. Kurt, Y.D., Uelkueseven, B., Guener, S., and Koeseogul, Y. (2007) *Transition Met. Chem. (Dordrecht)*, **32**, 494.
401. Traa, Y., Murphy, D.M., Farley, R.D., and Hutchings, G.H. (2001) *Phys. Chem. Chem. Phys.*, **3**, 107.
402. Kumar, G.N.H., Parthasarathy, G., and Rao, J.L. (2011) *Am. Mineral.*, **96**, 654.
403. KovalaDemertzi, D., Skrzypek, D., Szymanska, B., Galani, A., and Demertzis, M.A. (2005) *Inorg. Chim. Acta*, **358**, 186.
404. Jezierska, J., Gtowiak, T., Oiarowski, A., Yablokov, Y.V., and Rmczyliska, Z. (1998) *Inorg. Chim. Acta*, **275**, 28.
405. Sakata, K., Odamura, T., Kanbara, Y., Nibu, T., Hashimoto, M., Tsuge, A., and Moriguchi, Y. (1998) *Polyhedron*, **17**, 1463.
406. Kripal, R. and Misra, S. (2005) *J. Magn. Magn. Mater.*, **294**, 72.
407. Kripal, R. and Misra, S. (2001) *J. Phys. Soc. Jpn.*, **70**, 2158.
408. Ucar, I., Karabulut, B., Pasaoglu, H., Bueyuekguengoer, O., and Bulut, A. (2006) *J. Mol. Struct.*, **787**, 38.
409. Bozkurt, E., Döner, A., Uçar, I., and Karabulut, B. (2011) *Spectrochim. Acta, Part A*, **79**, 1829.
410. Mithira, S., Natarajan, B., Deepa, S., Ravikumar, R.V.S.S.N., and Rao, P.S. (2007) *J. Mol. Struct.*, **839**, 2.
411. Parthipan, K. and Rao, P.S. (2010) *J. Mol. Struct.*, **977**, 130.
412. Kripal, R. and Misra, M.G. (2013) *Appl. Magn. Reson.*, **44**, 759.
413. Bozdog, C., Chow, K.H., Watkins, G.D., Sunakawa, H., Kuroda, N., and Usui, A. (2000) *Phys. Rev. B.*, **62**, 12923.
414. Pogni, R., Baratto, M.C., Busi, E., and Basosi, R. (1999) *J. Inorg. Biochem.*, **73**, 157.
415. Hoffmann, S.K., Goslar, J., and Lijewski, S. (2012) *J. Magn. Reson.*, **221**, 120.
416. Ravikumar, R.V.S.S.N., Komatsu, R., Ikeda, K., Chandrasekhar, A.V., Ramamoorthy, L., Reddy, B.J., Reddy, Y.P., and Rao, P.S. (2003) *Physica B*, **334**, 398.
417. Hoffmann, S.K., Goslar, J., and Tadyszak, K. (2010) *J. Magn. Reson.*, **205**, 293.
418. Narasimhulu, K.V., Sunandana, C.S., and Rao, J.L. (2000) *Phys. Status Solidi A*, **217**, 991.
419. Narasimhulu, K.V., Sunandana, C.S., and Rao, J.L. (2000) *J. Phys. Chem. Solids*, **61**, 1209.
420. Maurya, B.P., Punnoose, A., Umar, M., and Singh, R.J. (1995) *Spectrochim. Acta*, **51A**, 661.
421. Kripal, R., Misra, M.G., and Dwivedi, P. (2012) *Appl. Magn. Reson.*, **42**, 251.
422. Demir, D., Köksal, F., Kazak, C., and Köseoglu, R. (2009) *Z. Naturforsch.*, **64a**, 123.
423. Ostrowski, A., Bednarski, W., and Waplak, S. (2003) *Acta Phys. Pol. A*, **104**, 549.

424. Ostrowski, A., Bednarsky, W., and Szczepanska, L. (2003) *Mol. Phys. Rep.*, **37**, 117.
425. Khan, S., Ikram, M., Singh, A., and Singh, R.J. (1997) *Physica C*, **281**, 143.
426. Kripal, R., Bajpai, M., Maurya, M., and Govind, H. (2008) *Physica B*, **403**, 3693.
427. Ravikumar, R.V.S.S.N., Komatsu, R., Ikeda, K., Chandrasekhar, A.V., Reddy, B.J., Reddy, Y.P., and Rao, P.S. (2003) *J. Phys. Chem. Solids*, **64**, 261.
428. Ravikumar, R.V.S.S.N., Komatsu, R., Reddy, B.J., and Ikeda, K. (2003) *Spectrochim. Acta, Part A*, **59**, 3321.
429. Dwivedi, P., Kripal, R., and Misra, M.G. (2010) *J. Alloys Compd.*, **499**, 17.
430. Rao, A.C., Mithira, S., Natarajan, B., Bavikumar, R.V.S.S.N., Anandhalakshmin, H., and Rao, P.S. (2007) *J. Phys. Chem. Solids*, **68**, 305.
431. Karabulut, B., Bulut, A., and Tapramaz, R. (1999) *Spectrosc. Lett.*, **32**, 571.
432. Selvakumar, P.N., Boobalan, S., Sambasiva Rao, P., and Subramanian, P. (2011) *Spectrosc. Lett.*, **44**, 285.
433. Kripal, R. and Shukla, S. (2011) *Appl. Magn. Reson.*, **41**, 95.
434. Kripal, R. and Bajpai, M. (2009) *Spectrochim. Acta, Part A*, **72**, 528.
435. Hoffmann, S.K., Hilczer, W., and Goslar, J. (1996) *J. Magn. Reson., Ser. A*, **122**, 37.
436. Sano, W. and Di Mauro, E. (1997) *J. Phys. Chem. Solids*, **58**, 391.
437. Di Mauro, E. and Sano, W. (1994) *J. Phys. Condens. Matter*, **6**, L81.
438. Mauro, E.D. and Domiciano, S.M. (1999) *J. Phys. Chem. Solids*, **60**, 1849.
439. Mauro, E.D. and Domiciano, S.M. (2001) *Physica B*, **304**, 398.
440. Kripal, R., Shukla, S., and Dwivedi, P. (2012) *Physica B*, **407**, 656.
441. Kripal, R. and Misra, S. (2004) *J. Phys. Soc. Jpn.*, **73**, 1334.
442. Keeble, D.J., Li, Z., and Harmatz, M. (1996) *J. Phys. Chem. Solids*, **57**, 1513.
443. Lingappa, Y., Rao, S.S., Ravikumar, R.V.S.S.N., and Rao, P.S. (2007) *Radiat. Eff. Defects Solids*, **162**, 11.
444. Ulanov, V.A., Krupski, M., Hoffmann, S.K., and Zaripov, M.M. (2003) *J. Phys. Condens. Matter*, **15**, 1081.
445. Brant, A.T., Yang, S., Giles, N.C., Zafar Iqbal, M., Manivannan, A., and Halliburton, L.E. (2011) *J. Appl. Phys.*, **109**, 073711.
446. Mashkovtsev, R.I., Smirnov, S.Z., and Shigley, J.E. (2006) *J. Struct. Chem.*, **47**, 252.
447. Ravi, S. and Subramanian, P. (2007) *Solid State Commun.*, **143**, 277.
448. Kripal, R., Yadev, M.P., and Shukla, S. (2011) *Spectrochim. Acta, Part A*, **78**, 354.
449. Vorotynov, A.M., Petrakovski, G.A., Sablina, K.A., Bovina, A.F., and Vasil'ev, A.D. (2010) *Phys. Solid State*, **52**, 2415.
450. Boobalan, S. and Rao, P.S. (2010) *J. Phys. Chem. Solids*, **71**, 1527.
451. Rao, P.S., Rajendiran, T.M., Venkatesan, R., Madhu, N., Chandrasekhar, A.V., Reddy, B.J., Reddy, Y.P., and Ravikumar, R.V.S.S.N. (2001) *Spectrochim. Acta, Part A*, **57**, 2781.
452. Boobalan, S. and Rao, P.S. (2010) *Appl. Magn. Reson.*, **38**, 25.
453. Poonguzhali, E., Srinivasan, R., Venkatesan, R., Ravikumar, R.V.S.S.N., and Rao, P.S. (2003) *J. Phys. Chem. Solids*, **64**, 1139.
454. Naidu, K., Shiyamala, C., Mithira, S., Natarajan, B., Venkatesan, R., and Rao, P. (2005) *Radiat. Eff. Defects Solids*, **160**, 225.
455. Yerli, Y. (2007) *Spectrochim. Acta, Part A*, **66A**, 1288.
456. Kripal, R. and Misra, M.G. (2012) *Phys. Scr.*, **85**, 025701.
457. Udayachandran Thampy, U.S., Rama Krishna, C., Venkata Reddy, C., Babu, B., Reddy, Y.P., Rao, P.S., and Ravikumar, R.V.S.S.N. (2011) *Appl. Magn. Reson.*, **41**, 69.
458. Karabulut, B., Tapramaz, R., and Karadag, A. (2008) *Appl. Magn. Reson.*, **35**, 239.
459. Karabulut, B., Duezguen, F., Keser, C., and Heren, Z. (2007) *Physica B (Amsterdam)*, **396**, 8.
460. Al-Sufi, A.R., Bulka, G.R., Vinokurov, V.M., Kurkin, I.N., Nizamutdinov, N.M. and Salikhov, I.K. (1993) *Russian Physics Journal*. **36**, 555.
461. Vasyukov, V.N., Prokhorov, A.D., D'yakonov, V.P., and Szymczak, H. (2004) *J. Exp. Theor. Phys.*, **99**, 189.

462. Kumari, J.L., Padmini, V.N., Kumar, J.S., and Cole, S. (2012) *Optoelectron. Adv. Mater. Rapid Commun.*, **6**, 807.
463. Murthy, K.S.N., Murty, P.N., Rao, P.S., and Ravikumar, R.V.S.S.N. (2009) *Optoelectron. Adv. Mater. Rapid Commun.*, **3**, 954.
464. Kiczka, S., Hoffmann, S.K., Goslar, J., and Szczepanska, L. (2004) *Phys. Chem. Chem. Phys.*, **6**, 64.
465. Sumalatha, B., Omkaram, I., Rao, T.R., and Raju, C.L. (2011) *J. Non Cryst. Solids*, **357**, 3143.
466. Idziak, S., Goslar, J., and Hoffmann, S.K. (2004) *Mol. Phys.*, **102**, 55.
467. Bale, S. and Rahman, S. (2009) *J. Non-Cryst. Solids*, **355**, 2127.
468. Bale, S. and Rahman, S. (2012) *ISRN Spectrosc.*, **5**, 58305.
469. Suresh, S., Babu, J.C., and Chandramouli, V. (2005) *Phys. Chem. Glasses*, **46**, 27.
470. Chakradhar, R.P.S., Murali, A., and Lakshmana Rao, J. (1998) *J. Alloys Compd.*, **265**, 29.
471. Rao, J.L., Sivaramaiah, G., and Gopal, N.O. (2004) *Physica B*, **349**, 206.
472. Krishna, R.M., Andre, J.J., Seth, V.P., Khasa, S., and Gupta, S.K. (1999) *Opt. Mater.*, **12**, 47.
473. Rao, N.S., Purnima, M., Bale, S., Kumar, K.S., and Rahman, S. (2006) *Bull. Mater. Sci.*, **29**, 365.
474. Chakradhar, R.P.S., Yasoda, B., Rao, J.L., and Gopal, N.O. (2006) *J. Non Cryst. Solids*, **352**, 3864.
475. Giridhar, G., Rangacharyulu, M., Ravikumar, R.V.S.S.N., and Rao, P.S. (2009) *J. Mater. Sci. Technol.*, **25**, 531.
476. Krishna, R.M., Andre, J.J., Pant, R.P., and Seth, V.P. (1998) *J. Non-Cryst. Solids*, **232**, 509.
477. Upender, G., Babu, J.C., and Mouli, V.C. (2012) *Spectrochim. Acta, Part A*, **89**, 39.
478. Ramadevudu, G., Shareefuddin, N., Bai, S., Lakshmipathi Rao, M., and Narasimha Chary, M. (2000) *J. Non-Cryst. Solids*, **278**, 205.
479. Rao, T.G.V.M., Kumar, A.R., Chakravarthi, C.K., Reddy, M.R., and Veeraiah, N. (2012) *Physica B*, **407**, 593.
480. Suresh, S.C. and Chandramouli, V. (2004) *Indian J. Pure Appl. Phys.*, **42**, 560.
481. Shareefuddin, M., Jamal, M., and Chary, M.N. (1996) *J. Non-Cryst. Solids*, **201**, 95.
482. Bagratashvili, V.N., Bogomolova, L.D., Jachkin, V.A., Krasil'nikova, N.A., Rybaltovskii, A.O., Tsypina, S.I., and Chutko, E.A. (2004) *Glass Phys. Chem.*, **30**, 500.
483. Srinivasulu, K., Omkaram, I., Obeid, H., SureshKumar, A., and Rao, J.L. (2012) *Physica B*, **407**, 4741.
484. Prakash, P.G. and Rao, J.L. (2004) *J. Mater. Sci.*, **39**, 193.
485. Chakradhar, R.P.S., Ramesh, K.P., Rao, J.L., and Ramakrishna, J. (2003) *J. Phys. Condens. Matter*, **15**, 1469.
486. Andronenko, S.I., Andronenko, R.R., Vasil'ev, A.V., and Zagrebel'nyi, O.A. (2004) *Glass Phys. Chem.*, **30**, 230.
487. Babu, J.C., Suresh, S., and Mouli, V.C. (2005) *Indian J. Pure Appl. Phys.*, **43**, 833.
488. Naga Padmini, V., Lakshmi Kumari, J., Santhan Kumar, J., and Cole, S. (2011) *Phys. Chem. Glasses: Eur. J. Glass Sci. Technol. B*, **52** (4), 167–170.
489. Upender, G., Sameera Devi, C., Kamalaker, V., and Chandra Mouli, V. (2011) *J. Alloys Compd.*, **509**, 5887.
490. Suresh, S., Prasad, M., Upender, G., Kamalaker, V., and Mouli, V.C. (2009) *Indian J. Pure Appl. Phys.*, **47**, 163.
491. Suresh, S., Pavani, P.G., and Mouli, V.C. (2012) *Mater. Res. Bull.*, **47**, 724.
492. Upender, G., Prasad, M., and Chandra Mouli, V. (2011) *J. Non Cryst. Solids*, **357**, 903.
493. Upender, G. and Chandra Mouli, V. (2009) *Phys. Chem. Glasses: Eur. J. Glass Sci. Technol. B*, **50** (6), 399–406.
494. Dhanuskodi, S. and Manikandan, S. (1999) *Ferroelectrics*, **234**, 183.
495. Raghavendra Rao, T., Rama Krishna, C., Udayachandran Thampy, U.S., Venkata Reddy, C., Reddy, Y.P., Sambasiva Rao, P., and Ravikumar, R.V.S.S.N. (2011) *Appl. Magn. Reson.*, **40**, 339.
496. Stevens, K.T., Garces, N.Y., Bai, L., Giles, N.C., Halliburton, L.E., Setzler, S.D., Schunemann, P.G., Pollak, T.M.,

Route, R.K., and Feigelson, R.S. (2004) *J. Phys. Condens. Matter*, **16**, 2593.

497. Moreno, M., Barriuso, M.T., and Aramburu, J.A. (1994) *Phys. Rev. B*, **49**, 1039.

498. Das, B.B. and Ambika, R. (2003) *Chem. Phys. Lett.*, **370**, 670.

499. Kool, T.W. (2010) *Arch. Condens. Matter*, **1003**.

500. Laguta, V.V., Slipenyuk, A.M., Rosa, J., Nikl, M., Vedda, A., Nejezchleb, K., and Blazek, K. (2004) *Radiat. Meas.*, **38**, 735.

501. Bravo, D., Martin, M.J., Gavalda, J., Diaz, F., Zaldo, C., and Lopez, F.J. (1994) *Phys. Rev. B*, **50**, 16224.

502. Vrielinck, H., Callens, F., Zdravkova, M., and Matthys, P. (1998) *J. Chem. Soc., Faraday Trans.*, **94**, 2999.

503. Badalyan, A.G., Polak, K., and Rosa, J. (2003) *Radiat. Eff. Defects Solids*, **158**, 191.

504. Yamaga, M., Honda, M., Shimamura, K., Fukuda, T., and Yosida, T. (2000) *J. Phys. Condens. Matter*, **12**, 5917.

505. Yamaga, M., Hattori, K., Kodama, N., Ishizawa, N., Honda, M., Shimamura, K., and Fukuda, T. (2001) *J. Phys. Condens. Matter*, **13**, 10811.

506. Yamaga, M., Kodama, N., Yosida, T., Henderson, B., and Kindo, K. (1997) *J. Phys. Condens. Matter*, **9**, 9639.

507. Rosa, J., Asatryan, H.R., and Nikl, M. (1996) *Phys. Status Solidi A*, **158**, 573.

508. Loncke, F., Zverev, D., Vrielinck, H., Khaidukov, N.M., Matthys, P., and Callens, F. (2007) *Phys. Rev. B: Condens. Matter Mater. Phys*, **75**, 144427/1.

509. Yamaga, M., Yosida, T., Lee, D., Han, T.P.J., Gallagher, H.G., Henderson, B., and Kindo, K. (1997) *Modern Applications of EPR/ESR*, Springer, New York, p. 400.

510. Yamaga, M., Lee, D., Henderson, B., Han, T.P.J., Gallagher, H.G., and Yosida, T. (1998) *J. Phys. Condens. Matter*, **10**, 3223.

511. Guedesa, K.J., Krambrocka, K., and Gesland, J.Y. (2002) *J. Alloys Compd.*, **344**, 251.

512. Asatryan, G.R., Badikov, V.V., Kramushchenko, D.D., and Khramtsov, V.A. (2012) *Phys. Solid State*, **54**, 2057.

513. Popescu, F.F., Bercu, V., Barascu, J.N., Martinelli, M., Massa, C.A., Pardi, L.A., Stefan, M., Nistor, S.V., and Nikl, M. (2010) *Opt. Mater.*, **32**, 570.

514. Yeom, T.H. and Lee, S.H. (2004) *J. Korean Phys. Soc.*, **45**, 1052.

515. Asatryan, H.R., Rosa, J., and Mares, J.A. (1997) *Solid State Commun.*, **104**, 5.

516. Watterich, A., Aleshkevych, P., Borowiec, M.T., Zayarnyuk, T., Szymczak, H., Beregi, E., and Kovacs, L. (2003) *J. Phys. Condens. Matter*, **15**, 3323.

517. Misra, S.K. and Isber, S. (1998) *Physica B*, **253**, 111.

518. Hinatsu, Y. and Edelstein, N. (1997) *J. Alloys Compd.*, **250**, 400.

519. Abdulsabirov, R.Y., Falin, M.L., Vazlizhanov, I.I., Kasakov, B.N., Korableva, S.L., Ibragimov, I.R., Safiullin, G.M., and Yakovleva, Z.S. (1993) *Appl. Magn. Reson.*, **5**, 377.

520. Mehta, V., Gourier, D., Mansingh, A., and Dawar, A.L. (1999) *Solid State Commun.*, **109**, 513.

521. Rosa, J., Asatryan, H.R. and Nikl, M. (1996) *Phys. Status Solidi A*. **158**, 573.

522. Park, I.W., Choh, S.H., Kim, S.S., Kang, K., and Choi, D. (2002) *EPR in the 21st Century*, Elsevier B.V, p. 288.

523. GuillotNoel, O., Mehta, V., Viana, B., and Gourier, D. (2000) *Phys. Rev. B*, **61**, 15338.

524. Popescu, F.F., Bercu, V., Barascu, J.N., Martinelli, M., Massa, C.A. et al. (2009) *J. Chem. Phys.*, **131**, 034505.

525. Falin, M., Bill, H., and Lovy, D. (2004) *J. Phys. Condens. Matter*, **16**, 1293.

526. Yamaga, M., Honda, M., Wells, J.P.R., Han, T.P.J., and Gallagher, H. (2000) *J. Phys. Condens. Matter*, **12**, 8727.

527. Rey, J.M., Bill, H., Lovy, D., and Kubel, F. (1999) *J. Phys. Condens. Matter*, **11**, 7301.

528. Gratens, X., Isber, S., Charar, S., Fau, C., Averous, M., Misra, S.K., Golacki, Z., Farhat, M., and Tedenac, J.C. (1997) *Phys. Rev. B*, **55**, 8075.

529. Yeom, T.H., Kim, I.G., Lee, S.H., Choh, S.H., Kim, T.H., and Ro, J.H. (2000) *J. Appl. Phys.*, **87**, 1424.

530. Misra, S.K. and Andronenko, S.I. (2000) *J. Phys. Chem. Solids*, **61**, 1913.

531. Boldu, J.L., Gleason, R.J., Quintanar, C., and Munoz, E. (1996) *J. Phys. Chem. Solids*, **57**, 267.
532. Matsuoka, H., Ozawa, N., Kodama, T., Nishikawa, H., Ikemoto, I., Kikuchi, K., Furukawa, K., Sato, K., Shiomi, D., Takui, T., and Kato, T. (2004) *J. Phys. Chem. B*, **108**, 13972.
533. Gratens, X., Isber, S., and Charar, S. (2007) *Phys. Rev.*, **B76**, 035203.
534. Misra, S.K., Chang, Y., Petkov, V., Isber, S., Charar, S., Fau, C., Averous, M., and Golacki, Z. (1995) *J. Phys. Condens. Matter*, **7**, 9897.
535. Isber, S., Charar, S., Mathet, V., Fau, C., Averous, M., and Galacki, Z. (1995) *Phys. Rev.*, **B52**, 1678.
536. Isber, S., Misra, S.K., Charar, S., Gratens, X., Averous, M., and Golacki, Z. (1997) *Phys. Rev. B*, **56**, 13724.
537. Gratens, X., Arauzo, A.B., Breton, G., Charar, S., Averous, M., and Isber, S. (1998) *Phys. Rev. B*, **58**, 877.
538. Kaiya, K., Takahashi, N., Nakamura, T., Matsuzawa, T., Smith, G.M., and Riedi, P.C. (2000) *J. Lumin.*, **8789**, 1073.
539. Matsuoka, H., Furukawa, K., Sato, K., Shiomi, D., Kojima, Y., Hirotsu, K., Furuno, N., Kato, T., and Takui, T. (2003) *J. Phys. Chem. A*, **107**, 11539.
540. Neilo, G.N., Antonyak, T., and Prokhorov, A.D. (2001) *Phys. Solid State*, **43**, 652.
541. Zayachuk, D., Polyhach, Y., Slynko, E., Khandozhko, O., and Rudowicz, C. (2002) *Physica B*, **322**, 270.
542. Yeom, T.H., Choh, S.H., and Rudowicz, C. (1997) in *Modern Applications of EPR/ESR* (ed. C.Z. Rudowicz), Springer, Hong Kong, p. 546.
543. Yeom, T.H. (2002) *J. Korean Phys. Soc.*, **41**, 371.
544. Yeom, T.H., Choh, S.H., Rudowicz, C., and Jang, M.S. (1999) *Appl. Magn. Reson.*, **16**, 23.
545. Chen, N., Pan, Y., and Weil, J.A. (2002) *Am. Mineral.*, **87**, 37.
546. Chen, N., Pan, Y., Weil, J.A., and Nilges, M.J. (2002) *Am. Mineral.*, **87**, 47.
547. Cherney, N.V., Nadolinny, V.A., and Pavlyuk, A.A. (2008) *Appl. Magn. Reson.*, **33**, 45.
548. Arakawa, M., Ebisu, H., and Takeuchi, H. (1997) *J. Phys. Condens. Matter*, **9**, 5193.
549. Arakawa, M., Ebisu, H., and Takeuchi, H. (1996) *J. Phys. Condens. Matter*, **8**, 11299.
550. Gleason, R.J., Boldu, J.L., and Munoz, P. (1999) *J. Phys. Chem. Solids*, **60**, 929.
551. Prokhorov, A.D., Prokhorov, A.A., Chernysh, L.F., Dyakonov, V.P., and Szymczak, H. (2011) *J. Magn. Magn. Mater.*, **323**, 1546.
552. Leniec, G., Kaczmarek, S.M., Typek, J., Kolodziej, B., Grch, E., and Schlif, W. (2006) *J. Phys. Condens. Matter*, **18**, 9871.
553. Szyczewski, A., Lis, S., Kruczynski, Z., But, S., Elbanowski, M., and Pietrzak, J. (1998) *J. Alloys Compd.*, **275–277**, 349.
554. Zapart, M.B., Zapart, W., Reng, P., and Otko, A.I. (2002) *Ferroelectrics*, **272**, 187.
555. Borowiec, M.T., Dyakonov, V., Kamenev, V., Krygin, I., Piechota, S., Prokhorov, A., and Szymczak, H. (1998) *Phys. Status Solidi B*, **209**, 443.
556. Cherney, N.V., Nadolinnyi, V.A., and Pavlyuk, A.A. (2005) *J. Struct. Chem.*, **46**, 619.
557. DebaudMinorel, A.M., Mortier, M., Buzaré, J.Y., and Gesland, J.Y. (1995) *Solid State Commun.*, **95**, 167.
558. Gaitey, J.M., Izotovy, V.V., Bulkaz, G.R., Khasanovaz, N.M., Galeev, A.A., Nizamutdinovz, N.M., and Vinokurov, V.M. (1998) *J. Phys. Condens. Matter*, **10**, 7609.
559. Misra, S.K., Andronenko, S.I., and Chemekova, T.Y. (2003) *Phys. Rev. B*, **67**, 214411.
560. Misra, S.K. and Andronenko, S.I. (2007) *Appl. Magn. Reson.*, **32**, 377.
561. Antonova, I.I., Nizamutdinov, I.N., Abdulsabirov, R.Y., Korableva, S.L., Khasanova, N.M., Galeev, A.A., Stepanov, V.G., and Nizamutdinov, N.M. (1997) *Appl. Magn. Reson.*, **13**, 579.
562. de Biasi, R.S. and Grillo, M.L.N. (2004) *J. Phys. Chem. Solids*, **65**, 1207.
563. Nistor, S.V., Stefan, M., Goovaerts, E., Nikl, M., and Bohacek, P. (2006) *J. Phys. Condens. Matter*, **18**, 719.

564. Pdungsap, L., Booneyun, S., Winotai, P., Udomkan, N., and Limsuwan, P. (2005) *Eur. Phys. J. B*, **48**, 367.
565. Misra, S.K., Li, L., and Isber, S. (1999) *Physica B*, **269**, 278.
566. Misra, S.K. and Misiak, L.E. (1997) *Phys. Rev. B*, **56**, 2391.
567. Vazhenin, V.A., Potapov, A.P., Ivachev, A.N., Artyomov, M.Y., and Guseva, V.B. (2012) *Phys. Solid State*, **54**, 1245.
568. Arakawa, M., Mirayama, F., Ebisu, H., and Takeuchi, H. (2006) *J. Phys. Condens. Matter*, **18**, 7427.
569. Arakawa, M., Nakano, T., Ebisu, H., and Takeuchi, H. (2003) *J. Phys. Condens. Matter*, **15**, 3779.
570. Prokhorov, A.D., Krygin, I.N., Prokhorov, A.A., Chernush, L.F., Aleshkevich, P., Dyakonov, V., and Szymczak, H. (2009) *Phys. Status Solidi A*, **206**, 2617.
571. Vazhenin, V.A., Potapov, A.P., Gorlov, A.D., Chernyshev, V.A., Kazanskii, S.A., and Ryskin, A.I. (2006) *Phys. Solid State*, **48**, 686.
572. Kumar, B.V., Velchuri, R., Devi, V.R., Prasad, G., Sreedhar, B., Bansal, C., and Vithal, M. (2010) *J. Appl. Phys.*, **108**, 044906.
573. Ryadun, A.A., Galashov, E.N., Nadolinny, V.A., and Shlegel, V.N. (2012) *J. Struct. Chem.*, **53**, 685.
574. Mungchamnankit, A., Limsuwan, P., Thongcham, K., and Meejoo, S. (2008) *J. Magn. Magn. Mater.*, **320**, 479.
575. Bravoy, D., Kaminskiiz, A.A., and López, F.J. (1998) *J. Phys. Condens. Matter*, **10**, 3261.
576. Vernier, N. and Bellessa, G. (2003) *J. Phys. Condens. Matter*, **15**, 3417.
577. Asatryan, G.R., Badikov, V.V., Kramushchenko, D.D., and Khramtsov, V.A. (2012) *Phys. Solid State*, **54**, 1245.
578. Merino, R.I., Orera, V.M., Povill, O., Assmus, W., and Lomonova, E.E. (1997) *J. Phys. Chem. Solids*, **58**, 1579.
579. Gaister, A.V., Zharikov, E.V., Konovalov, A.A., Subbotin, K.A., and Tarasov, V.F. (2003) *JETP Lett.*, **77**, 753.
580. Astryan, H.R., Zakharchenya, R.I., Kutsenko, A.B., Babunts, R.A., and Baranov, P.G. (2007) *Phys. Solid State*, **49**, 1074.
581. Kim, Y.H., Yeom, T.H., Eguchi, H., and Seidel, G.M. (2007) *J. Magn. Magn. Mater.*, **310**, 1703.
582. Borowiec, M.T., Prochorov, A.A., Prochorov, A.D., Dyakonov, V.P., and Szymczak, H. (2003) *J. Phys. Condens. Matter*, **15**, 5113.
583. Kaczmarek, S.M. and Bodziony, T. (2008) *J. Non-Cryst. Solids*, **354**, 4202.
584. Milori, D.M.B.P., Moraes, I.J., Hernandes, A.C., de Souza, R.R., Siu Li, M., and Terrile, M.C. (1995) *Phys. Rev. B*, **51**, 3206.
585. Baranov, P.G., Ilyin, I.V., and Mokhov, E.N. (1997) *Solid State Commun.*, **103**, 291.
586. Wells, J.P.R., Yamaga, M., Mosses, R.W., Han, T.P.J., Gallagher, H.G., and Yosida, T. (2000) *Phys. Rev. B: Condens. Mater Mater. Phys.*, **61**, 3905.
587. GuillotNoel, O., Goldner, P., Le Du, Y., Baldit, E., Monnier, P., and Bencheikh, K. (2006) *Phys. Rev. B: Condens. Matter Mater. Phys.*, **74**, 214409.
588. GuillotNoël, O., Goldner, P., Le Du, Y., Baldit, E., Monnier, P., and Bencheikh, K. (2008) *J. Alloys Compd.*, **451**, 62.
589. Pashchenko, V.A., Jansen, A.G.M., Kobets, M.I., Khats'ko, E.N. and Wyder, P. (2000) *Phys. Rev. B*, **62**, 1197.
590. Kaczmarek, S.M., Leniec, G., and Boulon, G. (2008) *J. Alloys Compd.*, **451**, 116.
591. Kaczmarek, S.M., Macalik, L., Fuks, H., Leniec, G., Skibinski, T., and Hanuza, J. (2012) *Cent. Eur. J. Phys.*, **10**, 492.
592. Biodziony, T., Kaczmarek, S.M., and Hanuza, J. (2005) *Proc. SPIE*, **5958**, 59580A.
593. Brant, A.T., Halliburton, L.E., Basun, S.A., Grabar, A.A., Odoulov, S.G., Shumelyuk, A., Giles, N.C., and Evans, D.R. (2012) *Phys. Rev. B*, **86**, 134109.

# 5
# Compilation of Hyperfine Splittings and g-Factors for Aminoxyl (Nitroxide) Radicals

*Lawrence J. Berliner*

## 5.1
## Introduction

Knowledge of the anisotropic hyperfine and g-factors for aminoxyl (nitroxide) radicals is important in orientation determinations, host crystal structures, and in simulations of dynamics. For example, with knowledge of the anisotropic hyperfine values and the placement of the principal axes in a stable paramagnetic spin label that is specifically bound to a protein, analysis of the single crystal EPR spectra of the spin-labeled single crystal enables one to localize the label in the protein structure. On the other hand, one can analyze the tumbling dynamics of a spin-labeled macromolecule from EPR spectral simulations with precise knowledge of the anisotropic hyperfine values and g-factors. This is especially important in spin-labeled polymers, membranes, and oriented bilayers. To the best of this author's knowledge, no compilation had been published on this topic since the first *Spin Labeling: Theory and Applications* volume in 1976 [1]. In the past, anisotropic hyperfine values could be "read off" an X-band powder or frozen glass spectrum or determined with extensive analysis of the radical bound in a host crystal. With the advent of high-frequency (HF) W-band ($\sim 95$ GHz) EPR, all of the hyperfine parameters of this class of radicals are more precisely determined. Use of HF EPR has helped expand and update what we knew in the 1970s [2]. Also, these values have been measured by pulse EPR methods, such as high-field echo-detected EPR and nitrogen electron-spin-echo-envelope-modulation (ESEEM). The following data, although not thorough, cover a range of these radical types at several frequencies, solvent environments, and hosts. In some selected cases where the data were readily available, we have included parameters in several host environments, solvents, and other states as polarity affects both the hyperfine and g-values.

Historically, these types of molecules were commonly called *nitroxides*. In the early years of synthetic work on this class of compounds, the terms *iminoxyl* or *nitroxyl* were also used. The use of the term, *aminoxyl* or *aminoxyl radical*, crops up so infrequently that the typical reader would have considered it as inappropriate or incorrect. Yet IUPAC RNRI Rule RC-81.2.4.D defines aminoxyl compounds with the structure $R_2NO^\bullet$ as "radicals derived from hydroxylamines by removal of the hydrogen atom from the hydroxy group, and they are in many cases isolable." While Chemical Abstracts Service (CAS) uses nitroxide as the parent name for $H_2N-O^{\bullet}$, for example, $(ClCH_2)_2N-O^{\bullet}$ or bis(chloromethyl) nitroxide, the IUPAC name is bis(chloromethyl) aminoxyl. Hence, nitroxide should not be used as the name of a class of compounds that are, as per IUPAC, specifically and correctly aminoxyl radicals. The terms *iminooxy* or *iminoxyl radicals* have been incorrectly applied to alkylidene aminoxyl radicals (also called *iminoxyl radicals*, $R_2C=N-O^{\bullet}$). Has improper usage been strongly discouraged since then? The Sigma/Aldrich catalog calls the spin probe, TEMPO, 2,2,6,6-tetramethylpiperidine 1-oxyl, which is the correct name, yet probably should be appended with aminoxyl radical. Fortunately or unfortunately, common usage sometimes prevails and evolves into acceptable nomenclature. Statistically, nitroxides are cited the most, then nitroxyl about $\frac{1}{4}$ of the time, and then iminoxyl about 1/30th of the time. The correct term, *aminoxyl radical*, has been cited even less number of times than iminoxyl. The compilation given here attempts to include correct IUPAC (or occasionally CAS) names, common names, and acronyms so that the reader can easily refer to the literature.

## 5.2
## Tabulations

The compilation that follows includes the structure and common and IUPAC nomenclature. For hyperfine constants, values are in Gauss unless noted otherwise.

## 5.3
## Concluding Remarks

These listings are grouped into structural types of aminoxyl radicals, that is, the five-membered pyrrolinyl rings, the six-membered piperidinyl rings, and the fatty acid labels where an aminoxyl radical in a ring structure is fused to the fatty acid chain at various positions. The compilation lists both isotropic and anisotropic hyperfine and g-factors at several frequencies. As mentioned earlier, at W-band or higher, these values are easily read off the spectrum, while at X-band one needs a frozen powder spectrum. These compilations should be useful to the EPR community as reference values for researchers investigating oriented biological macromolecules and polymers, and especially in elucidating the anisotropic features of molecular tumbling.

| Common name | Name | IUPAC name | | |
|---|---|---|---|---|
| DTBN | Di-tert-butyl nitroxide | Di(tert-butyl)aminyl oxide | CAS registry number: 2406-25-9 | |

Structure

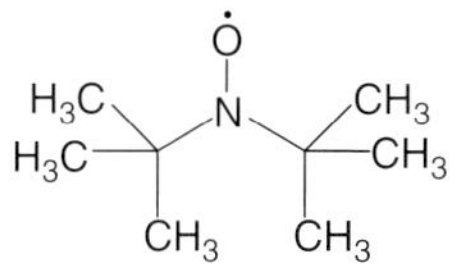

$H_3C$, $H_3C$–C(CH$_3$)–N(O•)–C(CH$_3$)–CH$_3$, CH$_3$

| Frequency (GHz) | $A_{iso}$ | $A_{xx}$ | $A_{yy}$ | $A_{zz}$ | $g_{xx}$ | $g_{yy}$ | $g_{zz}$ | $g_{iso}$ | Solvent/host | Comments | References |
|---|---|---|---|---|---|---|---|---|---|---|---|
| 9.5 | 15.1 | 7.6 | 6 | 31.8 | 2.0089 | 2.0061 | 2.0027 | 2.006 | Tetramethyl-1,3-cyclobutanedione | — | [3] |
| 9.5 | — | 7.1 | 5.6 | 32 | 2.0088 | 2.0062 | 2.0027 | — | Tetramethyl-1,3-cyclobutanedione | — | [4] |
| 9.5 | 15.8 | — | — | — | — | — | — | — | Adamantane | — | [5] |

| Common name | Name | IUPAC name |
|---|---|---|
| TEMPO | | 2,2,6,6-Tetramethylpiperidine-1-oxyl |

Structure

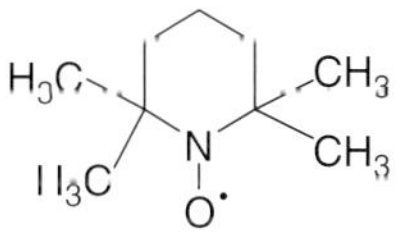

$H_3C$, $H_3C$–(piperidine ring with N–O•)–CH$_3$, CH$_3$

| Frequency (GHz) | $A_{iso}$ | $A_{xx}$ | $A_{yy}$ | $A_{zz}$ | $g_{xx}$ | $g_{yy}$ | $g_{zz}$ | $g_{iso}$ | Solvent/host | Comments | Reference |
|---|---|---|---|---|---|---|---|---|---|---|---|
| 9.5 | — | — | — | — | 2.003 | 2.0069 | 2.003 | — | Tetramethyl-1,3-Cyclobutanedione | — | [6] |

| Common name | Name | IUPAC name |
|---|---|---|
| Tempone | | 2,2,6,6-Tetramethylpiperidinone 1-oxyl |

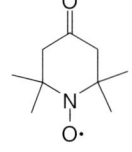

Structure:

| Frequency (GHz) | $A_{iso}$ | $A_{xx}$ | $A_{yy}$ | $A_{zz}$ | $g_{xx}$ | $g_{yy}$ | $g_{zz}$ | $g_{iso}$ | Solvent/host | Comments | References |
|---|---|---|---|---|---|---|---|---|---|---|---|
| 9.5 | 14 | 5.2 | 5.2 | 31 | 2.0104 | 2.0074 | 2.0026 | 2.006 | Tetramethyl-1, 3-cyclobutanedione | — | [4, 7] |
| 9.5 | — | 6.5 | 6.7 | 33 | — | — | — | — | — | — | [8] |
| 9.5 | 15 | 5.6 | 5.1 | 33.7 | 2.0099 | 2.0062 | 2.00215 | — | Liquid crystal | — | [9] |
| 15.6 | — | 6.5 | 6.7 | 33 | 2.0094 | 2.0061 | 2.0021 | — | Ascorbic acid | — | [10] |
| 250 | — | 4.1 | 5.1 | 33.6 | 2.00936 | 2.00633 | 2.00233 | 2.00602 | Toluene | Perdeuterated | [11] |
| 360 | — | 5.5 | 4.9 | 33.1 | 2.0099 | 2.0062 | 2.00215 | — | Calixarene | 150 K | [12] |

| Common name | Name | IUPAC name |
|---|---|---|
| TEMPOL | TANOL 4-Hydroxy-TEMPOL | 2,2,6,6-Tetramethyl-4-piperidinol-1-oxyl |

Structure:

| Frequency (GHz) | $A_{iso}$ | $A_{xx}$ | $A_{yy}$ | $A_{zz}$ | $g_{xx}$ | $g_{yy}$ | $g_{zz}$ | $g_{iso}$ | Solvent/host | Comments | References |
|---|---|---|---|---|---|---|---|---|---|---|---|
| 9.5 | — | — | — | — | 2.0095 | 2.0064 | 2.0027 | 2.0058 | — | — | [6] |
| 9.5 | — | 10.0 | 10.0 | 48.9 | 2.0096 | 2.0062 | 2.0022 | — | $^{15}$N perdeuterated in polycarbonate, 77 K | — | [13] |
| 9.5 | 22.2 | — | — | — | — | — | — | 2.0058 | $^{15}$N perdeuterated in polycarbonate, 520 K | — | [13] |

| Common name Name | IUPAC Name |
|---|---|
| Methoxy-TEMPO | 4-Methoxy-2,2,6,6-tetramethylpiperidine-N-oxyl |

Structure

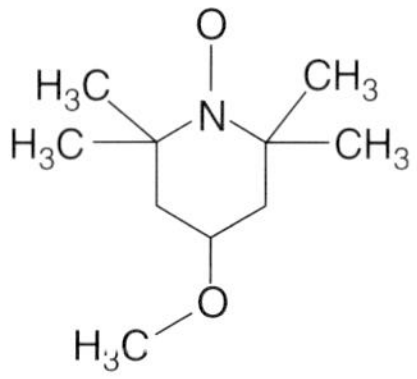

| Frequency (GHz) | $A_{iso}$ | $A_{xx}$ | $A_{yy}$ | $A_{zz}$ | $g_{xx}$ | $g_{yy}$ | $g_{zz}$ | $g_{iso}$ | Solvent/host | Comments | Reference |
|---|---|---|---|---|---|---|---|---|---|---|---|
| 9.5 | — | 7.1 | 7.1 | 33.5 | 2.01015 | 2.0062 | 2.00212 | — | p-Hexanoyl calix[4]arene | — | [12] |

| Common name Name | IUPAC name |
|---|---|
| MTSSL/MTSL | (1-Oxyl-2,2,5,5-tetramethyl-Δ3-pyrroline-3-methyl) Methanethiosulfonate |

Structure

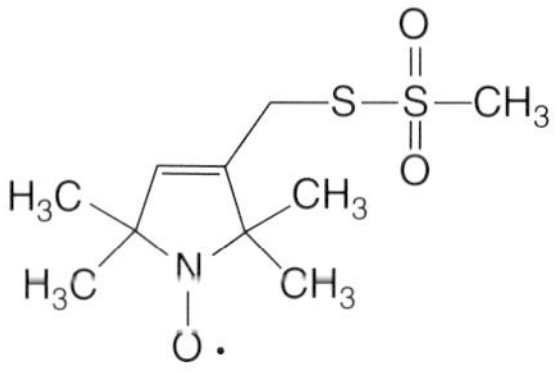

| Frequency (GHz) | $A_{iso}$ | $A_{xx}$ | $A_{yy}$ | $A_{zz}$ | $g_{xx}$ | $g_{yy}$ | $g_{zz}$ | $g_{iso}$ | Solvent/host | References |
|---|---|---|---|---|---|---|---|---|---|---|
| 9.5 | — | — | — | 33.9 | — | — | — | — | Toluene | [14] |
| 95 | 14.29 | — | — | 33.78 | 2.00869 | 2.0059 | 2.00198 | 2.00577 | Toluene | [14] |
| 9.5 | — | — | — | 35.18 | — | — | — | — | Ethanol | [14] |
| 95 | 15.02 | — | — | 35.1 | 2.0083 | 2.0059 | 2.00199 | 2.00574 | Ethanol | [14] |
| 9.5 | — | — | — | 35.56 | — | — | — | — | Methanol | [14] |
| 95 | 15.11 | — | — | 35.9 | 2.00812 | 2.00578 | 2.00189 | 2.00551 | Methanol | [14] |
| 9.5 | — | — | — | 36.88 | — | — | — | — | Water | [14] |
| 95 | 16.12 | — | — | 37.5 | 2.008 | 2.00586 | 2.00199 | 2.00594 | Water | [14] |

| Common name | Name | IUPAC name |
|---|---|---|
| Carbomyl-TEMPYRO | | 3-Carbomyl-2,2,5,5-tetramethyl-3-pyrrolin-1-oxyl |

**Structure**

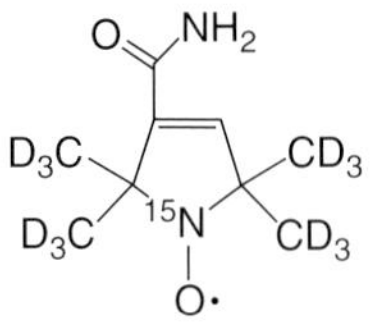

(structure: 3-carbomyl-2,2,5,5-tetramethyl-3-pyrrolin-1-oxyl with $D_3C$, $CD_3$ groups, $^{15}N$, and $NH_2$ carboxamide)

| Frequency (GHz) | $A_{iso}$ | $A_{xx}$ | $A_{yy}$ | $A_{zz}$ | $g_{xx}$ | $g_{yy}$ | $g_{zz}$ | $g_{iso}$ | Solvent/host | Comments | References |
|---|---|---|---|---|---|---|---|---|---|---|---|
| 9.5 | — | 16.7 MHz | 16.7 MHz | 129.9 MHz | 2.0092 | 2.0062 | 2.0024 | — | o-Terphenyl | Echo detected | [15] |
| 9.5 | — | 16.5 MHz | 16.5 MHz | 131.7 MHz | 2.0093 | 2.0069 | 2.0037 | — | 4-Pentyl-4′-cyanobiphenyl | Echo detected | [16] |

| Common name | Name | IUPAC name |
|---|---|---|
| DANO | Dianisyl-nitroxide | |

**Structure**

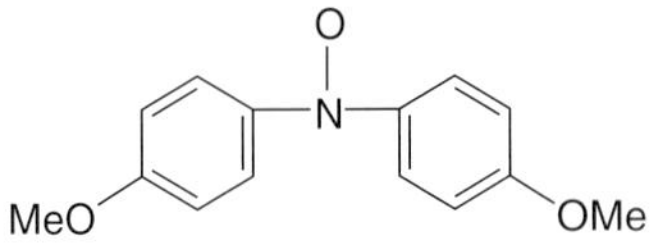

(structure: MeO–C$_6$H$_4$–N(O)–C$_6$H$_4$–OMe)

| Frequency (GHz) | $A_{iso}$ | $A_{xx}$ | $A_{yy}$ | $A_{zz}$ | $g_{xx}$ | $g_{yy}$ | $g_{zz}$ | $g_{iso}$ | Solvent/host | Comments | Reference |
|---|---|---|---|---|---|---|---|---|---|---|---|
| 95 | — | 7.0 MHz | 8.5 MHz | 2.5 MHz | 2.0091 | 2.0053 | 2.0023 | — | Dimethoxy benzophenone | ESEEM | [17] |

| Common name | Name | IUPAC name |
|---|---|---|
| DOXYL | Oxazolidine spin label | N-oxyl-4′,4-methyloxadolidine |

**Structure**

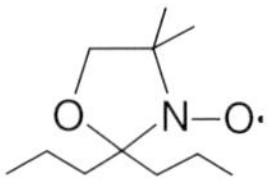

| Frequency (GHz) | $A_{iso}$ | $A_{xx}$ | $A_{yy}$ | $A_{zz}$ | $g_{xx}$ | $g_{yy}$ | $g_{zz}$ | $g_{iso}$ | Solvent/host | Comments | Reference |
|---|---|---|---|---|---|---|---|---|---|---|---|
| 9.5 | — | 4.7 | 4.7 | 31 | 2.0068 | 2.0058 | 2.0022 | — | — | — | [18] |

| Common name | Name | IUPAC name |
|---|---|---|
| CSL | 3-Doxyl-17-hydroxy-5α-androstane | 17β-Hydroxy-4′,4′-dimethylspiro(5α-androstane-3,2′-oxazolidin)-3′-yloxy |

**Structure**

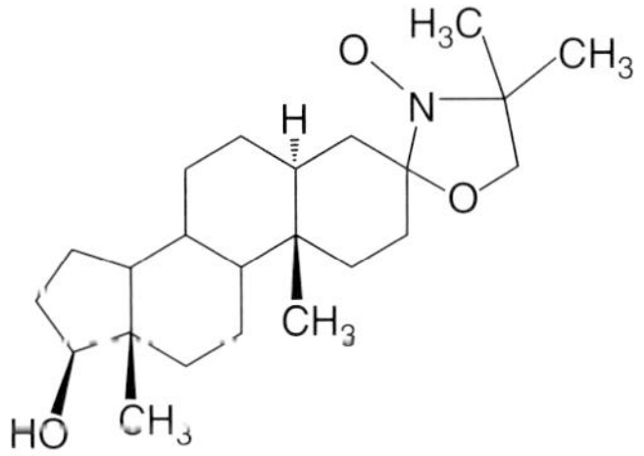

| Frequency (GHz) | $A_{iso}$ | $A_{xx}$ | $A_{yy}$ | $A_{zz}$ | $g_{xx}$ | $g_{yy}$ | $g_{zz}$ | $g_{iso}$ | Solvent/host | Comments | Reference |
|---|---|---|---|---|---|---|---|---|---|---|---|
| 94.3 | 14.3 | 5.3 | 4.9 | 32.4 | 2.00913 | 2.0061 | 2.00231 | 2.00584 | o-Xylene | −135 °C | [19] |

| Common name | Name | IUPAC name |
|---|---|---|
| n-PCSL | Labeled on sn-2 chain {1-acyl-2-[n-(4,4-dimethyloxazolidine-N-oxyl)stearoyl]-sn-glycero-3-phosphocholine | |
| n-SASL | Oxazolidine-N-oxyl stearic acid | |

**Structure Example 16-SASL**

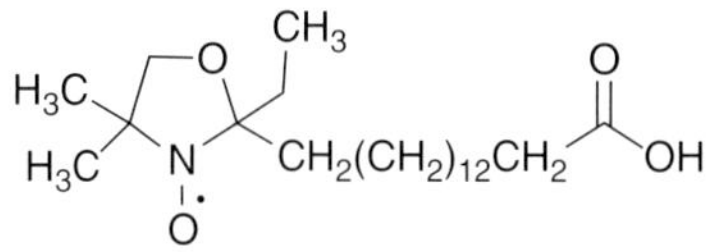

H$_3$C, H$_3$C—C—N(•O)—O—C(CH$_3$)—CH$_2$(CH$_2$)$_{12}$CH$_2$—C(=O)OH

| Frequency (GHz) | $A_{iso}$ | $A_{xx}$ | $A_{yy}$ | $A_{zz}$ | $g_{xx}$ | $g_{yy}$ | $g_{zz}$ | $g_{iso}$ | Solvent/host | Comments | References |
|---|---|---|---|---|---|---|---|---|---|---|---|
| 94.3 | — | 6.4 | 5.7 | 33.3 | 2.008 | 2.0063 | 2.0026 | — | Dimyristoyl phosphatidyl-choline (DMPC)/ dimyristoyl glycerol (DMG) | 0–10 °C | [20] |
| 94.3 | — | 6.4 | 5.7 | 32.5 | 2.008 | 2.0063 | 2.0026 | — | DMPC/DMG | 20–30 °C | [20] |
| 94.3 | — | 5.9 | 5.3 | 32.9 | 2.0076 | 2.0072 | 2.0021 | — | DMPC/DMG | 60 °C | [20] |
| 250 | — | — | 6.5 | 33.6 | 2.0084 | 2.0064 | 2.0031 | — | DMPC/cholesterol | 5-PCSL | [21] |
| 250 | — | — | 7 | 31.8 | 2.00840 | 2.00634 | 2.0031 | — | DMPC/cholesterol | 8-PCSL | [21] |
| 250 | — | — | — | 28.8 | 2.0082 | 2.00696 | 2.00362 | — | DMPC/cholesterol | 12-PCSL | [21] |
| 250 | — | — | — | 21.9 | 2.00745 | 2.00745 | 2.00446 | — | DMPC/myristic acid | 14-PCSL | [21] |
| 250 | — | — | 6.6 | 33.7 | 2.00895 | 2.00617 | 2.00257 | — | DMPC/cholesterol | 5-SASL | [21] |

**References**

1. Berliner, L.J. (ed.) (1976) *Spin Labeling: Theory and Applications*, Academic Press, New York.

2. Grinberg, O. and Berliner, L.J. (eds) (2004) *Very High Field (VHF) ESR/EPR*, Biological Magnetic Resonance, Vol. 22,

Kluwer Academic/Plenum Publishing Corp, New York.

3. Libertini, L.J. and Griffith, O.H. (1970) Orientation dependence of the electron spin resonance spectrum of di-t-butyl nitroxide. *J. Chem. Phys.*, **53**, 1359–1367.
4. Griffith, O.H., Cornell, D.W., and McConnell, H.M. (1965) Nitrogen hyperfine tensor and g tensor of nitroxide radicals. *J. Chem. Phys.*, **43**, 2909–2910.
5. Berman, J.A., Schwartz, R.N., and Bales, B.L. (1974) Electron paramagnetic resonance studies of Di-tert-Butyl nitroxide in adamantane. *Mol. Cryst. Liq. Cryst.*, **28**, 51 61.
6. Bordeaux, D., Lajzerowicz, J., Briere, R., Lemaire, H., and Rassat, A. (1973) Determination des axes propres et des valeurs principales du tenseur g dans deux radicaux libres nitroxydes par etude de monocristaux. *Org. Magn. Res.*, **5**(1), 47–52.
7. Capiomont, A., Chion, B., Lajzerowicz, J., and Lemaire, H. (1974) Interpretation and utilization for crystal structure determination of nitroxide free radicals in single crystals. *J. Chem. Phys.*, **60**, 2530–2535.
8. Snipes, N., Cupp, J., Cohn, G., and Keith, A. (1974) Analysis of the nitroxide spin label 2,2,6,6-tetramethylpiperidone-N-oxyl (TEMPONE) in single crystals of the reduced TEMPONE matrix. *Biophys. J.*, **14**, 20–32.
9. Lin, W.-J. and Freed, J.H. (1979) Electron spin resonance studies of anisotropic ordering, spin relaxation, and slow tumbling in liquid crystalline solvents. 3. Smectic. *J. Phys. Chem.*, **83**(3), 379.
10. Snipes, W., Cupp, J., Cohn, G., and Keith, A. (1974) Electron spin resonance analysis of the nitroxide spin label 2, 2, 6, 6-tetramethylpiperidone-N-Oxyl (TEMPONE) in single crystals of the reduced TEMPONE matrix. *Biophys. J.*, **14**, 20–32.
11. Earle, K.A., Budil, D.E., and Freed, J.H. (1993) 250-GHz EPR of nitroxides in the slow-motional regime: models of rotational diffusion. *J. Phys. Chem.*, **97**, 13289–13297.
12. Polovyanenko, D.N., Bagryanskaya, E.G., Schnegg, A., Möbius, K., Coleman, A.W., Ananchenko, G.S., Udachin, K.A., and Ripmeester, J.A. (2008) Inclusion of 4-methoxy-2,2,6,6-tetramethylpiperidine-N-oxyl in a calixarene nanocapsule in the solid state. *Phys. Chem. Chem. Phys.*, **10**, 5299–5307.
13. (a) Lee, S. and Ames, D.P. (1984) Temperature-dependent ESR hyperfine constants for nitroxides and orientational correlation time determination). *J. Chem. Phys.*, **81**, 4206–4209. (b) Brustolon, M., Maniero, A.L., and Corvaja, C. (1984) E.P.R. and ENDOR investigation of TEMPONE nitroxide radical in a single crystal of tetramethyl-1,3-cyclobutanedione. *Mol. Phys.*, **51**(5).
14. Owenius, R. (2001) M. Engstro1m, and M. Lindgren, Influence of solvent polarity and hydrogen bonding on the EPR parameters of a nitroxide spin label studied by 9-GHz and 95-GHz EPR spectroscopy and DFT calculations. *J. Phys. Chem. A*, **105**, 10967–10977.
15. Kulik, L.V., Rapatsky, L.L., Pivtsov, A.V., Surovtsev, N.V., Adichtchev, S.V., Grigor'ev, I.A., and Dzuba, S.A. (2009) Electron-nuclear double resonance study of molecular librations of nitroxides in molecular glasses: quantum effects at low temperatures, comparison with low-frequency Raman scattering. *J. Chem. Phys.*, **131**(6), 64505–6.
16. Pivtsov, A.V., Kulik, L.V., Surovtsev, N.V., Adichtchev, S.V., Kirilyuk, I.A., Grigor'ev, I.A., Fedin, M.V., and Dzuba, S.A. (2011) Temperature dependence of hyperfine interaction for $^{15}$N nitroxide in a glassy matrix at 10–210 K. *Appl. Magn. Reson.*, **41**, 411–429.
17. Bloeß, A., Möbius, K., and Prisner, T.F. (1998) High-frequency/high-field electron spin echo envelope modulation study of nitrogen hyperfine and quadrupole interactions on a disordered powder sample. *J. Magn. Reson.*, **134**, 30–35.
18. Jost, P., Libertini, L.J., Hebert, V.C., and Griffith, O.H. (1971) Lipid spin labels in lecithin multilayers. A study of motion along fatty acid chains. *J. Mol. Biol.*, **59**, 77–98.
19. Smirnova, T.I., Smirnov, A.I., Clarkson, R.B., and Belford, R.L. (1995) Wband (95 GHz) EPR spectroscopy of nitroxide

radicals with complex proton hyperfine structure: fast motion. *J. Phys. Chem.*, **99**(22), 9008–9016.

20. Schorn, K. and Marsh, D. (1996) Lipid chain dynamics in diacylglycerol-phosphatidylcholine mixtures studied by slow motional simulations of spin label EPR spectra. *Chem. Phys. Lipids*, **82**, 7–14.

21. Gaffney, B.J. and Marsh, D. (1998) High-frequency, spin-label EPR of nonaxial lipid ordering and motion in cholesterol-containing membranes. *Proc. Natl. Acad. Sci. U.S.A.*, **95**(22), 12940–12943.

# Index

## a

absorption spectrum  7
adiabatic rapid passage  4
alternating linewidths  98
aminoxyl (nitroxide) radicals  287
amorphous hydrogenated silicon  17, 56–57
amplifier, coil driver
– *see also* scan driver  9
– heating  22, 28
– push-pull  25
amplitude, scan rate dependence  9–11
anticrossing  86
AWG (arbitrary waveform generator)  11, 31
axial Maier–Saupe potential  81

## b

$B_1$  9, 10, 20, 55
background  28–33
– removal  37–41
bandwidth  10, 41
– *see also* resonator bandwidth  10
– signal  43, 45
basis  75
basis choices in Liouville space  76
BDPA (α,γ-bisdiphenylene-β-phenylallyl)  17, 20, 31
Bloch equations  20, 76
– Bloch sphere  100
Bloch magnetization vector propagation  97
BMPO-OOH  40, 42, 52–54

## c

$C_{60}$  57
3-carbomyl-2,2,5,5-tetramethyl-3-pyrrolin-1-oxyl (Carbomyl-TEMPYRO)  292
carbomyl-TEMPYRO  292
$Ce^{3+}(4f^1)$, SH Parameters  246
Chebyshev approximation  102

Chebyshev polynomials  86
chemical exchange  94
– electron transfer  98
– general exchange theory  99
– intermolecular exchange  98
– intramolecular exchange  98
– multisite exchange  99
– mutual exchange  98
chirp  11
choke coils  30
Clebsch–Gordan series  76
$Co^{2+}(3d^7)$, SH parameters  213–215
$Co^{3+}(3d^6)$, SH parameters  211
continuous wave  3
correlation NMR  4
coupling between two electron spins  74
$Cr^+(3d^5)$, SH parameters  189
$Cr^{2+}(3d^4)$, SH parameters  185
$Cr^{3+}(3d^3)$, SH parameters  176–183
$Cr^{3+}(3d^3)$ glasses, SH parameters  183
$Cr^{4+}(3d^1)$, SH parameters  174
$Cr^{5+}(3d^1)$, SH parameters  142
crystal field theory  92
crystals  80
CSL  293–294
CTPO (3-carbamoyl-2,2,5,5-tetramethyl-3-pyrrolinyl-1-oxyl)  11

## d

DANO  292
data format
– BES$^3$T  114
– ESP  114
data formats  114
– JCAMP-DX  114
deconvolution  6, 17–20
*density matrix*  75

## Index

density operator  75
diamond defects  17, 58
dianisyl-nitroxide (DANO)  292
direct detection  5, 62
direction cosines matrix  79
distance distribution  108
– MtsslWizard  108
– PRONOX  108
double-balanced mixer detection  7
double-quantum coherence  103
doubly nitroxide-labeled  96
down-conversion  15
DOXYL  293
3-doxyl-17-hydroxy-5α-androstane (CSL)  293–294
DPPH  4, 20
DTBN, di-tert-butyl nitroxide  289
$D$ tensor  92
$Dy^{3+}(4f^9)$, SH parameters  266

### e

$E$  92
$E^{prime}$ defect in quartz  7, 16, 55
eddy currents  28, 30, 32, 35
eigenfield  84
*eigenframe* or *principal-axes frame*  78
electron spin coupling
– Heisenberg–Dirac–van Vleck Hamiltonian  74
– biquadratic exchange  74
– coupling matrix  74
– dipole–dipole coupling  74
– Dzyaloshinskii–Moriya interaction  74
– isotropic exchange coupling  74
electron spin relaxation times  7, 17, 20
electron spin resonance software database (ESDB)  71
ENDOR
– $^1$H ENDOR  105
– $^{55}$Mn ENDOR  105
– coils  28, 30, 31
– Davies ENDOR  106
– detection envelopes  107
– high-spin pulse ENDOR  107
– hyperfine enhancement  106
– inversion pulse  106
– resonator  32
– selectivity  106
ENDOR/transition intensity
– Boltzmann polarization  106
energy level crossings  86
energy level diagram  84
EPR simulation programs  70
equations of motion  76
$Er^{3+}(4f^{11})$, SH parameters  266
Ernst equation  58
error analysis  109
ESEEM
– rf-driven  103
$Eu^{2+}(4f^7)$, SH parameters  251–254
Euler angles  79
exchange
– Heisenberg exchange  94
exchange interactions  21

### f

fast-motion limit  94
fast-motion regime  94
$Fe^{2+}(3d^6)$, SH parameters  211, 212
$Fe^{3+}(3d^5)$, SH parameters  189–197
$Fe^{5+}(3d^3)$, SH parameters  184
Fermi contact  73
field modulation  78
field-swept echo detection  57–58
filter
– Butterworth  20
– comb  15
– low-pass  20
filtering  114
– analog  114
– Savitzky–Golay  114
finite-element methods  104
Floquet theory  103
Fourier transform  41
Fourier transformation  93
frames  78
Frank–Condon factor  88
free induction decay  3
Fremy's salt  20
frequency, 250 MHz  12, 13, 19, 35
frequency scans, microwave  5, 11, 64
frequency-sweep  93
frequency-to-field conversion factor  84, 88
frozen solution  79, 81
FT-EPR  57–58

### g

$Gd^{3+}(4f^7)$, SH parameters  254–264
glasses  81
global dynamics  97
g-matrix  73
g-matrix/g-factor
– nuclear g-factor  73
gradient smoothing  93

### h

Hall probe  21
Hellmann–Feynman theorem  91

high-field limit   98
high-spin systems   102
Hilbert space   75
hindered rotation   99
Ho$^{3+}$(4f$^{10}$), SH parameters   266
homogeneous   90
homogeneous broadening   20, 21
hybrid models   88
17β-hydroxy-4,4-dimethylspiro
   (5α-androstane-3,2′-oxazolidin)-3′-yloxy
   293
4-hydroxy-TEMPOL   290
hyperfine (hf) interaction   73
hyperfine coupling matrix $A$   73
HYSCORE   101

*i*

image reconstruction
– filtered back projection   60
– maximum entropy method   61
– regularized optimization   61
inhomogeneous   90
inhomogeneous broadening   6
interpolation schemes   83
ions, detected by EPR   139, 140
irradiated tooth   5

*j*

Jacobi diagonalization   86
Jahn–Teller pseudorotation   99

*k*

Kramers-Kronig relation   42

*l*

Lanczos methods   104
Larmor precession   76
least-squares fitting   108
– bacterial foraging   111
– error function   110
– fitting algorithms   110
   Gauss Newton   110
– genetic (evolutionary) algorithm   111
– Hooke and Jeeves pattern search   111
– Levenberg-Marquardt   111
– metropolis algorithm   111
– Nelder–Mead   111
– neural network models   111
– nonlinear least-squares methods   112
– parameter values   110
– particle-swarm optimization   111
– Powell's conjugate gradient method   111
– principal component analysis   112
– systematic grid searches   112

line broadening
– dipolar broadening   91
– Pake broadening   91
– strain
– – strain broadenings   91
line broadenings   90
linear response regime   18
lineshape
– *Voigt function*   92
– *pseudo-Voigt profile*   92
– convolution   92
– Gaussian function   92
– Holtzmark   92
– lineshapes   92
– Stoneham   92
– Tsallis distribution   92
Liouville basis
– basis of irreducible spherical tensor
   operators   76
– Cartesian Zeeman product operator basis
   76
Liouville space   75
Liouville–von Neumann (LvN) equation   77
LiPc (lithium phthalocyanine)   8, 9, 18, 20,
   60, 61
Litz wire   22, 23, 28
looping transitions   86
Lorentzian function   92
Lorentzian lineshape   44

*m*

magic angle spinning   103
magnet
– air-core   23
– four-coil   21
– Helmholtz   21
– iron-core   23
magnetic dipole–dipole coupling   73
magnetic field
– gradients   60
– homogeneity   21
– modulation   3
magnetic isotopes   77
– structural and kinetic isotope effects   78
magnetically concentrated   20
matrix diagonalization   85
– extrapolation   85
– Hermite cubic spline interpolation   85
– homotopy   85
– least-squares fitting   85
– root-finding algorithms   85
mechanical vibrations   30, 33

methoxy-TEMPO  291
4-methoxy-2,2,6,6-tetramethylpiperidine-
   N-oxyl  291
mHCTPO  13, 47–50
250 MHz  3, 12, 13, 19, 35
microphonics  33
microwave power  50
Mims matrix  101
$Mn^{2+}(3d^5)$, SH parameters  198–209
$Mn^{2+}(3d^5)$ glasses, SH parameters  210
$Mn^{3+}(3d^4)$, SH parameters  186–188
$Mn^{4+}(3d^3)$, SH parameters  184
$Mn^{5+}(3d^2)$, SH parameters  174
$Mo^{3+}(4d^3)$, SH parameters  245
$Mo^{5+}(4d^1)$, SH parameters  244
modulation broadening  7
modulation coils, Bruker  22, 46
molecular dynamics  95
MOMD  96
Monte Carlo  99
*mosaic misorientation linewidth model*  83
MTSSL/MTSL  291

*n*

$N@C_{60}$  57
NARS (nonadiabatic rapid sweep)  5, 62
$Nd^{3+}(4f^3)$, SH parameters  247–250
Newton–Raphson method  90
$Ni^+(3d^9)$, SH parameters  244
$Ni^{2+}(3d^8)$, SH parameters  215–217
nitroxide (aminoxyl) radicals  287
NMR  103
noise  17
– reduction  15
*n*-SASL, oxazolidine-*N*-oxyl stearic acid  294
nuclear electric quadrupole moment  73

*o*

objective function  109
organic radicals  109
oscillations (see also wiggles)  6, 22
oxazolidine spin label
   N-oxyl-4′,4-methyloxadolidine  293
oxazolidine-*N*-oxyl stearic acid  294
(1-Oxyl-2,2,5,5-tetramethyl-Δ3-pyrroline-3-
   methyl) methanethiosulfonate
   (MTSSL/MTSL)  291

*p*

Padé approximation  102
parallelization  103
parameter space  91
passage effects  4
path integrals  82

$Pd^+(4d^9)$, SH parameters  246
periodic signal  15, 16
perturbation theory  87
– Scalar perturbational expressions  87
powder sample  79
power saturation  9, 16
projection techniques  93
propagator matrices  102
pseudomodulation  20
pulse EPR  81
– integration window  103
– multifrequency  103
– product rules  103
– pulse DEER  108
pulsed EPR  3, 45, 54–58

*q*

quadrature detection  41
– mixer  41
– nonorthogonality  40
quantitative EPR  13
quaternions  79

*r*

rapid scan  3–64
– acquisition parameters  42–45
– background  28, 32, 37–42
– bridge  41
– coil driver  24–29
– coils  22, 24, 28, 46–50
– deconvolution  18–20, 64
– digital  11, 13
– driving function  18
– Fourier transform NMR spectroscopy  3
– imaging  59–62
– multifrequency  46
– post-acquisition treatment  17–20
– scan rate selection  45
– signal bandwidth  43, 45, 46
– signal-to-noise advantage  14, 15, 17,
   53–58, 64
– simulation  8–10, 20
– sinusoidal  11, 19, 24, 28, 29, 34, 40–41, 64
– $T_2$ measurement  51, 52
– trapezoidal  21
– triangular  11, 17–18, 25–28, 37–38, 64
Redfield–Wangsness–Bloch relaxation theory
   95
reduced spin system  106
reference frame
– lab frame  79
– *molecular frame*  79
references (chapter 4)  270–286
relaxation times  45, 54, 58

resistance, AC   23
– $T_1$   46
– $T_2$   6, 21, 44, 46, 50, 52
– $T_2^*$   3, 6, 43–45
resonator
– 250 MHz   35, 36
– bandwidth   10, 43
– cross-loop   33, 42
– design   32–41
– dielectric   33–35, 46, 47
– isolation   33
– Q   11, 33, 35, 43–46
– rectangular   33, 34
– shield   30
– split-ring   33, 34
$Rh^+(4d^8)$, SH parameters   245
$Rh^{2+}(4d^7)$, SH parameters   245
*rigid limit*   80
rigidly attached labels   97
rotational dynamics   79, 94
– correlation time $\tau_c$   94

**s**
saturation-transfer EPR   4, 89
$Sb^{2+}(5s_25p_1)$, SH parameters   269
scan coils   21–24, 46–49
scan driver
– sinusoidal   6, 27–29
– triangular   6, 25–28
Schrödinger equation   76
second-harmonic out-of-phase detection   5
segmental acquisition of spectra   4, 63
semiquinones   13
signal-to-noise advantage   3
– *see also* rapid scan   3
simulation noise   93
simulations software
– DDPOW   71
– DIPFIT   71
– E-SpiReS   71
– EasySpin   70
   EPRNMR   71
– EPRSIM-C   71
– EPRsim32   71
– EWVoigt   71
– GENDOR   71
– HYSCORE   71
– MAGRES   71
– Molecular Sophe   71
– OPTESIM   71
– QPOW   71
– Sim   71
– SimBud   71
– SimFonia   71

– SIMPOW6   71
– SIMPSON   72
– Sophe   71
– SPIN   71
– Spinach   72
– SpinCount   71
– SPINEVOLUTION   72
– WinMOMD   71
– WinSIM   71
– Xemr   71
– XSophe   71
slow and intermediate motion
   regime   94
slow-scan spectrum   6, 18
$Sm^{3+}(4f^5)$, SH parameters   250
smoothing   114
sodium in liquid ammonia   4
solid-state cw EPR   75
Sparse matrix methods   104
spherical grid   81
– igloo grid   82
– analytical   82
– Delaunay triangles   82
– Fibonacci grid   83
– icosahedral   83
– iteratively generated grid   83
– Lebedev grids   83
– numerically optimized   82
– octahedral   83
– randomly generated   82
– Voronoi cells   82
– Zaremba–Conroy–Wolfsberg scheme   83
spin basis
– coupled basis   75
– eigenbasis   75
– uncoupled Zeeman   75
spin echo   100
spin Hamiltonian   72
– isotropic   89
spin Hamiltonian, non-zero coefficients of
   spin operators for various symmetries   139
spin quantitation   113
spin relaxation   76
spin system   72
spin trapping   40, 52–54
spin–orbit coupling   72
spin-relaxation   94
SRLS   96
stochastic dynamics   95
– Brownian rotational diffusion   95
– jump diffusion   95
strain
– *A* strains   92
– *D* strain   92

strain (contd.)
– g strain   91
– spin–strain tensor   92
strains   91
sum of squared deviations   109
superhyperfine (shf) interactions   90
superoperator   75
– Liouville superoperator   77
– chemical exchange superoperator   77
– stochastic relaxation superoperator   77
– super-Hamiltonian   84
– Zeeman superoperator   84
– zero-field superoperator   84
symmetry   80
– Laue classes   80
– molecular symmetry group   80
– space group   80

**t**
TANOL   290
TCNQ   5, 62
TEMPO   291
TEMPOL   290
TEMPONE   290
tempone-$d_{16}$   19, 50, 51
2,2,6,6-tetramethyl-4-piperidinol-1-oxyl (TEMPOL)   290
2,2,6,6-tetramethylpiperidine-1-oxy (TEMPO)   291
2,2,6,6-tetramethylpiperidinone 1-oxyl (TEMPONE)   290
$Ti^{2+}(3d^2)$, SH parameters   175
$Ti^{3+}(3d^1)$), SH parameters   142, 143
Tikhonov regularization   108
time domain   94
$Tm^{3+}(4f^{13})$, SH parameters   268
transition field   84
transition frequencies   105
transition intensity
– Fermi's Golden Rule   88
– nuclear overlap matrix   88
– spin transition moment   88
– thermal polarization   89
– transition probability   88

transition preselection   86
transition metal complexes   92
transition probability   81
triangle projection   83
triplet spectra   82
trityl radical   13, 44, 60
tunneling   99

**u**
uncertainty analysis   14

**v**
$V^{2+}(3d^3)$, SH parameters   184
$V^{3+}(3d^2)$, SH parameters   175, 176
$V^{4+}(3d^1)$ glasses and solutions, SH parameters   143–146
$V^{4+}(3d^1)$, SH parameters   146–148
vanadyl porphyrin   7
$VO^{2+}(3d^1)$, SH parameters   148–160
$VO^{2+}(3d^1)$ glasses, SH parameters   161–173

**w**
W-band   11
wiggles   4, 6
Wigner functions   96
Wigner rotation matrices   79

**y**
$Yb^{3+}(4f^{13})$, SH parameters   269

**z**
Zeeman interaction   72
– electron   72
– higher order electron Zeeman terms   73
– nuclear Zeeman interaction   73
zero-field spin-Hamiltonian parameters(ZFSHP), common notations   140, 141
zero-field splitting   74
– $B_{kq}$   74
– $O_{kq}$   74
zero-field tensor $D$   74